CHRIS KRAFT

FLIGHT

MY LIFE IN MISSION CONTROL

A PLUME BOOK

PLUME
Published by the Penguin Group
Penguin Putnam Inc., 375 Hudson Street,
New York, New York 10014, U.S.A.
Penguin Books Ltd, 80 Strand,
London WC2R 0RL, England
Penguin Books Australia Ltd, Ringwood,
Victoria, Australia
Penguin Books Canada Ltd, 10 Alcorn Avenue,
Toronto, Ontario, Canada M4V 3B2
Penguin Books (N.Z.) Ltd, 182–190 Wairau Road,
Auckland 10, New Zealand

Penguin Books Ltd, Registered Offices:
Harmondsworth, Middlesex, England

Published by Plume, a member of Penguin Putnam Inc.
Previously published in a Dutton edition.

First Plume Printing, March 2002
10 9 8 7 6 5 4 3

The Library of Congress has catalogued the Dutton edition as follows:

Kraft, Christopher C.
    Flight : my life in mission control / Christopher C. Kraft with James L. Schefter.
        p.   cm.
    ISBN 0-525-94571-7 (hc.)
    ISBN 0-452-28304-3 (pbk.)
    1. Kraft, Christopher C. 2. Aerospace engineers—United States—Biography.
3. Astronautics—United States—History. 4. United States. National Aeronautics and Space
Administration—History. I. Schefter, James L. II. Title.
TL789.85.K7.A3   2001
629.4'.092—dc21
[B]   00-057291

Printed in the United States of America
Original hardcover design by Eve L. Kirch

Acclaim for Chris Kraft's *Flight*

"*Flight* allowed me to relive many amazing and trying incidents.
Those readers who saw our accomplishments in the Mercury,
Gemini, and Apollo programs only in news stories and on television
will be caught up and captivated by the behind-the-scenes
truthfulness of Chris Kraft's *Flight*. It is a must-read for all who
would understand our role in human space flight."
—Charles A. Berry, M.D., M.P.H.,
former director of medical operations, NASA

"Chris Kraft is one of those unheralded heroes without whose
commitment and dedication America's space program would
have never become a reality. Now he divulges what heretofore
many of 'us' never knew."
—Eugene A. Cernan, astronaut and author of *Last Man on the Moon*

"Chris Kraft was indeed a giant in America's race to beat the
Russians to the moon. He was a great leader as well as an
outstanding engineer. He understood operations and he made
things happen. Now, at last, he tells his story. He chronicles one
of those proud periods in American history when we
undertook the impossible and did it."
—Frank Borman, commander of Apollo 8 and author of *Countdown*

CHRIS KRAFT joined the Langley Aeronautical Laboratory of the National Advisory Committee for Aeronautics (NACA) in 1945 after graduating from Virginia Polytechnic Institute with a B.S. in aeronautical engineering. In 1958, he was selected as one of the original members of the Space Task Group, the organization established to manage Project Mercury, the original manned space mission. He went on to become the first flight director for NASA and personally served as flight director for the Mercury missions and many of the Gemini missions. Following that, Chris Kraft went on to become the director of the NASA Lyndon B. Johnson Space Center in Houston, Texas, from 1972 to 1982. Since his retirement from federal service, Dr. Kraft has served as an aerospace consultant to Rockwell International, IBM, and a number of other companies. He has two grown children and lives in Houston with his wife, Elizabeth.

All of the author's funds received from this publication are to be given to the Scholarship Fund for children of the employees at the NASA Johnson Space Center in Houston, Texas.

To the thousands of dedicated people in the industry, academe, and government who contributed to the success of one of man's greatest adventures

CONTENTS

Section III: The Gemini Missions

Section IV: The Apollo Missions

ACKNOWLEDGMENTS

Writing a book is a different experience for an engineer who was used to presenting technical facts in both written and verbal form. I recall that my professor in technical English in college would chastise us by saying "one of these days we will graduate educated men as well as engineers from this university." His point was that most engineers pay little attention to the art of communications. During my tenure in the space program, speaking to my peers and the public became an integral part of my job. I grew to enjoy it because it was a pleasure to talk about the passion I had for what we were doing. The keen interest by the public added to this pleasure. This passion and my desire to tell the story of manned space flight from the perspective of one who participated in the management of these programs is what prompted this book.

As I began to write, I realized I needed someone to evaluate the quality of my stories, to assure they would be understood by the reader and to help formulate my words. A close and dear friend, Mary Ann Moore, had a keen interest in writing and had recently been taking courses in writing novels at Rice University. Mary Ann volunteered her services and became a tremendous asset during my struggles to write my life experiences. She offered many suggestions, became an indispensable critic, and gave me the encouragement I needed to continue.

The Johnson Space Center helped me in many ways. They supplied technical reports, searched their archives, and answered many questions necessary to the book. The library technicians and Abby Cassell in the director's office were indispensable.

After completing the first manuscript, I found a fine literary agent, Dominick Abel, to aid in the publishing process. He worked closely with me and recommended I obtain an accomplished writer to put the manuscript in a more suitable form. He recommended James Schefter. I had known Jim as a writer for *Time* magazine and as a news reporter. Jim was a fortunate choice because he was knowledgeable in almost every aspect of my career. Jim did a marvelous job of making my writing come alive and creating a compelling memoir. I hope our readers agree. We became close associates and I came to admire his ability to interpret my description of events. James Schefter died of pulminary fibrosis in February 2001 while awaiting a lung transplant.

Brian Tart, the editor in chief at Dutton who opted to publish the book, has been great to work with and very helpful in what turned out to be a rewarding process.

Finally, my wife, Betty Anne, and my daughter, Kristi-Anne DuPont, were my in-house editors and critics. They provided important input and showed marvelous recall of many of the events that took place.

Thanks, of course, to all of my associates in manned space flight over the years who made my life such a wonderful experience. They are the ones who landed men on the moon and brought them safely back to earth.

Full many a gem of purest ray serene,
The dark unfathom'd caves of ocean bear:
Full many a flower is born to blush unseen,
And waste its sweetness on the desert air.

—from Thomas Gray's "Elegy Written in a Country Churchyard"

PREFACE

My name is Christopher Columbus Kraft Jr. My gut's got a knot in it, but for the next few minutes there's nothing I can do. I'm in a room that I conceived in my mind, then invented, it seems, almost overnight. Some of the men who helped me are here now, as quiet and grave as I am.

We're waiting for news.

I'm thirty-six years old on this day, January 31, 1961. Exactly three years ago other men worked in a dingy room only a few miles from here, and in the dark before midnight, they made history. One of them flipped a toggle switch. Not far from that firing room, a Jupiter-C rocket spit flame and soared into the night sky. It carried a little thirty-one-pound package of instruments with the grand name of Explorer. A few minutes later, *Explorer 1* was a new satellite in orbit around the earth.

America, frightened and confused by the two Sputniks sent into orbit by our Cold War enemy the Soviet Union, had finally joined the space race.

Now I'm standing here mute in Mercury mission control, wanting to curse the silence in my headset, wanting to curse the Redstone rocket that was a Jupiter-C's closest relative, wanting to curse the damned arrogant

German who promised this wouldn't happen. *I should have punched him when I had the chance,* I grumble to myself.

But if I had, I probably wouldn't be here today. And somebody else would be making the decisions that could mean life or death to an astronaut in space.

It isn't an astronaut out there today. It's a chimpanzee named Ham. No matter. We've all learned something today, beyond the lessons laid out so carefully in our mission plan. We learned on this flight, and will repeat the lesson on the many flights yet to come, that our first concern is for the crew. We've known this instinctively, of course, from the beginning of America's program to put men into space. The crew comes first. But today, when things were going wrong, we learned just how visceral those instincts are.

I'm the flight director for this mission, Mercury-Redstone 2, the first mission in Project Mercury to put a living thing into space. Ham was the living thing, but we never thought of him as anything but *crew.*

We all have monikers. They call me Flight on the mission control intercom loop. The doctor—he and his brethren have given us fits for years—is Surgeon. The engineer responsible for getting the capsule down, for monitoring and calculating its retrofire systems, is Retro. Flight dynamics, the infant science of trajectories and propulsion, is the domain of FIDO. There are others, too. The voice link between mission control and the capsule is Capcom, short for "capsule communicator."

Eventually, an astronaut would be Capcom, but today the console is manned by an engineer. Alan Shepard is nearby, in the launch blockhouse. He has a personal interest in today's events. If the Redstone rocket and the Mercury capsule work well, and if Ham does his job on board that capsule, and if we do ours on the ground, Al will be next. He'll be the world's first man in space.

There's only one flight director. From the moment the mission starts until the moment the crew is safe on board a recovery ship, I'm in charge. I ask. I listen. I make decisions. No one can overrule me. Not my immediate boss in the still-young National Aeronautics and Space Administration, the mission director, Walt Williams. Not his boss, a man I respect and revere, the guiding light of America's manned space program, Bob Gilruth. Not even Jack Kennedy, the president of the United States, who's only had his job for ten days or so. They can fire me after it's over. But while the mission is under way, I'm Flight. And Flight is God.

I don't feel so godly right now, I muse.

The problems on the launch pad weren't so bad. I'd held the countdown when an electrical unit overheated and Ham started to get warm. He was strapped into a form-fitted couch and sealed in a small pressurized capsule where an astronaut would normally sit. If it had been Al up there, I could have asked the Capcom to inquire about his comfort. Knowing Al, he would have said something like "No sweat," making a joke of the heating problem. And he would have changed settings on his environmental control unit to cool down a bit.

Ham and I didn't have that option. Outer space was new territory for exploration, and nobody knew much about it. A lot of doctors were predicting that zero gravity would have dire consequences for the human body. Most of us, including test pilots and astronauts, didn't believe them. But the only way to make our point was to substitute monkeys or chimpanzees for men, then see what happened. Al would gladly have traded places with Ham on that January day. He had supreme confidence in what rigors the human body—especially his own—could handle. The decision wasn't his to make, so a trained chimp was out there on the pad.

Instead of passing an order to an astronaut, I told EECOM, on the environment, electrical, and communications console, to turn off Ham's unit. While we waited for it to cool down, I asked Surgeon to evaluate Ham's comfort level.

"He's go, Flight."

I took Surgeon at his word. In mission control, nobody lies to Flight. They tell what they know, or they tell me their best informed guess. There's only one other option: "I don't know, Flight." Anybody who gives me that answer more than a few times will be looking for a new job.

We picked up the count and had the same problem. I held us for an hour this time, letting the electrical unit and Ham get comfortable again. But it was getting late; we wanted as much daylight as possible in the recovery area in case things went wrong. We should have launched at 9:30 A.M. Now it was after eleven. Then the elevator at the pad stuck. Another hold while a technician fixed the problem, and then the pad was clear.

By now, I'd stopped thinking of Ham as a monkey. Some of the jokesters were calling this a monkey flight, and the phrase had been picked up by the press. So had a one-liner from some nightclub comic who was pointing out that Ham was paving the way for Al Shepard. "First the chimp, then the chump," that's what some of them were saying. Al wasn't amused, and neither was I.

In my mind as the countdown headed toward zero, Ham was a real as-

tronaut, he was crew, and we were treating every moment just the way we
would when it was Al up there on that skinny little, black-and-white rocket
built by Wernher von Braun and the same Germans who'd bombed London. I heard the numbers in my headset.

"...three, two, one...lift off..."

Ham was on his way at 11:55 A.M. The Redstone was supposed to fire for
about 140 seconds, reaching 4,400 miles per hour, and then letting Ham
and his Mercury capsule coast to 115 miles high. Ham would get about five
minutes of weightlessness as he arced over the top and started back toward
a splashdown on the Atlantic Missile Range beyond the Bahamas.

All the while, he was supposed to convince the doctors that it was
safe for Al Shepard to go next. He'd do this by watching colored lights
and hitting levers in the right sequence. If he did it right, he'd get a banana pellet. We figured that by now, Ham was getting pretty hungry. If
he did it wrong, he'd get an electrical shock to his foot, and the doctors
would use that to bolster their fears that man couldn't function in zero
gravity.

Gilruth's design team had included an escape tower on top of the Mercury capsule. It had a solid rocket with three nozzles, and if something
went wrong with the Redstone, the tower would fire and pull the capsule
safely away. If nothing went wrong, we'd just jettison the tower. One of our
preflight questions was that timing. We didn't want to let it go too early.
And if our timing was too late, there could be real problems. So we asked
von Braun and his experts for advice.

"We've never had a Redstone burn for less than 139 seconds," von
Braun said. "Never."

"So if we disarm the tower at 137.5 seconds, we should be all right?" my
planners asked.

"*Ja, ja,* that should give you plenty of time."

Within sixty seconds of launch, it started to go wrong.

"Flight, FIDO."

"Go, FIDO."

"One degree high."

Redstone was not following the precise path we'd anticipated.

"Are you go?"

"Go, Flight."

Okay, we'd stick with it. The range safety officer, the RSO, had his finger ready to hit the abort button. If he did, the Redstone would blow into
a million pieces and Ham would be dragged to safety by the escape tower.

We hoped. RSO was monitoring the comm loop and did nothing. Then we got more bad news.

"Flight, FIDO."

"Go, FIDO."

"LOX depletion in twenty seconds." .

Damn it! We've got a hot Redstone. It's burning liquid oxygen too fast and it must be overthrusting. That's why it's higher than the flight path. And if I can add, it's going to run out of fuel a few seconds early.

"Go or no-go, FIDO?" *If he says no-go, I have to tell RSO to hit the button. And God help Ham because it'll be out of my hands.*

"Go, Flight."

"Surgeon?"

"Go, Flight." *So Ham's handling this pretty well. Okay, we let it ride.*

I watched the clock, hoping that the Redstone would keep burning until after 137.51 seconds. Then the tower would be safed and we could release it in the normal fashion. At one hundred thirty-seven and a fraction, my headset crackled.

"Flight, FIDO. LOX depletion. Engine shutdown. Abort initiated."

Damn it! Von Braun promised this wouldn't happen. So much for his Teutonic perfection . . .

Ham had just gotten a 17-g kick in the ass—something a few degrees worse than being blindsided hard by a pro football linebacker—when the escape tower fired and pulled his Mercury capsule away from a perfectly safe Redstone. The extra rocket power was sending him through 115 miles and all the way up to 157 miles. He was getting at least seven minutes of weightlessness, and if the readings Surgeon was getting were accurate, he was one pissed-off monkey. It didn't last long. With the extra speed, the high-g impulse, and the extra time in space, he'd missed a couple of cues with his right hand and gotten some shocks. Then he settled down and kept trying to do his job. He was giving Al a shot at immortality.

"Flight, EECOM."

"Go, EECOM."

"Cabin pressure's down to one psi." *Damn! Now what? Had the cabin wall ruptured? Mercury was losing oxygen fast.*

"How's the suit circuit?" Ham was inside his sealed couch. He should have been getting oxygen at about 5.5 psi just as if he were Al Shephard in his space suit. If he wasn't . . .

"Holding at five point five, Flight." *Good. Probably a valve popped open.*

Capsule integrity is still good. Ham's okay. I didn't bother to answer. I had another problem to deal with.

"Flight, FIDO, we're going to overshoot."

"How far, FIDO?"

"Maybe one hundred miles, Flight." *He'll have better data when the parachute opens and the SARAH beacon activates. But we better get the recovery ships moving to the east.*

I glance at the clock and visualize Ham plummeting through the atmosphere. We had tracking data and telemetry. In another few minutes, the drogue chute would deploy and pull out the main recovery parachute. "Flight, Recovery. Looks like about one hundred fifteen miles long."

A few minutes later, the SARAH beacon aboard the capsule activates and we're getting good fixes on its location. Ham is indeed 115 miles farther downrange than we'd planned. The Navy has planes on the way and the *Donner* is steaming at flank speed toward the impact point. All we can do is wait for news.

I look around mission control. Every face is somber. While we wait, I think about how I'd gotten here from my old neighborhood in the tough Tidewater town of Phoebus, Virginia. It had been an interesting ride.

And if today turned out all right, it was going to get a whole lot better.

SECTION I

A Boy from Phoebus

1

Raised in a Town That No Longer Exists

By any other name, Phoebus was still a tough town. It did have another name, spoken only by the locals with a mixed measure of quiet pride and quieter concern: Little Chicago, because Phoebus sat at the end of the Chesapeake & Ohio Railroad, the old C&O, and if the Chicago end of the line could be tough, so could the Tidewater end. Phoebus hadn't yet been engulfed by Newport News and Hampton, but its unique gentle hardness would one day disappear into the urban sprawl. You won't find Phoebus on many maps today. It's been swallowed by Hampton. But back then, there were still open fields and minor countryside on the north and west of Phoebus.

To the east and south was water, Chesapeake Bay and the Atlantic. The biggest business in town was the Newcome Seafood Company. But the biggest employer was the federal government. Phoebus was neighbor with Forts Monroe and Eustis, with the Newport News Shipbuilding and Drydock Company, with Norfolk Naval Base, and a veterans' hospital, and most important in my life, with Langley Field, where the great journey really began.

That was the world I came into and the world that gave me shape and direction.

My grandfather August was a saloonkeeper living in midtown New York City, in that boisterous era of the Gay Nineties when the theater district was still below Eleventh Street. He died young—I don't know how—but not before siring six children. My father was the fourth, born just before Columbus Day in 1892. The circle a few blocks away at Eighth Avenue and Fifty-ninth Street had just been renamed in honor of the four hundredth anniversary of Columbus's discovery of America. So they called him Christopher Columbus Kraft, a fittingly mixed name for the son of Bavarian immigrants born into nineteenth-century polyglot America. When Grandmother Clara—I always called her MaMa—was widowed, she wasted little time in finding a new husband. His name was Albright and he took the family to Phoebus in 1904. Sometime later, he disappeared. I never knew the details.

Clara ran a bakery in the Bronx, but she knew that the real money was in booze, not bread. Phoebus men were working class and good drinkers, so the saloons she opened in the century's early years made money. It didn't last. The Eighteenth Amendment was ratified in January 1919, and by the end of the year, her saloons were empty and my grandmother was broke. My father was back from army duty. Somewhere along the line, he'd had a nervous breakdown and didn't get sent to Europe. His problems would get worse in later years, and his influence on me would always be more shadowy than substantive. I saw strong father-son relationships among my friends, but I didn't miss it at home. I was a child, with a child's understanding, and I accepted life as it was without wondering if it should be different.

When I was young, my father's symptoms of depression were masked. But he was feeling low, and his family was closing in on poverty. He found himself working in the finance department at the veterans' hospital just outside of town. It was a good thing for me that he got that job.

The hospital's patients included men who'd fought in the Spanish-American War and World War I. Several hundred old veterans of the Civil War still survived and were getting help there. A lot of patients were drunks, burying their memories under piles of empty bottles. My father didn't pay much attention to them. He'd spotted a pretty nurse and he focused on her. They started going out and in 1922 took a trip together to New York City. There in the old neighborhood that he'd left eighteen years before, Christopher Columbus Kraft proposed to Vanda Olivia Suddreth, and she said yes.

They were married a few days later in the Church of the Transfiguration, the fabled "Little Church Around the Corner" on Twenty-ninth Street between Fifth Avenue and Madison. Though he'd been raised Catholic, my father quickly embraced the Episcopal faith, and Vanda did, too. Their little sneak-away vacation trip became their honeymoon, and they celebrated in part by going to the Hippodrome Theater, where they saw live elephants performing. New York City was an amazement to the new Mrs. Christopher Kraft. "After seeing the streets of New York City," she would say for the rest of her life, "every other street in the United States looked like Main Street." She dreamed of going back to see Broadway and Fifth Avenue again, but she never made it. The newlyweds returned to little Phoebus, Virginia, and settled down for the long haul and some hard times ahead.

I missed being a leap year baby by one day. My mother went into labor on February 28, 1924, and Grandmother Clara sent for a doctor. There were only two in town, and Dr. Charles Vanderslice got the call. At one point in the delivery, he handed Clara a bottle of ether and told her to get it ready "just in case." In her excitement, Clara spilled ether on my mother's face. When she came to, Dr. Vanderslice was holding a baby boy.

"Chris Junior, has arrived," were the first words my mother remembers, "and he's a fine healthy boy."

Of the names they'd discussed, Christopher Columbus Kraft Jr. was nowhere on the list. My father considered his name onerous; it had forced him to survive teasing as a boy and an occasional gibe as an adult. But in that glowing moment, it hit my mother just right. I was only minutes old and now the burden was mine, too. Can a name influence the course of a life? I've had most of a century to ponder that question. I think with a name like Christopher Columbus Kraft Jr. some of my life's direction was settled from the start.

I was born at home. The building next door was one of Clara's empty saloons, and it was a marvelous playhouse. The great barroom was perfect for my electric trains. The storeroom and the little office offered hiding places. I had no feel for the sorrows that had been drowned here, or the joys that had been toasted, or the simple tensions that had drifted away from a working man's shoulders while he sipped a beer or downed a Scotch. I only knew that I was luckier than most kids in town because I had a whole building to play in while they were stuck with a shared bedroom or a living room overseen by a keen-eyed mother ready to yell if the play got too boisterous.

I knew about such things because our house was crowded. Mother, Father, Grandmother, and I lived on the first floor. Uncle August Kraft, Aunt Bertie, and my four cousins lived in an upstairs apartment. There were only two bedrooms on our floor. Grandmother got the bigger one. I slept in a crib in the little bedroom with my parents, and that lasted until I was seven or eight. When I began to understand something about the noises that came from my parents' bed, it was time to make other arrangements. I was moved to a daybed in the dining room, and I stayed there until high school.

We had a big yard, with fruit and nut trees, a vegetable garden, and a chicken house. I can still see my mother catching one of those chickens, wringing its neck with her bare hands, dipping it into a tub of hot water, and then plucking it clean. She was quite a woman. I don't remember ever going hungry. I do remember understanding that some things die so that others may live. It wasn't quite the same lesson that farm boys got, or the life-and-death lessons that helped mold the great American character along the westward trails, but it was vivid to me, and it stuck.

There was something else outside that had an extraordinary impact on my life. It bordered our yard and I can't forget it: the town dump.

This is my first indelible memory of childhood. I was three. Trash was burning in the dump and I took my peanut-butter-and-jelly sandwich to the back of our yard to watch. Somehow in my thrall with the flames and smoke, I dropped the sandwich. I reached for it and fell full into the hot coals and ash. It hurt and I screamed and managed to get to my feet. I remember running past the grape arbor, screaming at the top of my lungs, and watching the skin on my right hand swell and bloat with grotesque pain. Then mother had me in her arms.

She was a nurse before a mother and immediately salved me with burn ointment from her medicine cabinet. Then she rushed me to Dr. Vander-slice.

I was burned on both hands and knees, but the right hand was awful. I remember crying and screaming while he treated me. My left hand and knees weren't bad. They'd be scarred, but otherwise unaffected. It was the pain in my right hand each time he touched it that I couldn't fight off. But there was no choice. It was 1927 and a small town. Burn wards and modern treatments were off somewhere in the future. For the next three months, I was in and out of Dr. Vanderslice's office, being painfully unbandaged and rebandaged until finally he and some doctors at the veterans' hospital decided to try a skin graft. It's a memory filled with a small boy's terror at being strapped to the operating table, with the acrid smell of ether and the red spiral into unconsciousness, and the nightmares of being under anesthesia. They peeled patches of living skin from the back of my right hand and grafted them to the worst burns on the palm. When they were done, they told my mother, all that could be said was that they had done the best they could. It was like trying to make a dress without a pattern, they said.

I was in the hospital for two weeks. Each day the bandages came off, ointments were applied, and new bandages went on. And each day, it hurt a little less. The rest of the day was paradise. I was the center of lavish attention from nurses, from my parents, and even from many of the old soldiers who were patients there. I had new toys and new friends, and finally one day the doctors smiled and said I could go home.

Even with a little boy's regenerative powers, healing was slow. But my hand did heal. I squeezed and I gripped, and if the hand never regained all of its powers and all of its strength, it was still a useful right hand. I never considered becoming a lefty. Right-handed I was born, and right-handed I stayed. It wasn't a perfect hand, but only once in my long life did it ever stop me from doing something I wanted to do. And that decision wasn't mine to make.

There was one school in Phoebus, first through ninth grade. My mother had put me into a private kindergarten and I entered school in the 1B class, skipping most of first grade. Grades eight and nine in Phoebus were considered high school, equal to freshman and sophomore, so when it was time to head over to Hampton High School, Phoebus kids enrolled as juniors. We all got a decent education, and I made some lifelong friends starting from the first day. Maybe it came from the saloonkeepers in my background, or maybe from seeing how much my father enjoyed being a member of the fire department and the American Legion. He'd sit at his typewriter in the dining room, a cigarette hanging on his lip until it burned out from his saliva, working on the minutes of fire department meetings. He had one of the few notary public licenses in our areas, and if he wasn't typing minutes, he'd be helping people prepare some kind of document. It all made me a gregarious kid. I had to be in every school activity, from sports to music to acting, and I was pretty much a straight-A student, including Latin. I liked being "on" and was comfortable when other kids felt the stress; that helped me handle a lot of things I couldn't even imagine in the 1930s.

The Depression touched every adult in Phoebus, but to us kids, it was hardly a factor. Mathew Carli's family ran the grocery store down the block, and like me, he was a baseball nut and we spent much of our childhood rounding up all the guys and playing. My father bought me a bat and ball, and eventually a catcher's mitt and mask. By the time we moved on to high school, I was pretty good. My right hand didn't bother me. But two of my heroes, Babe Ruth and Lou Gehrig, were lefties, so I batted left, threw right.

Washington, D.C., and Griffith Stadium were an eight-hour car ride away. A few times each season, we went to games, usually to see the Yankees play the Senators. My mother was a big Yankee fan, and my father loved the game almost as much. I desperately wanted to be a professional ballplayer, and the trips only made me want it more. We always got to the park early, and I usually brought a baseball along in hopes of getting an autograph. It didn't happen often. But twice was enough. The first time was when Babe Ruth started talking to kids along the third-base foul line. I ran down, and sure enough, he signed my ball.

The next season, 1938, I brought the same ball, carefully wrapped, to the park again. Lou Gehrig was taking batting practice and I couldn't resist. I jumped over the guard rail and shoved the ball to him right there at home plate. "Will you sign it please?"

Gehrig laughed. "Sure, kid. Gimme a pen."

My heart stopped. I'd forgotten to bring one. "Just a minute," I yelled, and ran back to the stands. My father held out his pen and I grabbed it, just hoping to get back to Gehrig before the ushers grabbed me and threw me out. I made it. Gehrig was still chuckling at my discomfort when he signed the same ball that Babe Ruth had signed a year earlier. Most of our childhood treasures disappear along the way. Not this one. I still have that ball.

Hampton High didn't have a baseball team. But we got up our own and convinced the Newport News newspaper to sponsor a lower-peninsula league in which we challenged other teams in the area. We were motivated and we got the kind of unstructured hands-on learning experiences that helped all of us as we grew toward adulthood.

We did, however, have pretty good teachers. The kids from Phoebus were the ones from the wrong side of the tracks. There were no school buses and we'd each get fifteen cents a day from our parents for the streetcar. We'd hitchhike to school and use the money for extra treats or drinks at lunch. When we couldn't catch a hitch, we'd chip in for the fifty-cent cab ride to school. The Phoebus kids arriving in a taxi, usually late, gave us an added reputation as a strange bunch. But the teachers liked us. I was lucky, too, that somehow I always had a mentor to guide me. My mother showed me that human relationships are as important as acquiring knowledge, and when a teacher took interest in me, I responded by trying harder.

Our math teacher at Hampton High was Mrs. Marguerite Stevens. She looked out for the Phoebus kids and she pushed me to succeed. "You can do it," she'd tell me, "just think about what you want in your life and you can do it." She said it so often that I believed her. Our physics teacher was Stephen Acierno, and he's the guy who got me started thinking about engineering. "When you really appreciate these things," he'd say, "you won't see a car going down the road, you'll see the pistons and other moving parts working together to make it go." He was right. Whether I'm looking at a car or a spacecraft, I always see what's under the skin as clearly as if the skin were transparent.

All of the Phoebus kids worked. We had to. Work of any kind was considered a noble calling in Phoebus. I tried unloading freight cars for a while, then clerking in a grocery store, until my high school French teacher got me a job in her family's steam-laundry business. Before long, I was handling the business finances and opening up each morning. After school, I'd work until closing, and on Saturdays, I was there from 7 A.M. to 10 P.M. At just over a buck an hour, I was making good money for a high school kid. I learned about business, but I also learned that this wasn't the life I wanted for myself. There was more to be had out there, and I wanted to be part of it. I just wasn't sure what "it" was.

I'd learned to play trumpet in the Phoebus band, but figured I had enough going on in my life and didn't try out for the Hampton High band. My father was a "joiner," a vestryman, treasurer, and secretary in his Episcopal church, but he was happiest in the American Legion Post No. 36. When it formed a drum-and-bugle corps, with the dream of competing in the state, regional, and even national championships, he got excited. "Join up," he said, "it'll be good for you." So I did and it was. Suddenly I was exposed to military-like training, precise marching, following orders, wearing a uniform, and being part of a bigger team. I liked it. Before long, we were getting invited to march in various parades and celebrations around Virginia. We even made it to the state drum-and-bugle finals, but our corps came in second. I did a little better on my own; I was the state champion bugler. I came away with a lot of lessons; discipline in an organization had to be most important. I made friends and acquaintances, including some Hampton kids who usually shunned the Phoebus crowd, and I'd run into them and share memories, long into the future and in places or circumstances that none of us could have dreamed.

I met someone else there, too. Betty Anne Turnbull was in my homeroom. She was petite and smart and, I thought, downright gorgeous. For her part, she thought I was loud and brash, just another kid from that awful Phoebus. She was right, too. But that didn't stop me from asking her for a date. When she said yes, I was floored. We went to a movie and that got us started. Betty Anne and I would date, then not date, and be on and off for years. Eventually we'd settle the questions about our relationship. But not just yet.

I knew that I'd go to college. I just didn't know where. Or what I should study. Duke University looked awfully good. We'd stop there on the way to visit my aunt, my mother's sister, in Lenoir, North Carolina, and I was awed by the campus architecture and the school's reputation. Then in the winter of 1940, I checked on the tuition, and that ended my thoughts of going to Duke.

Virginia has many fine schools. One that didn't appeal to me at all was in Blacksburg, so far west in the Virginia hill country that even Roanoke was well east of it. And to Tidewater people, Roanoke seemed like the western frontier. After Roanoke came the wilderness. The school was Virginia Polytechnic Institute, VPI. It was a full-time military school with a good reputation for turning out engineers, scientists, and Army officers. My cousin August had gone there, and the stories he brought back about the first-year "rat" system spread terror among his listeners. VPI, it seemed, was a place for the crazies, the gung ho, or the poor kids who couldn't get into any other school.

I saw VPI for the first time in the spring of 1940 when I was the lower-peninsula representative to Boys' State. We were on campus for a week, learning about state and federal government, listening to speeches by area politicians, and getting some insight into public service in general. We were housed in the lower quadrangle, in buildings of gray fieldstone surrounded by a campus that was every bit as beautiful as Duke's. The school wasn't in full session, but the students who were on campus seemed normal. Nobody looked stressed or distressed. Just the opposite. Every student I saw looked disciplined and motivated, even satisfied. There were a few girls, too. VPI wasn't all military; it had agriculture, business, and engineering schools, plus a home economics department. About two hundred girls were on campus, staying in a residence nicknamed "the skirt barn." By midweek of Boys' State, I felt comfortable, even at home. In my free time, I read the bulletin boards and wandered through the library. I could see into classrooms, and if there was a military cast to some of the classes, there also was dedication. Students were listening and responding and asking questions. Before I went home to Phoebus, I knew. Virginia Poly was the place for me.

Some of my friends the following year applied to the University of Virginia, to Virginia Military Institute, even to the apprentice school at Newport News Shipyard. I sent my application to VPI in Blacksburg. I was accepted for the fall semester. The wars in Europe and the Far East were already two years old. We all had some concerns. But I was seventeen, the wars were a long way away, and we weren't involved. I set out to get my education in some kind of engineering.

It was September 1941.

Rat Year was intended to put stress on us. It wasn't as bad as Cousin Gus had made out, but it was bad enough. Rats learned to march (I already knew that from my drum-and-bugle days), to line up silently and walk with measured steps to and from the mess hall for every meal, and to accept confinement to our dormitory rooms after 7:30 P.M. The exceptions were those "crap" meetings three nights a week.

There was another four-letter word for those meetings, and we caught lots of it. The sophomores, still smarting from their own crap meetings of the year before, hazed us unmercifully. They yelled, they screamed, they demanded instant and carefully worded answers to questions about the military, about the school, about our personal lives. We were in full dress uniform, and they had us on the floor doing push-ups, or standing at attention and marching in place. It wasn't easy to take a sophomore screaming into my face and telling me I was nothing but a dumb rat who didn't deserve to wear the uniform. In the first few weeks, a lot of the twelve-hundred-member rat class had enough and resigned. I stayed. It didn't take long to realize that survival meant coping. The rats who flinched or talked back or broke only got it worse. I took what they gave me, and before long the hardest part of a crap meeting was staying at attention while a sophomore yelled at some other rat. Those who survived developed a strong camaraderie, supported each other, and came together as a unit. And that, of course, was the purpose of it all.

At Hampton High, I was in the academic top twenty. At Virginia Tech, I was close to being the dumbest kid in the class. At the end of the Depression, Virginia's education system ranked near the national bottom. I didn't know that until I got to Tech and was put into first-quarter booster courses in English, math, and science. I had enough background from high school to get through them and to realize that I still had a hell of a lot to learn. In that first quarter, I made a 1.78 grade point average. On the 3 scale, call it a B-minus. Not bad, but not good enough.

While I was wrestling with college and was adapting to this new military-style life, my parents were wrestling with their own problems. My father's mental condition, more severe than simple depression, came howling back as full-blown schizophrenia. They called it dementia praecox, and the only known treatment was electric shock therapy. It wasn't helping. I knew only a little of the troubles at home. My memories were of his love for his cars, always a Ford or a Studebaker; of the Sunday drives we took to explore Virginia or Washington, D.C.; and of the firemen or Legion conventions he took me to. The conventions were a mixed experience. We didn't always have enough money for our own hotel room, so we would stay in a room with one of his friends. As I grew older, I saw that my father was as embarrassed by the necessity as I was.

My mother told me just enough to keep me informed, and not enough to cause me much worry. I remember the trip from Phoebus to Blacksburg when I enrolled as a rat. There was something different about my father, sort of a detachment. But I was so overcome by the excitement of college that I didn't dwell on it. I finally figured out that my mother was protecting me. She understood that my future was out there somewhere, and that I shouldn't be anchored to my father's incurable problem. So while she worried over my father, I was free to let Virginia Tech take over my life.

I was surrounded by smart kids, especially the upperclassmen, and even though I wanted to be an engineer, it didn't take long to figure out that I had no idea what an engineer did. My nominal major was mechanical engineering. But in the freshman and sophomore years, all engineers took the same set of classes.

I wasn't alone in the engineering wilderness, though. So several of us signed up for a second-quarter course called Introduction to Engineering. Maybe an intro course would have some clues that would help clear things up. It did. Each engineering department spent two or three hours, once a week, telling us about their field, about what this kind of engineer or that kind of engineer did in the real world. They outlined the mandatory courses and the electives, and they told us something critically important: "It's not fatal to change your mind as you discover more about your own interests."

It's not fatal to change your mind. How many times would that lesson come back during the next four decades as I confronted problems and searched for solutions that were beyond anything I had dreamed at Virginia Tech? I was seventeen when they opened that door for me. It became part of my core philosophy, part of the package that is me.

For the time being, I stuck with mechanical engineering. I thought that I wanted to design automobile engines. But over the coming quarters, I picked electives that would give a deeper look at other engineering fields, too. One of them, in my sophomore year, was aerodynamics.

On December 7, 1941, my life, along with every other American's life, changed. Japan attacked Pearl Harbor and America was at war. The Corps of Cadets formed on the quadrangle and marched to the mess hall singing patriotic songs and chanting slogans about defeating Tokyo. My voice was as loud as any of them. We knew that our world had changed while the bombs were falling and the *Arizona* was sinking.

I went home for Christmas and my world changed again. Betty Anne Turnbull had come to VPI for a prom, but our romance was fading. Distance makes the heart grow fonder of someone else, and we had drifted apart. So I had a date with Betty Verell during the holidays. Wendell Fuller was driving my father's car, with Betty and me also in the front seat. Wendell was a great football player at Hampton High, but a bit of a troublemaker. He'd put firecrackers in the air vents, with a lit cigarette near the fuses. Then he'd sit at his desk looking innocent until the thing went off, scaring the hell out of everybody. And he wasn't an experienced driver. Near Williamsburg, he was going 80 mph over a rise that dropped down to a bridge and we rolled three times through a ditch and into a field. Betty was thrown out the door. I went through the windshield. Wendell was still behind the wheel, dazed and calling to me. I had cuts and abrasions and a concussion, and I could hear him, but I couldn't make my voice work or even move. *I guess I'm dead*, I thought. Then a doctor who just happened by was giving me first aid. Betty Verell had a broken collarbone and Wendell was mostly unhurt. In my dazed state, I remembered that we had a bottle of whiskey in the glove compartment. Somehow I got to it and tossed it away.

The doctor got us to a hospital. Betty's broken collarbone was fairly serious, but my injuries amounted to bruises and a lot of cuts. While a nurse was tending me, a state trooper walked in holding the whiskey bottle. "Nice try, kid," he said, "but you didn't have to throw it away. The seal isn't broken, so I know you guys weren't drinking." He walked out with the bottle and I suspect that it went home with him that night.

I got home the next day, bandaged and bruised. My mother fainted when she saw me. The car didn't make it home at all. It was totaled. (And Wendell Fuller didn't make it home from World War II. He enlisted soon after in the Army Air Corps and was killed in England in a B-25 training crash.)

Maybe my accident contributed to my father's downward spiral. My father was getting worse, suffering hallucinations and delusions. For the first time, I was hit by the changes in him. It was obvious that he needed hospitalization. But he was still home when I went back to school in January. I stayed in Blacksburg a month, then got a weekend pass to go home again before my father went into the VA hospital. It wasn't a happy homecoming, and my sore throat only made me feel worse. By the time I hitchhiked back to school, I was hot and flushed and felt terrible. It was scarlet fever. Virginia Tech was in the beginning of an epidemic. I was in a quarantine ward for three weeks, with a fever spiking at 105, and enduring the kind of hallucinations that I thought, in my rational moments, had to be similar to what my father was suffering. I remember that small glimmer of understanding and how I wondered why I could be cured and he couldn't. The one thought I never had was about my own future state. We knew little about inherited illnesses, but we seldom saw mental problems passed from generation to generation. My hallucinations were strictly from scarlet fever. My father's were something else entirely. By the time I was released from quarantine at Virginia Tech, he'd entered the VA hospital at Phoebus.

He was in a ward with other demented souls, and he'd improve enough to come home for a short visit, then sink even deeper into his dementia. He'd try to be "up," but he couldn't make it last. I was home for one of these visits and it was such a bad scene—he was berating himself uncontrollably—that I wanted only to get away. Later I worried that I should have tried to be with him more, to somehow help him. But even the VA hospital couldn't do that. They transferred him to Roanoke.

I was still shaky and weak from the aftereffects of scarlet fever when I talked to my course adviser. I'd missed too much, he said. Instead of sticking it out and failing the finals, I should go home and plan on taking the second quarter over again from the beginning. I was a few days short of my eighteenth birthday, the war news wasn't good and I'd probably get drafted, the emotional and financial stresses at home were taking a big toll on my mother, and I'd just lost a full quarter of my rat year at Virginia Tech. Suddenly I was at one of life's major crossroads. It was enough to give me a depression of my own.

* * *

The Corps of Cadets used the artillery table of organization. We weren't assigned to companies, like the infantry or armor, but to batteries. They were essentially the same thing. Two of the upperclassmen in my battery were a shirttail cousin, Nelson Fuller, and the cocaptain of the football team, Elmer Wilson. It didn't take long before they were knocking on the door in Phoebus to make sure I'd be back in school at the start of the spring quarter. "Listen to them," my mother said. "You can't sit around and mope. You have to go back to school." When I worried about the financial burdens of having my father hospitalized and me away at college, she dismissed it. Whatever it took, she said, the money could be found. But still, I hesitated. My books alone would cost almost $100 a year.

Finally Fuller and Wilson mentioned baseball and I perked right up. They knew that I wanted to try out for the freshman team. "You can make the team," they told me. "But not if you stay in Phoebus." It's good to have family and friends who care. I went back to Blacksburg, got more help from my adviser in taking courses to finish my freshman year, and then I headed for the baseball diamond. I was five-eight and weighed 128 at the start of the school year. Now I'd lost a lot of weight in my bout with scarlet fever, was almost emaciated, but I was motivated. And it didn't hurt that I was a pretty good catcher. I made the team. From that moment on, I also was seated at the athletes' training table in the mess hell. The food was only a little different from the regular mess, but it was enough to make me thrive. I gained weight during the spring quarter, all the way to a new high of 142. I also felt terrific about myself and about Virginia Tech. We played a six-game schedule in the spring of 1942 and won them all. But while we were winning on the baseball field, our country was trying to figure out how to win a war.

In the original scheme of things, I looked forward to being a member of the class of '45. Then in the fall of 1942, the Army called up our entire class of '43. They were two quarters short of graduating and left almost overnight for officer candidate school. When they got their gold second-lieutenant bars, they weren't quite "ninety-day wonders," because of their Corps experience. But they were cannon fodder for the coming battles, and they distinguished themselves and our school in the way they led and fought and died.

The class of '42 left for the service immediately after graduating. In a matter of months, they'd be the young second lieutenants leading squads of still younger men, and fighting and dying in North Africa and the Pacific.

At the same time, Tech canceled summer vacation. The school went to a twelve-month schedule to expedite the training of engineers, who were suddenly in short supply. I looked ahead and wondered about my future. I was only eighteen and not likely to be drafted immediately. But suddenly there were other opportunities and other obligations. By summer's end, the Army had assigned a large contingent of new students to Tech under its Army Special Training Program, ASTP. And other services were offering new programs, particularly in aviation, to entice bright young men into uniform. Our Corps of Cadets was still important, but it was being whittled away.

Instructors were disappearing, too, many of them called up or enlisting or reactivating their old Reserve commissions. By Thanksgiving of 1942, the Virginia Tech campus was being depleted of men. Those of us left behind were torn between our duty to country and or desire to finish school. We argued our options inside our batteries, and many of us made the same choice. We would apply to become aviation cadets in the new Navy V-7 and V-12 programs. *Flying for the Navy!* It had a glamorous ring to it. The first step was to get a physical examination.

We reported in a group to the medical detachment in Richmond, about 150 miles away, in a state of high excitement. This was commitment taken to a new level for an eighteen-year-old, and I could hardly wait to get on with it. A few hours later, we felt like cattle. Hundreds of young men were in line that day, draftees, volunteers, kids like us hoping to become pilots. I had never seen such a mass of seminaked bodies, or such a mess. The corpsmen drew blood— and some strong-looking boys passed out when the needle hit. They took urine specimens, checked blood pressures, took our temperatures. Doctors looked into every orifice, and the only way to survive the embarrassment was to remember that we weren't alone. Finally, near the end of the day, we were formed into groups of twenty in front of a stern-looking Marine colonel with the medical caduceus on his collar. He put us through a series of exercises, watching each of us carefully. I thought I was doing fine until he ordered, "Halt!"

"What's wrong with your hand, boy?" he barked at me.

"Nothing."

He stared at me for a long moment, then put us back to doing exercises. Finally it was over. He turned and marched away. A few minutes later, a corpsman tapped me on the shoulder. "The office down the hall," he said. "Go in there and wait."

It seemed like forever. Finally the Marine colonel showed up. His demeanor had changed. He was cordial and kindly as he questioned me about my right hand. He examined it while I told him about the fire when I was three, about the skin grafts, and to make sure that he understood that the hand was just a regular old hand, about being the catcher on the Virginia Tech baseball team. My hand was fine, I insisted. It had never stopped me from doing anything.

"No, it isn't fine," he said, not in a harsh way. "And it's going to stop you from being a Navy pilot. In fact, it'll probably keep you from being drafted. I'm recommending a classification of 1A-L. That's 'limited service.' Your best bet, young man, is to go back to Blacksburg and put everything you've got into studying engineering. The country needs engineers, too, not just pilots or soldiers."

He shook my hand, my right hand, wished me luck, and sent me home. I'd be lying if I didn't admit to some held-back tears. It was the first time in years that I'd even thought about my hand. Now it was a determining factor in where my life would go. I made a fist, as close as I could come to a fist, and went back to school.

One of the joys of being a sophomore is not being a rat. My rat year was still a vivid memory, and though I had to take on the role as a hazer of the new rats, I kept it toned down. There were enough other sophomores who needed to prove themselves to be the loudest or the meanest. I had other things on my mind.

I plunged into engineering with a new intensity. I had to take such subjects as physics, statics, dynamics, operational calculus, and thermodynamics, and I was doing well. But that elective course in basic aerodynamics grabbed my attention. The department head, Leon Seltzer, was excited about aviation; it had only been forty years since the Wright brothers flew and aviation advances seemed to be occurring almost every day. The Aero Department instructors all shared the same enthusiasm, so different from the plodding faculty in Mechanical Engineering. It was infectious. I finished the last quarter of my sophomore year in early 1943 and declared aeronautical engineering as my formal major. I was never one of those kids who knew from the beginning what he wanted to do in life. But what I lacked in brilliance, I made up for in luck, and I found my niche.

When I was a twenty-year-old senior, I was elected president of the Corps of Cadets, the youngest cadet officer ever to hold that position. I not only had more administrative duties and the heavy class load of my senior year, but I was now responsible for bringing charges under the honor code.

Most of the offenses called for probation. But now and then a serious of-
fense such as cheating came before us. When the evidence was solid, we
offered the offenders a chance to resign. They always did. The biggest problem I faced was that the code considered you as guilty for not reporting a
violation as it did the violator himself. It was difficult to administer, but I
learned to walk that line and to live with it.

From my secure vantage point in the twenty-first century, I can see what
the Corps of Cadets, and particularly that senior year, did for me. It gave
me my first, and almost only, training in leadership. I've seen too often in
the business world where professional skill leads to a management promotion, and the new manager is woefully unprepared to lead. He's suddenly
responsible for production, or overseeing the work of many others, or even
training them in their jobs. But he hasn't had the training or the experience
himself. I know beyond doubt that my own leadership skills were honed
by the direction, example, and practices of the Corps of Cadets. Experience is a great teacher and the Corps gave me those experiences when I was
still young and impressionable. To this day, I encourage young men to recognize and take advantage of this aspect of military training. Some won't
be suited for it. But many others will, and the military experience will become an invaluable part of their core person through whatever career or
profession they follow. Times have indeed changed and so has the nature
of military training at land-grant universities such as Virginia Tech or
Texas A&M, and at old-line military schools such as The Citadel and Virginia Military Institute. Change is a good thing, but so are the fundamental lessons of military life. Our nation will lose much if it misplaces those
lessons.

By the fall of 1944, it was time to graduate. I would get one of the first
degrees in aeronautical engineering granted by VPI, and my B+ average
was good, if not spectacular. I was a kid, but I felt ready to be a man. Somehow I knew that the jewel of a college education wasn't in the knowledge
I'd gained, but in the process I'd gone through. We all were lucky on that
campus. The process was delivered by professors who understood the difference between teaching and learning. We learned the formulas and the
theories. But we learned, too, that it all can become obsolete overnight and
that we have to be ready to encounter new technologies and to discover
new truths. We had teachers, particularly in aeronautics, who knew how to
measure us, how to quiz us and stroke us and confront us and sometimes
even to bully us into thinking through a problem. That kind of teaching is
a real art. That kind of education—the Corps, the dance weekends, base-

ball, and learning not just facts, but learning how to learn—was a jewel that I hold forever precious. Virginia Tech, the school I hadn't even wanted to consider, took this boy from Phoebus and made him into the man I would become.

The recruiters came to campus. Teams from Chance Vought, Curtis Wright, North American, Douglas, Martin, and all the other big airplane companies looked at our records, interviewed us, enticed us with stories about the airplanes they were building in the war effort and about the job opportunities that seem endless in 1944. So did the National Advisory Committee for Aeronautics (NACA), based on Langley Field, only seven miles from my house in Phoebus. NACA was the government's research arm in aeronautics. They flew experimental planes there and tried new designs for wings and fuselages. With the war raging on two fronts, Langley people were finding and fixing the flaws in the bombers built by Martin or Douglas, or the fighter planes built by North American or Chance Vought. It was an exciting and important place to be during World War II. Orville Wright and Charles Lindbergh were members of the committee that oversaw the research and the flying at Langley. We'd studied some of the NACA work and read the technical papers they'd produced. One of the Langley men, Robert R. Gilruth, was becoming a legend in aeronautics. If his papers read like dry sand, they were filled with revelations and discoveries about the art and science of flying.

I didn't want to work at Langley. I thought it was too close to home. So I applied to Chance Vought. Then in an afterthought, I sent an application to NACA anyway. I already knew the value of having a backup plan.

There were no graduation ceremonies at VPI in 1944. This school that had encased us in a cocoon and put us through a metamorphosis was casting us loose. I had a heavy heart when I went to the registrar's office in mid-December to pick up my diploma. Then I walked back to my battery, took off my uniform for the last time, and walked out to face the real world. I had a job-offer letter from Chance Vought in my pocket. And in a little over two months, I'd be twenty-one years old.

2

The Boy Becomes a Man

For a young fellow who had never traveled farther from Phoebus than Bristol, Virginia, or Washington, D.C., Connecticut seemed to be a long way to go for a job. But Chance Vought wanted me and was giving me a chance to work on some of the nation's best warplanes in the first month of 1945.

The big news over Christmas was the Battle of the Bulge. My cousin Allison Kraft had left Virginia Tech with the class of '44 to become an infantry officer. He was sent to Belgium in late 1944 and was in the battle. My uncle August and I kept track of his unit, and then the telegram came. Allison was hit by shrapnel near Bastogne, was wounded in both legs, and was in an Army hospital. He'd survived, his recovery would be long and slow, and there was both relief and pride in knowing how bravely he'd served his country. Now with Chance Vought waiting for me, I could finally make my own contribution to winning World War II. I could hardly wait.

I took the train to New York City, arriving at Pennsylvania Station and walking crosstown on a Sunday afternoon in lightly falling snow to Grand Central Station for the train to Bridgeport. My suitcase was light with a few new clothes just purchased to replace the uniforms I'd been wearing for three years. My mother was right: New York was a big city. I promised myself that I'd be back to explore it at the first opportunity. It turned out to be a while.

In Bridgeport, the snow had become a blizzard. I spent the rest of the day in the hotel room furnished by Chance Vought. The next morning, I reported for work and discovered bureaucracy.

"Birth certificate?" The clerk was officious.

"I don't have it with me. Nobody told me."

"Can't enter the plant without one. Security regulations."

"What do I do now?"

It took a round of telephone calls and then a telegram sent to the right office in Richmond to request a copy of that vital document. My future rested in the hands of clerks, and all I could do was wait. The certificate should have arrived by special delivery on Wednesday. It didn't. "Come back Friday," the clerk said.

But it still wasn't there. I'd moved to migrant-worker quarters, was bored with the nothingness of waiting, and was getting a little angry.

"I'd like to see the men who interviewed me at VPI," I told the clerk. "Maybe they can help."

"Nope, can't do that."

"Why not?"

"Security. They can't talk to you until you complete the paperwork."

"But I need their help to—"

"Sorry. Regulations."

I stalked back to my quarters with dark thoughts about futility and stupid rules and what-do-I-do-now? By the time I got there, I knew the answer. I called the employment office at Langley Field, the National Advisory Committee for Aeronautics, and asked if their job offer was still valid. I was put through to a manager and heard the magic words: "It sure is! How soon can you be here?"

"Next week," I said.

"Make it Thursday. See you then."

It took two minutes to pack my suitcase, another five to write a letter to Chance Vought explaining my decision. If they still wanted me, they could contact me at Langley. I never heard from them again.

So there I was, a college graduate with a degree in aeronautical engineering, still six weeks short of being old enough to vote or drink legally, and living back in the house where I was born with my mother. That was, in truth, a pretty good deal. She had always been my strongest supporter and my closest confidant. Suddenly I realized how much she meant to me. At the same time, I worried about my father in the hospital in Roanoke. Today's drugs for treating mental disorders didn't exist, and he was living a life of hell among the insane. I will always regret that I couldn't help him. What I couldn't see at the time, and even now only vaguely understand, is what I drew from him. His absence as a strong figure in my life forced me to find strength in myself, and maybe to be more open to guidance and advice from the strong mentors I was so lucky to encounter. If this much is true, and the insights come late in life, I owe my father more than I suspected. But like all father-son relationships, there were differences between us. My father was never a leader. Yet even as a boy, before I had any understanding about such things, I wanted to lead and I became good at it. I don't know what my father wanted. If he had dreams, he never discussed them. For my part, our relationship was founded on duty as much as love. My duty was to do what I could. But it was beyond me to do more than visit him now and then and hope that he drew something from a son's love. Whatever the legacy of our relationship, I know that the idea of blaming my father for anything never occurred. He was a victim of a tragic disease.

Langley Field was just seven miles from Phoebus. When I passed through the main gate, I entered the world of real aviation, where some of the world's best wind tunnels were used to find airplane problems and to discover new facts about flying, where test pilots and planes flew through the sky, and where some of the country's top aeronautical experts came to work every day. I didn't know what I'd be doing. I just knew that it felt important. Mel Butler in personnel did me a real favor. After a few minutes of talk, he saw that doing wind-tunnel work or statistical analysis weren't my strengths. He sent me to the Flight Research Division (FRD), where they actually flew planes and found ways to make them better. The boss over there was Floyd Thompson, and Robert R. Gilruth was chief of research. I'd read some of Gilruth's technical papers in college and knew that he had a good reputation as a scientist/engineer and as a research manager. What I didn't know was that Gilruth's reputation had been understated at VPI. He wasn't just good. He was held in awe. He'd found and fixed some flaws in the British Spitfire early in the war, and the Brits had assigned a team to him during the rest of World War II. Every time he made a discovery or developed an innovation, it was carried back to Britain and found its way into the warplanes being built and flown over there. On the U.S. side, he worked on American bombers and fighters, always making them better, and his skills were becoming legendary.

The NACA was founded in 1915 and was the mecca for aeronautical science and engineering in the United States, and in fact, the world. If there was an aviation advance, it probably had its roots at Langley Field—everything from the shape of wings to the design of flight controls to the master book on the flying qualities of airplanes. The latter was a Gilruth innovation and with the in-flight help of chief test pilot Mel Gough became the bible of aircraft designers. Gilruth had arrived at Langley in 1937, fresh out of the University of Minnesota. He quickly became a national treasure in time of war.

I was assigned to FRD's Stability and Control Branch. That put me far below the line of sight of either Floyd Thompson or Bob Gilruth, at least for a while. My base pay was $2,000 a year, and with overtime for Saturday work, the war years norm, I made $2,300. My boss was William Hewitt Phillips, and he, too, was a jewel, one of the few true geniuses I've known in a life of working with brilliant people. His people called him Hewitt. He made me his protégé, shaped my engineering abilities, and taught me this amazing profession. He was my postgraduate school. When I finally figured out what he'd done for me, I decided that I'd pass along the favor whenever I could.

My first assignment was to help analyze airplane structures. That meant doing such things as calculating the loads that went into the main structure of the airplane when changes were made to the flight controls or other components. If we changed a flap, for instance, the forces going into the main spar inside the wing changed. That kind of information is important because an overloaded spar can break, or in the long term, metal fatigue can have the same result. I was fresh out of college, where the newest techniques were taught. So it didn't take long before some of the older engineers, guys twenty-five to thirty-five, were asking for my help. The first of them to hand me an assignment was Walter C. Williams, only a few years my senior. It was the beginning of a working relationship that would last for two decades and take us both to strange places and undreamt-of heights.

Hewitt Phillips introduced me to Gilruth's "flying qualities" and to the flight-test business. Flying qualities was a comprehensive set of criteria to judge an airplane's flight characteristics. Gilruth developed the yardsticks by riding along with Mel Gough on test flights, sometimes with both of them crammed into an airplane's single seat, and recording the various forces, stresses, and other things that made the plane easy or difficult to fly. For instance, Gilruth figured out that a fighter pilot should put so many pounds of force on his control stick to get the right "feel" in a tight turn, and to maneuver safely. The force varied with the tightness of the turn, and Gilruth calculated it perfectly. That was just one of many flying qualities that Gilruth discovered and documented through years of taking to the air with Mel Gough. By the early forties, the Civil Aeronautics Agency and the military services adopted Gilruth's flying qualities as the standards to be used in all aircraft design. With only a few changes to accommodate modern high-performance airplanes, Gilruth's work is still being used at the beginning of the twenty-first century.

I knew this only vaguely when Walt Williams dropped the assignment on my drawing table. He was measuring the flying qualities of the Navy F6F fighter-bomber and told me to make an accurate model of the aileron's cross-section. An aileron is that piece on the outer trailing edge of a wing that moves up and down when a pilot moves his stick left or right. This causes the airplane to bank left or right, but how much depends on speed and altitude, and the changes don't follow a smooth curve on a graph. The F6F aileron had a fabric skin over metal ribs. Williams suspected that it was being distorted during some maneuvers and that Navy pilots were being put in danger. But to prove it, he needed to know the aileron's exact shape at the beginning of a maneuver. My job was to make the perfect model for his tests, and it helped to prove that Williams was right. It only took minor design changes to fix the problem.

Before long, I was loaded with warplane assignments and loving it. In those first few months of 1945, I found myself working on versions of the Army's P-47, on the Navy SB2-C torpedo bomber, and the Navy F6F. We had all of those planes in the Langley squadron, usually getting one of the first to come off the assembly lines. This wasn't classroom theory. This was real. With an airplane's operating manual in hand, I'd climb over and into these new machines, figuring out what instruments we'd need to test them on the ground and in flight, where to attach the instruments, how to calibrate them before the test, then collect the readings after the test. It was exciting, and a day's work didn't mean going home at 4:30 or 5:00 P.M. We all stayed into the evening almost every day.

Working late put a crimp in my social life. My old girlfriend, Betty Anne Turnbull, had worked at a local bank since high school, and I assumed she'd be awed by the return of this handsome young engineer with a degree from VPI and a job at the most prestigious aviation center in the world. She wasn't. She was dating some old high school boyfriends, then met an Army Air Force officer at a USO dance. There was no place for me, so I looked around the Flight Research Division at Langley and spotted several attractive and eligible ladies. Finding the time to date them was a problem, but I managed to ask some of them out once or twice a week. One was a beautiful and blond mathematician from Lynchburg, Virginia, named Katherine. I called her Kakki. She'd graduated from Randolph-Macon Woman's College, loved aviation, and for a while I even convinced myself that she loved me. I didn't count on the air of superiority that Randolph-Macon and Lynchburg high society bred into their chosen few.

On the job, I met all the pilots and was on a first-name basis from the beginning. Pilots don't put up with much ceremony or protocol. What counted was getting the job done. I'd help prepare the plans and procedures for each flight test, then go over them with the pilots. They were smart and experienced. When they had suggestions about how to do something better, or why some maneuver or test should be added to a flight, I listened. And so they were part of my ongoing education. Test pilots weren't just going through the motions; they had insights and skills that added immeasurably to the efforts of us ground-bound aeronautical engineers. I learned a kind of respect for test pilots that would serve me well almost every day of my career.

World War II was ending. I'd had only four months to help defeat Germany, but I knew that I'd made some small contributions. I also learned to detest the Nazi war machine, and not just in general. At Langley, we had names and faces to put on some of Hitler's key people. We heard stories from the British aeronautical team assigned to Langley, particularly about the German V-1 and V-2 rocket attacks on England. The havoc and indiscriminate destruction being caused by a German named Wernher von Braun focused our hatred on the Nazi rocket program. Von Braun was the engineering genius who developed those rockets, and if his technology had our curiosity and respect, we wished nothing but ill on the man himself. His rockets could not be aimed, except in the most general sense. At least, we told ourselves, our fighters and bombers went after specific targets such as factories and railheads. We managed to avoid thinking about some of the Allied firebomb raids that were every bit as indiscriminate as von Braun's rockets.

Still, we had a keen interest in what those German rocket men and their counterparts in aviation were doing. We saw classified reports on German experiments with swept-wing aircraft, jet engines, and liquid-fueled rockets. We heard that a German jet fighter had engaged Allied bombers and escaped without a scratch after shooting several planes down. On other subjects, we heard a mix of rumor and fact. We didn't have intimate details of von Braun's operation at Peenemünde along the Baltic coast, or the rocket factory he now ran under a mountain in central Germany, but we knew enough for it to generate strong emotions, bordering on loathing. Wernher von Braun was the name of the enemy. I wasn't alone in those feelings; everyone in the Flight Research Division felt the same way. Our memories are long. The day would come when many of FRD's people would have to deal directly with von Braun and his people, and the events of World War II would bitterly taint those relationships.

The war's end only added to our workload. Wars push technology to new horizons, and now we were moving into the era of jets and rockets. What we knew at the time wouldn't fill an engineer's notebook. Our job at NACA was to take the lead in researching what happens to airplanes as they approach Mach 1, the speed of sound. (Mach is a relative number. The name came from Ernst Mach, the Austrian physicist who pioneered the study of ballistics. Sound travels at different speeds, depending on the density and temperature of the air. So an airplane flying at Mach 1 close to the ground is going faster than an airplane flying Mach 1 at forty thousand feet.) A few World War II fighters such as the P-51 Mustang flew fast enough to give us a taste of the problems we faced. The faster an airplane flies, the more it compresses the air in front. Transonic speeds just before reaching Mach 1 were theoretically the most dangerous. But since nobody had flown at the speed of sound, we didn't know for sure.

What we did know was that at some point, the airplane would run into compressibility problems. Airplanes don't only fly through the air; they push some of that air on ahead, and it collides with, and compresses, the air that's already there. Airplanes fly because air flow goes faster over the top of a curved wing than it does under the bottom. The difference creates lift. But the faster a plane flies, the more compression there is. Theory said that when an airplane approached the speed of sound, the airflow would separate from the wings and other control surfaces. The plane would lose its lift and probably become unstable.

But we didn't have a mathematical formula that could predict exactly what would happen. In the past, the math came from wind tunnel tests, but now the tunnels were no help. At the very point in a tunnel—the throat—where you mounted a model, transonic air turned turbulent from bouncing off the tunnel walls. The tunnel airflow "choked" and the data were unusable. So we had no math, no test data, and all we really knew was this: Since it's airflow that provides lift and keeps a plane flying, transonic separation was dangerous and could be fatal.

Bob Gilruth was working on the problem in the war's last months. Gilruth's mind grasped concepts that ordinary engineers couldn't conceive, and he already had two programs under way. One of his best engineers, Charles Mathews, asked me to work on one of them. Every few weeks, we designed and built different five-hundred-pound models of various aerodynamic shapes, attached them like a bomb under the wing of a B-17 (later a B-29), and dropped them from thirty thousand to thirty-five thousand feet, near the Virginia shore. Instruments measured the forces on the wings and body as the model accelerated, and it would hit transonic, then supersonic speeds on the way down. But how to get our data? We could have recorded it on film inside the model (this was before the days of tape, or even wire, recorders), but when it hit the ground, everything would probably be destroyed.

Our solution was one of the first uses of radio telemetry. A continuous stream of measurements was radioed to a receiver on the ground. At the same time, we tracked the falling model with radar to get its speed. When the test was over, we put the telemetry and radar data together to get accurate readings of what happened when the model went transonic, then supersonic. The tests were a breakthrough in aerodynamic research and produced a catalog of transonic and supersonic information for the country's airplane designers. We didn't know it at the time, but those early experiments in telemetry would have far-reaching implications. Sending detailed data by radio became a key to future space programs, and then to collecting and transferring the most complex industrial and commercial information as well.

Gilruth had another team working on his second idea to get transonic data. It was nothing short of brilliant. The basic principle of lift is that air flows faster over the top of a wing than the bottom. But at a wing's maximum thickness, the airspeed is relatively constant over a short distance. Gilruth designed a glove about three feet wide and several inches thick along the entire wing chord so that with the airplane traveling at about Mach 0.75, the air flowing over about a two-foot length of the glove was at transonic speeds. Then if a model airplane or model wing was attached, he could find out how it acted in that dangerous regime. It worked. I did some of the grunt work in building the models and making sure that they were instrumented and attached to the Mustang properly. We ran tests for months, comparing the data to our findings during the drop test. But before the testing was well under way, Gilruth turned the project over to Harold Johnson and set to work on his third idea. I would have gone with him if he'd asked. He didn't and I stayed behind to work on drop tests and P-51 flights.

All of us in NACA were airplane people. But again, Gilruth's vision took him to a new place in our profession. With World War II over, supply depots were filled with excess munitions. Thousands of solid rockets, the kind fired from airplanes and from the Army's mobile rocket launchers, were to be destroyed. "Why not requisition some of those rockets and attach airplane models to their tips?" Gilruth asked. The sleek rockets accelerated quickly to supersonic speeds. Here was the best way yet to gather information on what happened to wings and fuselages in the unknown beyond Mach 1.

Gilruth quickly got permission to establish a new division with NACA. He was barely thirty-three, but was one of the most senior men in the Flight Research Division. Now he would have a division of his own. The question quickly arose about what to call it. Gilruth knew that it wouldn't be politic to mention rockets; Langley's heritage was airplanes, and rocket science was still an alien technology. So with a wry bow to his own background, and to the rest of NACA, too, the Pilotless Aircraft Research Division (PARD) was born. The men he took with him to form the nucleus of PARD would be among his inner circle, or close to it, for the rest of his career.

They got to work quickly and set up a rocket firing range at Wallops Island off the eastern shore of Virginia. Almost immediately, they discovered something that flew in the face of conventional wisdom. Engineers believed that the way to fly through transonic turbulence and to survive the forces of supersonic speed was to make the airplane stronger. Thick wings, the theorists said, were the answer. Gilruth and his team tried models with thick wings and models with thin wings. To everybody's surprise, thin wings sliced through the air with the least stress. This breakthrough led Bell Aircraft to redesign the wing on its X-1 rocket plane and opened the way for jet aircraft to rule the skies.

While Gilruth and his crack team combined their talents to write more new chapters in the history of rocket flight, my path in the postwar years moved in a different direction. I became a young flight-test engineer working for Mel Gough, the chief test pilot. He took over as chief of the Flight Research Division and set about guiding my career. Mel had been a test pilot himself for too many years; he knew all the tricks and suspected that his pilots weren't telling him everything. Because I had a close rapport with FRD's test pilots and they had seen me work and knew that I was there to keep them alive by making sure that their planes met reasonable safety standards, we reached an unspoken compromise. The pilots assumed that anything I reported to Mel would be technically honest, and that on non-technical subjects I'd keep my mouth shut. It was a workable system. Everybody understood that I was only doing my job to ensure the safety of our operation.

Mel never questioned me. Instead, he adopted me off work and on. I never thought of him as a father figure, but as a brilliant man from whom I could learn a great deal. He was a golf fanatic and we played often. As our friendship grew, Mel made me his aide, driving him to speaking engagements around the state. He was renowned at the podium, regaling audiences with stories of the early days of flight and his experiences as a test pilot. As we drove, he taught me volumes about aviation history, and his long discussions about the fine points of our profession made me a better engineer. In Mel Gough, I had a boss, a friend, and a private tutor. I made the most of it.

At the same time, I was courting Kakki. She took me home to meet her parents and suddenly the "boy from Phoebus" was mingling in Virginia high society. I liked it at first, but as I got to know her friends and spent more time in Lynchburg, it began to feel odd. More and more, I listened to people talking about superficial social activities that just didn't match up with the work I was doing at Langley, or with the life I'd grown up in. Sometimes at dinner or afternoon tea, I had to bite my tongue to hold back the urge to tell it like it is—or least, my view of how it is. More than anything, I thought that high society and its views were hilarious. But I was careful not to laugh out loud.

One Sunday afternoon, we were invited to a buffet at a "farm" outside of Lynchburg. This was not my idea of a farmhouse. It was a classic Virginia mansion, filled with elaborate antiques and overlooking the lush, rolling countryside. The buffet was superb, with more salads, meats, breads, and desserts than I'd ever seen assembled in one place. And through it all, with the lovely Kakki at my side, I had this desire, almost overwhelming, to pile peas on my knife and shovel them into my mouth. I looked around the lavish room and could imagine the reaction from those proper Virginians. That time, it was really hard to keep from laughing.

Maybe my growing disdain for the shallow life showed through. It wasn't long before Kakki took my hand on an evening date and told me she'd taken a job in Lynchburg. She was going home. That was that. I'd loved her, I thought, but I had my own concerns about our future. Now she was leaving and I didn't feel that bad. My strongest reaction was to conclude that it was her loss, not mine. A few months later, I ran into one of her friends. She was getting too serious about me, he said, and her parents ordered her to come home. An airplane engineer didn't fit their plans; she was expected to marry within her social class. I'm not sure it actually worked out that way, but then, my concept of social classes may be warped. Much later, I heard that she'd found a butcher with lots of money and lived happily ever after.

The X-1 was America's first rocket plane, a collaboration that began in 1946 between the Army, the Navy, and NACA. Bell Aircraft in Buffalo, New York, had the contract to build the plane, and Walt Williams landed the job of lead engineer for NACA. Hewitt Phillips assigned me to build a quarter-scale model of the plane that we'd use in drop tests from a B-29. This was early in the program, before the final design was locked in. The drop tests were supposed to tell us something about the X-1's stability and control at transonic speeds. We learned enough to give the X-1 designers some valuable input, and the test ranked among the most bizarre experiences of my young career

I spent weeks at my engineering design board, designing the model and getting it just right. Some of my toughest hours came in designing the plane's elevator—that horizontal control surface at the tail that deflects up or down to make a plane climb or dive—so that it could move during descent and give us information on the longitudinal (up and down) stability of the X-1. If the X-1's nose was going to bounce up and down, we needed to know about it. I made a few nonstandard additions, such as wood wedges on the wingtips, to act as deflected ailerons and thus make the descent trajectory predictable. Ailerons are the movable sections on the trailing edges of wings; they move in opposite directions to bank a plane left or right. I designed these particular "ailerons" to roll the model as it fell, so that the lift was always toward the centerline of the model's falling spiral. Then I had to set the elevator so that the model would dive from under the B-29 mother ship's wing and not slam into it. By the time the NACA machine shops finished building the model, it weighed about fifteen hundred pounds. This was no little kit airplane such as us kids used to hang by thread from our bedroom ceilings.

By now, I was a twenty-two-year-old veteran of drop tests. Whatever we were dropping, we didn't want it to come down in a populated area. So our aiming point was in the swamps off Poquoson, Virginia, and our tracking and telemetry station— really just a couple of trailers—was on the bank of the Back River about twenty miles from Langley Field. I usually manned the radar tracker, keeping the radar pointed precisely at the thing being dropped.

The big day finally came. The B-29 crossed back and forth over our drop zone, and I tracked a balloon with the radar to get wind speeds and directions. Finally when we were confident that we knew where the X-1 model would come down, the test conductor gave the B-29 the go-ahead to make the drop. The release was perfect. I tracked the rolling model through the radar's telescope and everything looked good. Then it happened.

At ten thousand feet, my perfectly designed model pulled out of its dive, leveled off, then went into a wings-level descent just as if a pilot were in control. It disappeared in the direction of Poquoson. Radar and telemetry readings showed that it had reached Mach 0.98. "Well," said one of my Flight Research friends who was watching the test, "at least you know that it can fly."

It could indeed. When we analyzed the data, it was clear that the X-1 design configuration was flyable and that the elevator was still effective in the transonic range. It was almost good enough to let a test pilot take it up. But before the full-size X-1 was built and shipped to the Flight Test Center at Muroc Dry Lake, California, some changes were made, including the thin wing recommended by Bob Gilruth. The next time we heard about the X-1, it was a flying machine and about to make history. And my fifteen-hundred-pound model? We never figured out why it decided to fly instead of fall. We did spot an impact crater far into the swamp. It was a dismal place, full of insects and snakes and other unpleasant creatures. If that's where my model ended its short test life, we let it lie.

My next job was a promotion—lead project engineer to conduct flying-qualities tests on the new P-51H. This latest Mustang had an advanced Rolls-Royce engine with more horsepower and a four-blade propeller. It also had the bubble canopy that has become synonymous with the P-51s that still fly air races today. The modifications gave the Mustang more speed, but the bigger engine meant lengthening the nacelle, the nose of the airplane. Airflow changes over the nacelle and the bubble canopy were giving pilots stability and control problems. The 51 had always been a handful to fly. Now it was getting dangerous, too. Our job was to determine the new flying qualities and then recommend fixes.

My test pilot was Jack Reeder. After a few easy familiarization flights, he was ready for the hard stuff. You don't determine all of an airplane's flying qualities by making nothing but routine flights. Some of the test flights involved putting the plane through its paces and looking for its limits. It didn't take long for Jack to run into trouble. In steep climbs and dives, something a fighter plane such as the P-51 had to do routinely in combat situations, the P-51H lost directional stability. A climbing or diving turn caused "rudder lock." With the wings tilted in a sideslip, the force Jack needed to move the rudder increased at first, then would decrease until the rudder was deflecting more than he wanted. It could become an out-of-control situation if he wasn't expecting forces to reverse themselves. At that point, all kinds of bad things could happen. With an inexperienced pilot at the controls, the plane might do a snap roll and go out of control. Or the vertical tail might rip off. Those aren't happy moments for the guy holding the stick.

I looked back at early P-51 models. The D had a similar problem that was solved by adding a dorsal fin—a rounded wedge running from the canopy to the tail. Our P-51H had the fin, but when I ran the numbers, it was obviously too small, given the plane's higher power and longer nose, to do the job. But a bigger dorsal wouldn't have much effect. After a lot of thinking, the answer came to me. The vertical tail needed to be taller, thus increasing the ratio between the tail's height and its chord, or width. I designed a tail extension, the shop produced the parts and we installed them.

Jack and I were in constant radio contact the day he took it up. After a few gentle maneuvers to get a feel for how the 51 handled with its bigger tail, the flight plan I'd written called for a full-power climb. Jack Reeder was a typical test pilot—laconic, given to short answers and short descriptions of what was happening while he was flying. He was given to the habit of negative reporting: If nothing was wrong or nothing unusual was happening, he kept his mouth shut. And like all test pilots then and now, he was completely calm in almost every circumstance.

I was learning to exercise that calmness myself, but it was a slow process, and on that day, I was feeling a bit nervous. This was the first time that my design of anything had been incorporated into a production airplane. Up there, climbing through twenty thousand feet, was my test pilot, operating according to my flight plan. If something went wrong . . .

I remember being frustrated. *What's happening up there?* I wanted to shout into the microphone. *Are you all right? Is the plane all right? What's happening?*

Finally, I couldn't take the silence. Keeping my voice as calm as I could—and feeling certain that every syllable betrayed my frazzled nerves—I asked Jack for a report. His answer sounded almost indifferent. "Nothing's wrong, Chris," I remember him saying. "Did everything in the plan and there's no rudder lock. She handles like a baby."

I sagged in my chair and felt relief flooding every muscle in my body. Then I started to grin. *It worked! Damn it, we did it!*

We recommended the tail extension to the Army Air Force and it went into production on the P-51H and later models. But it went further than that. Even into this new century, owners of rebuilt P-51Ds dig out our original report and are getting better performance by retrofitting the modification to their still-flying airplanes. I see tall-tail 51Ds all the time in photos and videos from air shows and air races.

While we were working on the 51, I let my social life slip a bit. Softball took up five or six nights a week, some of them coaching a junior team and some playing with the Fox Hill All Stars. We almost always played doubleheaders on Wednesdays and Saturdays, against some of the best teams up and down the East Coast, including the perennial world champs, the Clear Water Bombers. We held our own and won often enough to be respectable, but we never made it to that championship game.

I went on occasional dates, but after my experience with dating a high-society gal, I wasn't eager to get back into the fray. Then along came a happy circumstance. My cousin Allison had fully recovered from the wounds he'd suffered at the Battle of the Bulge, and he was dating Betty Anne Turnbull now and then. I saw her often enough to rekindle the feelings I'd had in high school. When I casually mentioned something to Allison, he did just what I hoped he'd do. "Ask her out," he said. "See what happens."

It took one evening. Our infatuation with each other picked up where it had fallen off five years earlier. She was petite, no more than five feet, with brunette hair, blue-green eyes, and a Miss America shape. She was, and is, the most beautiful woman I've ever seen. From the late summer of '47 on, we were inseparable. But I've always wondered about what Allison really felt. He and I were like brothers for the rest of his life, and he went from one beautiful woman to another, leaving a trail of broken hearts. He never did marry. He died in 1978 in a construction accident, and many of the tears at his funeral came from the ladies he'd loved.

Betty Anne solved the marriage question for me. She was the only woman I wanted, but the more I proposed, the more she'd smile and nod and say the words I hated to hear: "Not yet." She never knew her father. He died before she was one. Her mother brought up Betty Anne and her two brothers, and the older the kids got, the more dominant and possessive their mother got. I saw it in high school. The apron strings between Betty Anne and her mother were made of iron chain.

"Somehow," I told myself, "we have to break those chains, if it takes forever." It's a good thing that I was still young enough to have a poor concept of how long "forever" can be.

My new title was aeronautical research scientist, and I moved into 1948 feeling pretty good about myself and my work. When Hewitt Phillips dropped a new assignment in my lap involving automatic control systems, I was ready. We all thought that automation was inevitable in aviation's future. This was my chance to be part of it.

From the beginning, flying an airplane had been a manual job. Cables or chains ran from the control stick and the foot pedals to the ailerons, elevator, and rudder. On some bigger and more modern airplanes, the controls might have hydraulic augmentation. But it was still a muscle job requiring good reflexes, and sometimes great strength, from the pilot.

In turbulent air or gusty wind, an airplane would bounce up and down, tilt left and right, pitch forward nose down or nose up. Handling an airplane in those conditions was real work. A pilot could correct the plane's attitude, but he couldn't prevent the next jolt from doing it again. In crazy air conditions, the plane could be pitching, rolling, and bouncing all at the same time. If it was hard on the pilot, it was hell on the passengers. Anyone who's been airsick understands the torment passengers endure when an airplane spends much time in rough air. Under the worst conditions, the jolting motions could damage the airplane itself. Pieces of the tail or wing might break off. A lucky pilot got his plane safely on the ground. A lot of pilots weren't lucky.

My assignment from Phillips was called gust alleviation. Could I devise some system that would take much of the workload off the pilot and automatically keep a plane straight and level, or at least dampen out most of the bad effects of rough air? If I could, it would be a genuine breakthrough in aeronautics. Almost immediately, I found one of the first technical papers ever issued by NACA. It dealt with gusts and their probable effects on airplanes. So the problem had been understood even in the early days of powered flight, but the solution was still missing.

Boeing and Douglas Aircraft were trying systems that freed the ailerons on the wings to respond to the sudden lift changes caused by wind gusts. It helped a little with the up-down motions, but aggravated other motions such as pitching around the airplane's center of gravity. Instead of making things better for passengers, the changes to control systems were making things worse. The engineers had neglected the pitching movements—nose up, nose down—that came when the controls surfaces counteracted the gusts. The British tried a system that deflected the wing flaps to counteract lift changes in rough air. When they tried it over the African desert, the ride was so rough that the pilots refused to go up again. So they reversed the flap system. This time the airplane handled turbulence better, but the jolts from g forces were worse. It seemed that every time somebody thought he had a solution, he only made one thing better while another got worse.

I found a French aerodynamicist, René Hirsch, who'd designed and built a gust-alleviation airplane and was beginning to test it. We corresponded about our various plans and concerns and seemed to be in some agreement. Then he was injured when his airplane crashed. I never learned the cause of the accident. Gust alleviation was not only a mysterious quest, but now I knew it was dangerous as well.

A conversation with Hewitt Phillips pointed me in the right direction. All of the research programs I'd found, he said, had a common flaw. They were so focused on modifying the wings, and thus stabilizing lift, that they ignored the pitching motions of the nose jerking up or down. It was as if a jolt of electricity hit me. *Of course! We were the experts in stability and control, and we'd all missed the obvious.* It was a glaring omission. In that moment in Philips's office, I knew that any automated gust-alleviation system had to do both at the same time—alleviate the lift changes on the wings and damp out the up-down movements of the airplane's nose. I knew the solution. I just didn't have any idea of how to get there. It turned out to be a job of many years.

It was also my introduction to computers. Until I ran into the mathematical complexities of gust alleviation, I was like every other engineer of the day. For quick calculations, I used a slide rule. A slide rule in experienced hands is accurate enough for a lot of work, and it's portable. They were used well into the sixties; when you saw a fellow with a long, slender green or brown case dangling from his belt and slapping his thigh, it was a safe bet that he was an engineer. But a slide rule wasn't good enough for the refined calculations in designing aircraft parts. When we needed accuracy out to three or four decimal places, we used calculators. The good ones were electric and made a funny *chuga-chug* noise when you hit ENTER. That noise has disappeared from our twenty-first-century world. The less good calculators were strictly manual. You entered your numbers on the keypad, then pulled a handle—*kachunck!*—and a bunch of mechanical wheels inside the thing rolled around to add or subtract or whatever. It did the job, but slowly and only one step at a time. A long day of running calculations could leave you with a sore arm. So the whole idea of a computer that would take some of this workload was incredibly appealing.

NACA had one of the early computers, a Reeves electronic analog computer (REAC). *Analog* meant that it worked by reading differences or changes in voltages coming from or going through resistors, condensers, and potentiometers. So all of my standard formulas and equations had to be converted into electrical terms. I had some help from a couple of electronic engineers provided by the Reeves company, but the only time I could get on the computer itself was from midnight to 8:00 A.M. I worked the dead man's shift for months. How I got it all converted isn't important. But eventually I was ready for the next step: convincing myself that REAC could be trusted.

Before I'd trust my analytical work to that computer, I devised some test cases on those mechanical calculators. I'd do a step-by-step analysis of something on the calculator, then run the same problem on REAC. It was laborious and I spent nearly three months at it. But in the end, REAC gave me the right answers— and in a few minutes or hours per problem instead of the days I'd sometimes spent setting up the test. Finally I was convinced that REAC could do the job. It was already well into 1948, and I spent most of the next two years working on gust-alleviation calculations and solutions. (A good laptop or desktop computer could do the same job in a week today, and NASA's large computers could probably do it in a few seconds! But at the time, I was enthralled with REAC, and I was learning a lesson that computer junkies know today: Speed is addictive. After I'd run my numbers on REAC, I knew that I'd never go back to the old calculators again.)

Even with this technical marvel of a computer, the analytical work took months. Then I finally had enough to put it all down on paper. NACA Technical Note 2416 was issued in 1951. I'd done enough math work to conclude that it was possible to get better comfort for passengers when a plane flew through rough air. But it was only a paper study. I still had to get the budget to build it and prove it in the air.

The tech note circulated throughout NACA and got good marks from both our managers and the senior people who acted as NACA advisers. About one hundred men were in the senior staff, and I was called to brief them on my work. One of them was Bob Gilruth. In the question-and-answer period, nobody asked more questions, or harder ones, than he did. But this had been my life for more than three years, and I was ready. In some corner of his brain, I must have impressed him. The day would come when he was handed the best assignment of them all, and he asked for me by name to join up. The immediate result of my tech paper and presentation was that my gust-alleviation system went into the budget. I had the money to design it and build it, and then to modify a C-45 transport and install it for flight tests.

Now I was into the fun part, designing and building the hardware, then putting a test pilot behind the controls of a C-45 to take it up. The C-45 was made by Beechcraft and was used extensively as a multiengine trainer by the military. It wallowed like hell in rough air and was perfect for the kind of testing I needed to do. The project would consume most of my working hours through the first half of the fifties.

At work, the time flew as fast as the airplanes we tested. I wished the rest of my life had been moving as quickly. The three years that I courted Betty Anne, taking her everywhere as proud as I could be, were wonderful and full of love and agonizingly slow. Our friends, including Mel Gough and Sig Sjoberg, all expected us to get married at any moment. Sig and I had once dated the same girl, and when we found ourselves working at NACA, we began a lifelong friendship. He was always there when I needed advice or counsel, and his observations were usually on the mark. He was an avid golfer. I'd never tried the game, but he convinced me to play a round one Sunday afternoon. Fifty-odd years later, to the day he died, I didn't know whether to thank him or damn him. All I know for sure is that I got hooked on the game, have spent a fortune on greens fees and equipment, and have spent thousands of hours whacking a golf ball around courses in every part of the world. I still play several times a week, with a ten handicap. And now that I look carefully at the friends I've made and the good times I've had, I do know the words that apply to Sig: "Thanks, friend."

But Sig was no help when it came to Betty Anne. No matter how I pleaded about marriage, I kept hearing those awful words from her lips: "Not yet." Those iron-chain apron strings hadn't even rusted, much less broken. Brother Robert had gotten away to get married, but Betty Anne and brother Carter still lived under the Turnbull roof. Mrs. Turnbull's hold on her family was ferociously strong.

It was a strain. We all felt it. But Mrs. Turnbull wouldn't let go. Betty Anne couldn't fight her. It took one of those decisions that I didn't quite think through to make something happen.

"You know I've been saving money for our marriage," I said one summer night in 1950. "There's quite a bit in the bank."

Betty Anne looked at me and nodded.

"Well, I guess if we're not getting married, I should get us one of those new powder-blue Oldsmobile Rockets over at Mike Suttle's in Newport News. What do you think?"

Betty Anne swallowed hard and I saw a little mist in her eyes. "That's some kind of choice. Marry me or buy a new car?"

"Well, the money's just sitting there. But I'd sure rather get married than buy an Oldsmobile."

That brought a little smile back to her face. She kissed me on the cheek and said, "Don't go spending it yet. Let me think about it for a few days."

My head started buzzing when I realized what I'd done. I'd accidentally given her an ultimatum and she hadn't responded with those dreaded words "Not yet."

Two days later, Betty Anne Turnbull said yes.

Now the conspiracy started. "I don't know how Mother will take it," Betty Anne said, "so we're not going to tell her." We were twenty-six, we'd known each other since we were fifteen, and we had to act like a couple of eighteen-year-olds planning to elope in the middle of the night. But, no, this would be no hurry-up, runaway marriage by some justice of the peace. We wanted to be married in the Episcopal Church, and we wanted our families to be there. Except for Mrs. Turnbull.

We picked the first Saturday in September, just before Labor Day. That gave us barely five weeks to put the scheme together. First I talked to George Little, the acting rector at the Emmanuel Episcopal Church in Phoebus. He knew the background, and with a chuckle he agreed to get the Reverend T. V. Morrison to perform the ceremony. He also agreed to keep our plans quiet until after the ceremony. I breathed a sigh of relief.

Two weeks before the big day, I told my mother. She'd been my friend and counselor all along. Now she was ecstatic. I think she loved Betty Anne as much as I did. Robert Turnbull was the next to know. His grin said it all when Betty Anne asked him to give her away. Carter was living at home with his mother and we didn't tell him about our plan.

Next came the blood test. My friend Dr. Frank Kearney drew the samples. He held the results until the last minute, then filed the right papers at the courthouse just before it closed Friday evening. There was no chance for the gossip to start.

The night before the wedding, we asked our friends Mathew and Shirley Carli to be the best man and maid of honor. This was real-time mission planning all the way, and they were happy to be part of it. Betty Anne and Shirley went off to look at the new wedding clothes that Betty Anne had secretly been buying for weeks and storing at my house. I barely slept that night and was nervous and on edge on the morning of the wedding. But I still had more to do.

I told Allison Kraft and his mother, my aunt. With only a few hours warning, they were ready to be part of the audience. Allison looked at me keenly and knocked me on the shoulder. "You look nervous, Chris," he grinned. "Let's go swimming." An hour in the surf at the local beach had me relaxed and ready. The plan proceeded.

The courthouse was open for a half day on Saturday. Betty Anne and I got our license, went to my house to change clothes, and were ready for our three o'clock wedding.

She was radiant (isn't that the word for a bride?) in a green suit and pillbox hat. She carried the corsage I gave her. Matt and Shirley Carli stood up for us, Robert gave his sister away, the Reverend Morrison officiated, Allison and his mother grinned, and my mother cried. It was short and simple. We were married.

The hard part was still to come.

We drove to the Turnbulls' red-brick, two-story home and found Anne Turnbull sitting in her usual white rocking chair on the front porch. Carter was there, too. I helped Betty Anne out of the car, both of us still in our wedding garb on an early Saturday afternoon, and we approached the house under Mrs. Turnbull's studying gaze.

"What have you done?" she asked when we reached the steps.

"We got married," Betty Anne replied. Her grip on my arm tightened.

"Did you have to?"

"No."

Mrs. Turnbull stared at us with blank emotion until Betty Anne walked up the steps alone and went into the house to pack her clothes. I stayed rooted, silent and tense on the sidewalk. The past six weeks of deception hadn't been easy. Standing there, I realized that I'd hardly eaten in days; my appetite was gone. I wanted to be gone, too. Not another word was said, not even when Betty Anne came out with her suitcases and I carried them to the car. Betty Anne started to wave when we drove away, then dropped her hand and gripped mine. We were married, we were free, and we on our way to a ten-day honeymoon at Natural Bridge, Virginia. By the time we reached the edge of Hampton, we felt the tension begin to fade and we were saying "I love you." September 2, 1950, was a fine day for the newly minted Mr. and Mrs. Christopher Columbus Kraft Jr.

We were home the next weekend, going straight to the Turnbull house. Betty Anne was worried about her mother, but she needn't have been. My new mother-in-law looked me over, and her first words had to be rehearsed: "A lot of boys are trying to avoid the draft these days. I guess that's why you wanted to get married."

I almost laughed. I'd tried to get in during World War II and been rejected. Now with the Korean War just two months old, my mother-in-law thought I was a draft dodger. "No, Mrs. Turnbull," I said, "that's not it at all." But there was no point in explaining further. We let it lie and somehow called a truce from that day onward. It was never a comfortable relationship, but at least it was reasonably calm.

We set up housekeeping in an apartment on Elizabeth Road, close to our friends, and I went back to work on my gust-alleviation project. I was a traditional fifties male and insisted that Betty Anne quit her job at the bank. She did, even though they begged her to stay. All that meant was that she no longer worked for a paycheck. The apartment needed painting and other work, and in time we both got the house bug. We bought a lot in East Hampton and started making plans.

About this time, I realized that I'd become a "NACA Brainbuster." The title was invented by the locals to describe Langley engineers who could sometimes act a little fuzzy-headed by peninsula standards. A Brainbuster would quiz a merchant about how a washing machine worked and expect an explanation in mechanical terms. Or he'd want advance information on repairs and spare parts before he'd put down the money to buy it. Or a Brainbuster might walk around totally preoccupied with some problem from work and forget to say hello or even to nod at a neighbor. The Brainbusters had their own gripes about know-nothing locals and unfamiliar Tidewater traditions. Now I was both a local and a Brainbuster, defending one side or the other, depending on who was complaining. It reminded me of my days as a senior officer in the Corps at VPI where I'd learned to handle that kind of thing.

My Brainbuster side focused on work, and by 1952 we started flight tests on my gust-alleviation system. I'd designed a sleek, sharp-ended cone to fit over the nose of the airplane. A long probe with a vane extended from the cone. The vane would sense—or feel—turbulent gusts before they hit the fuselage and control surfaces. We tested it on a DC-3, and it worked. It measured what was about to happen to the rest of the airplane. So next we modified a C-45 for flight trials of the full system.

That was the simple part of my complex system. I'd spent more than a year designing modifications to the C-45's flaps, ailerons, and elevator, then interconnecting pieces of each so that response to a sudden gust or "air pocket" was automatic. At the same time, I had to give the pilot normal control over the airplane. That meant regearing his control column so that it felt right when he climbed, dived, or banked. The automatic parts of the system used hydraulic actuators that we installed in the passenger cabin and a standard Sperry autopilot system to stabilize the plane in roll and yaw.

Finally we were ready to fly. And fly. And fly. I rode along on almost every test flight, monitoring the system, taking notes and measurements, and figuring ways to fix this or that, or to make one part or another more or less responsive. It was a question of balancing the system, and the hard work was done in rough air. We flew for months, then more than a year. Whenever the weather turned bad, we were in the air. When the weather was good, I was at my engineering board calculating the ever-smaller fixes that would lead to a smoother flight through a storm.

It worked like this. When a wind gust hit the nose vane, it triggered a signal to the flaps. They'd deflect up or down as needed to balance out the sudden lift (or stomach-wrenching drop) from the gust. Simultaneously, parts of the elevator moved to keep the nose straight instead of pitching up or down. Finally, small interior sections of the flaps moved in the opposite direction of the main flap. That counteracted the gust just as it reached the airplane's tail.

There was more to it than that, of course. Endless calculations. Changes in how fast and how much each piece of the system reacted. Varying the gear ratios here and there as we got more experience with actual flight in bumpy air. In some ways, what I was doing was the highest form of aeronautical science. In other ways, it was pure trial and error.

During the initial design, I realized that the gust alleviation system would actually prevent the pilot from maneuvering the airplane. That is, when the pilot moved the airplane's nose up or down, the system acted the same as if a wind gust had hit it. It would counteract with the flaps, so even though the nose wasn't level, the airplane wouldn't get any lift. This could become an uncontrollable situation. I fixed it by connecting the control yoke directly to the system's flap controller and to the elevator. It solved the problem completely. The airplane did what it was supposed to do, and with even quicker response than normal. The flaps and elevator gave or took away lift immediately, and then the gust system neutralized the flaps. The pilots loved it.

Through those years, I always had two or three projects going, and as a section head I had engineers under me working on their own assignments. One of mine was to study wake turbulence. Pilots called it prop wash—the turbulence that lingers behind a moving plane. The bigger the plane, the bigger and stronger the prop wash. A small plane passing through a big plane's wake turbulence could be in big trouble.

It didn't take long to learn that prop wash was the stuff of legend. I used a P-51 as my "generator" airplane and discovered that wake turbulence has nothing to do with the propeller. It comes from the wingtips. Fast-moving air rolls off the wings into two little tornadoes, or vortices. The bigger the plane and the slower it flies, the stronger and more enduring are the vortices. (The worst case today would be flying behind a 747 or a C5A with its flaps down during a landing approach.)

I used smoke to make the wake turbulence visible, then had a P-80 jet fly through it. At first, the P-80 pilot had a hard time hitting the turbulence exactly. "Stretch a piece of gum across your windshield," I told him, "then line it up with the smoke and just fly it."

He gave me a funny look, but he tried it. After that, he hit the turbulence perfectly on every try. The results were dramatic. We discovered that wake turbulence starts during the takeoff roll and that the tornadoes can linger over the runway in calm conditions. If the wind is blowing, they move aside and dissipate quickly. At thirty-five thousand feet, in otherwise quiet air, the vortices lasted three to four minutes. Another plane coming along would never see these dangerous little tornadoes and could fly right into one. It happened often enough to give pilots a few scary moments.

I thought my report, issued by NACA, was a classic. In fact, it was so good that it was completely ignored. Years later, the Federal Aviation Administration did its own tests, came to the same conclusions, and used the results to set rules for how far apart commercial airliners must fly. They could have saved a lot of money by checking the NACA archives first.

By 1955, our gust-alleviation research was ending and we had the best alleviation system in the world. It worked to keep the plane flying smoothly in rough weather. We demonstrated it far and wide, and design engineers who rode along on our show-and-tell flights thought it was a great solution to an age-old problem. Unfortunately, by 1955, they didn't need it anymore. While I was turning a C-45 into a rough-weather smooth-flyer, the big airplane companies were designing the first generation of jet transports. The electronics companies were designing smaller and better radar systems. And the weather forecasters were getting better and better at predicting bad weather and clear-air turbulence.

So the big jets were coming. They could fly over most of the bad weather. They'd soon have onboard radar systems to spot the storms. And their preflight planning included pretty good weather forecasts so they'd plot a course that avoided the known areas of turbulence.

I wrote a report that was eventually released by NACA in 1956. By then, I was on a new assignment, project engineer to determine the flying qualities and to solve some serious problems with the Navy's new supersonic F8U jet fighter. One of the contentious fellows I had to deal with was a Marine major assigned to the old Navy Aeronautics Bureau, located about where the Vietnam Veterans Memorial is today.

His name was John Glenn.

3

Flight Tests and a Stubborn Marine

That lot in East Hampton kept calling to us. Betty Anne and I knew what kind of house we wanted. We also knew what we could afford. I started doing sketches that gradually turned into a complete design of a house that we both thought we'd like. When Betty Anne announced in August 1951 that she was pregnant, we decided to build it. I was my own general contractor, hiring crews and workers, and doing painting, insulating, and anything else I could manage to keep the costs down.

It wasn't enough. Our $12,000 house ended up costing $14,000, my first personal experience with an overrun. Two thousand dollars was a lot of money in 1952. My cousin Allison gave us a loan. We skimped for the next two years to pay him back. Years later, I would face cost overruns with numbers so huge that I couldn't imagine them in the early fifties. Building our first house showed me how easy it can be to exceed a budget. But it didn't make me tolerant. I still yelled at contractors who brought me that kind of bad news.

We moved in six months after Gordon Turnbull Kraft was born. He was an example of something that almost never occurred in my professional life, then or later. Not waiting for the mid-April countdown we had marked on our calendars, he arrived March 29, 1952, two weeks early.

We had our house, our son, then our daughter, Kristi-Anne, three years later. I finished up the gust-alleviation project, which took far longer than I ever thought when it started, and moved on in the world of testing airplanes. One of the first things I did was to take on the Navy brass over defects in their F8U fighter. We got the number three plane off the assembly line and went to work.

The Chance Vought F8U was a complicated beast. Like all Navy fighters, it was designed to be flown from aircraft carriers. I couldn't help remembering that I'd almost gone to work for Chance Vought. Maybe, in a different ending to the story, I would have helped to design this machine. Instead, I was the NACA project engineer charged with testing it and fixing its faults. The F8U's innovations, many of them coming straight from NACA research at Langley, made it one of the most advanced production aircraft in the world. Its design included some of the first automatic control systems, an all-moving tail that eliminated the standard horizontal elevator, and a wing that moved in flight. One wing position, slightly tilted up to make the plane fly slower, was used for takeoffs and landings. Then for high-speed flight, the wing tilted straight into the airflow and fit tighter against the fuselage. Flight-testing this plane was a challenge to both engineers and pilots.

It hadn't taken long for Navy test pilots to discover a problem with the wing movement. The hydraulic system that tilted the wings wasn't strong enough. It could only move them at speeds under 220 knots. A pilot needing to climb fast or fly fast couldn't make his move quickly enough to meet the Navy's combat standards. Naval Aviation asked us to fix it.

It didn't take a genius, or even an aeronautical research scientist, to see where the problem was. The wing strut activator was obviously too weak. The question was, by how much? I installed strain gauges on the strut and sent a pilot up to put the movable wing through its motions. Within a few days, I had flight data on both the strut and the hydraulic system. I recommended that the struts piston be increased in size. The change would deliver more force when the wing was lowered at a high speed. Chance Vought accepted my recommendation and quickly redesigned the strut system. It worked.

We settled back into the routine of testing the plane through its normal range of speed and maneuverability to write the book on its flying qualities. At the same time, the Navy was shipping the F8U to its fighter squadrons. Then more problems surfaced. Engines flamed out. There were crashes and some Navy pilots died. Other pilots reported problems in high-speed maneuvers. Some of them lost control of the plane and others barely recovered. It was happening with my test pilots, too, but none of them died.

Something was wrong, something that turned the F8U into a dangerous and maybe fatal machine.

In a bull session with Navy test pilots at Patuxent Naval Air Station, Jack Reeder mentioned that he'd felt some g-force changes during turns. It was purely subjective and he almost didn't bring it up. In a controlled turn, the g forces on a pilot should be constant. That was the law, in both physics and aerodynamics, and you don't mess with Father Physics. So he was surprised when one the Navy pilots set down his coffee cup and said offhandedly, "Oh, yeah, we get that all the time."

Back at Langley, we dug into the problem. The law wasn't wrong, so it had to be the airplane. It didn't take long to figure out that the F8U's elevator kept moving in a tight accelerating turn, even when Jack held the stick absolutely motionless. I scratched my head when I saw the data plots. "How the hell can that be happening?" I asked.

By now, we'd instrumented almost everything but the stick. So I put some instruments on the bell crank at the bottom of the stick and Jack went up again. He reported the same fluctuations in g force, and I could hardly wait to see the data. There it was: The bell crank was moving when Jack whipped into a tight turn. It was putting more elevator into the turn than the pilot was asking for.

"Damn!" I said, and Jack saw it, too. "If that crank's moving, the whole airplane is unstable."

"We've been lucky," Jack said. "So why's the crank moving?"

That was the question, and within a day, I had the answer. The F8U had a big air scoop designed into its nose, to suck air into its engines. In a fast turn, the loads were so heavy that the scoop actually bent. And when it did, it moved the bell crank. The stick didn't move, so the pilot got no tactile feedback. But he could feel the changes in g force on his body. It was obvious that the F8U instability was just a crash waiting to happen. We took the data to Washington.

The Navy's aviation bureau was still housed in some World War II wooden buildings, about where the Vietnam Veterans Memorial stands today. We had an appointment to see the Fighter Desk, which was manned by a Marine major with thinning reddish blond hair, tough blue eyes, and an attitude. We had our briefcases loaded with hard data and were on a mission of the just. It took about ten seconds for us to get off on the wrong foot.

"The F8U is dangerous," I said bluntly. We'd barely made introductions and I wasn't feeling like waltzing around with the usual small talk. "It's stick-fixed unstable and we want to show you this data."

Major John Glenn fixed me with a glare that could have stopped an F8U in midair. "I've flown that airplane," he snapped. "You don't know what you're talking about."

I backed off and started over. "There is a problem, Major. Will you look at our data?" It took some self-control to keep my voice quiet and reasonable, but I did it. Glenn nodded and we spread out our papers. A few minutes later, he was looking troubled.

"Just happens that there's a couple of Chance Vought pilots in the building today," he said. "Let's see what they have to say."

Before long, a group of pilots were listening to Jack Reeder talk about what he'd felt and me explaining what we'd found. "Yeah," one of the Chance Vought pilots said, "we see a little of that. The way you tell it, it sounds dangerous."

That was enough for John Glenn. He was a believer. He packed our data and our recommendations off to Chance Vought before the sun set that day. Fixing all the F8U's in service could have been a major and expensive job. But after looking at our data, the Chance engineers came up with a brilliant fix. They simply removed the bell crank and reinstalled it backward. Not only did it lock the crank solidly in place, but suddenly the F8U was more agile than ever in a quick turn. Jack Reeder confirmed the fix on a test flight, and I could hear the grin in his voice.

Unfortunately, the F8U's problems weren't over. It wasn't all the plane's fault. This was one of the first planes to have a yaw damper, which prevented the kind of fishtailing oscillations at high speed that could throw a plane out of control. We'd gotten some instructions from Chance Vought on modifications, did the work, and reinstalled the yaw-damper electronics. Jack Reeder took it up for a run.

The F8U was a supersonic fighter, but it only carried enough fuel for about six minutes of flight faster than Mach 1. I was listening to Jack and he was cool and happy. He kicked it through the sound barrier, didn't say much except that it felt good, then slowed down and headed home to Langley. At about five thousand feet, and maybe 220 knots, he ran into some clear-air turbulence. All hell broke loose.

"Hey!" he yelled. "This thing's a bitch. What's going on, I'm all over the place. . . . Damn, this plane is dangerous!"

The crew chief and I looked at each other and had the same thought: "The yaw damper . . . something's wrong with the yaw damper."

I keyed my mike while Jack was wrestling with a wild airplane. "Jack, turn off the yaw damper. Repeat, turn the yaw damper *off!*"

There was a moment of silence, then Jack's voice came back cool and easy. "Damn, that did it. It's solid now. We're okay."

We tore into the plane a few minutes after it landed and there it was. We'd installed the yaw-damper switch backward. ON was OFF and OFF was ON. The F8U book had a big bold warning about never flying supersonic with the damper off. Jack had just done it, with no problem. But the clear-air turbulence almost killed him.

About that time, more problems cropped up in Navy squadrons. The F8U was suffering compressor stalls—the engine quit—and they'd lost about six of them. Then came the tragedy that shook everyone to the core. You'll still see the film of it in some of those videos of wild airplane accidents. Next time you do, remember that the guy flying the F8U was killed.

It happened at the Chance Vought—later Ling Temco Vought—plant in Fort Worth. A bunch of air cadets was visiting the plant, and one of Vought's test pilots did a high-speed flyby over the field. An engineering camera caught every moment. No more than one hundred feet above the runway, flying fast and straight and level, the F8U's wing suddenly folded up and ripped off. There was no time to react. The plane hit the ground in a ball of fire.

A few days later, a courier brought a copy of the film to Langley. We could see in slow motion exactly what happened. The question was why. I mounted a camera in the fuselage of our F8U and aimed it at the wing. Our film was shocking. Every time a pilot did a maneuver, the wing would pulse up and down. It wasn't supposed to do that. I hadn't taken my strain gauge off the wing strut, though we hadn't been looking at the data much since that problem was fixed. Now I looked again and made a horrifying discovery: When fuel was low, that strut was under tension instead of compression. It was being pulled when it should have been compressed by the aerodynamic loads.

A few calculations showed that the center of gravity was moving as the fuel tanks emptied, and it was enough to reverse the direction that forces were being applied to the strut. I immediately called Major Glenn on the Fighter Desk.

"What altitude was this test?" he asked.

"Thirty-five thousand feet."

"Well, it shouldn't be happening. There's something wrong with your data. Try it at twenty-five thousand feet."

I bit my lip to keep from yelling at him. This damn upstart Marine major might be a great fighter jockey, but he was listening too much to the

Chance Vought engineers and accepting the company view of things. I was steaming when I told Jack Reeder to repeat the runs at twenty-five thousand feet. Reeder just shook his head and suited up. And damned if it wasn't worse at twenty-five thousand feet than thirty-five thousand feet. The tension loads on that strut went way up.

"I don't believe it," John Glenn said again when I called with the new evidence. "Try it at twenty thousand feet."

My next call was barely civil. "It's worse at twenty thousand feet, Major. Do you want us to keep trying or will you finally do something?"

"Go lower," Glenn ordered after more conferences with Chance Vought. "Your data's wrong."

Before I sent one of my pilots back up, I ran the full range of calculations. The stress would be much worse at lower altitudes, until somewhere close to sea level, the wing would rip off. Of course, that's exactly what happened at Fort Worth. We flew at fifteen thousand feet, then ten thousand feet, and the data exactly matched my predictions. That was low enough. I called a halt to the flights and tried Major Glenn one more time. If I could have reached through the phone and grabbed his throat, I would have done it.

"Your strut loads must be wrong." Glenn still wouldn't accept our data over the Chance Vought company line. That was enough for me. I wouldn't accept his gyrene stubbornness either. I showed the numbers to Mel Gough, who was running the Flight Research Division. He looked, he listened, he picked up the phone and called an admiral he knew in the office of the Chief of Naval Operations.

"We've got this data on why your F8 wing is coming off, and we're getting a runaround from Air Bureau," he said. "What are you going to do about it?"

"What's your advice?"

Mel knew the guy pretty well and he wasn't afraid to be blunt. "Ground the airplane. Nobody flies."

"We can't ground the whole fleet of F8s."

"Then get some flags ready for funerals. You're going to need them."

There was a long pause. "Okay, Mel. Come up and show us your data. I'll look at it myself."

Two days later, we made our presentation directly to the admiral in the CNO's office. I could see his face tighten as he listened. He knew truth when he saw it. He also understood the politics of grounding the Navy's most expensive fighter jet. "I'm sending your package to Vought and ordering them to recheck everything on this strut," he said. "I'm going to give them forty-eight hours."

A few days later, Vought issued a letter to the Navy. It was a classic in mealymouthed military-contract politics. They'd found an error in their data, Vought admitted, then added that upon talking to Kraft at Langley, he'd found some calibration errors in his data, too. If you read the letter carefully, it says there were mistakes all around. But the bottom line was that the plane was dangerous. That was all the admiral needed. He grounded the F8U effective immediately.

The problem was a U-shaped fastener, a clevis, attached to the hydraulic strut and the nose section of the wing. The clevis was too weak to sustain the tension load. When a pilot was in a low-fuel condition, he was fine at low speeds. But if he had to make an abrupt high-speed maneuver, the load on the clevis was too much. It pulled up and the wing ripped off. Company engineers quickly designed a stronger clevis and the problem went away.

That was my first big argument with a major contractor, and it taught me a lifetime lesson. When it hits the fan, a contractor is likely to tell me anything. Engineers in trouble have too often sold their souls to the company store. So I learned early to verify everything. I learned that everybody knows about normal operations. But what the hell caused the abnormal operation? John Glenn wasn't all wrong to question my data. Frequently in operations, it's the instrument that's wrong because you don't build the reliability into the instrument that you build into the system. But he was wrong not to follow up and pin down the culprit.

On the F8U, the system was flawed and the instrument was right. We'd both suffer through life-threatening moments in the future when we'd face the exact same question: Is it the system or the instrument? Give the wrong answer and a man might die.

By the end of 1956, I'd been working for NACA for almost twelve years. Sure, there was a lot of excitement and a lot of satisfaction in testing the world's best airplanes to find and fix their flaws. But with each new assignment, I kept thinking that I'd done this before. I started getting headaches; before long, they turned into severe migraines. I was smoking a pack a day of cigarettes, and I'd sometimes get so sick at the end of the day that I'd vomit, then crawl into bed to sleep off the migraine and the general damage I was doing to myself. Our second child, Kristi-Anne, didn't sleep through the night until she was three, so Betty Anne was feeling the wear and tear from both of us, and from handling an active four-year-old boy.

There were outside pressures, too. My father had a fatal heart attack on New Year's Day, 1957. He was sixty-four. His brother, August, had died in 1952 of the same thing, and at the same age. Suddenly long life for a male Kraft didn't seem likely. And I hate funerals. I prefer a good old-fashioned wake and all it entails. My father's funeral was open-casket, and the eulogies seemed to go on forever. He was buried in East Hampton's Oakland Cemetery, in the Kraft family plot. Funerals are still tough for me. But getting through my father's was the toughest I ever had to do.

A few days later, it all caved in on me. I was violently sick after dinner, throwing up blood and getting weaker by the minute. Betty Anne called Dr. Kearney. "I'd say you've got a pretty bad ulcer, Chris," he said, and while he was talking, I passed out. He rushed me to Dixie Hospital by ambulance and ordered an immediate blood transfusion. He sent me home the next day with some magical blue pills and a strong lecture about stress and the need for a lot of milk and a proper diet.

"And that smoking has to stop," he said in a severe voice. "It's like pouring gasoline on a fire."

I listened. The cigarettes went into the trash, I followed my new diet religiously, and I found something to occupy my mind away from work. Betty Anne and I talked my mother into letting me build an addition to our house in East Hampton. Mother's money had been freed up after my father died, and she was skeptical at first that I'd squander her fortune on construction. But the idea of living under the same roof with us and her grandchildren was just too good. I designed an apartment, and with the help of some great Flight Research Division mechanics and carpenters, we built an eighteen-hundred-square-foot addition in six months. Now Mother had total privacy when she wanted it—her own kitchen, living room, bedroom, bath, and attached garage. She loved the children, so we had a built-in baby-sitter, and it didn't hurt that Betty Anne and Vanda got along so well.

But I still felt a nagging unhappiness about my job as an aeronautical research scientist with NACA. The job was getting to be more theoretical with less in-flight action, and I was losing my edge. I'd always considered myself a doer more than a thinker. But now thinking was a bigger part of my job than doing. To make things worse, a lot of my friends had left for more exciting jobs at companies such as Douglas in California, or government agencies such as the Federal Aviation Administration in Washington.

They were moving on with their lives and I was still doing what I'd always done. I didn't know that Betty Anne was praying every day that I'd get another job.

I'll never know if it was Betty Anne's prayers, simple fate, or some combination of both. I do know that I woke up on Saturday morning, October 5, 1957, to discover that the whole world had changed the night before.

SECTION II

THE MERCURY MISSIONS

4

Reds and Red Tape

A Russian satellite called *Sputnik* was in orbit around the earth. Nothing would ever be the same again.

Few events in the twentieth century had the catalytic effect of the Russian *Sputnik,* and none sent such important and everlasting ripples across the face of humanity. With *Sputnik,* this curious human race of ours embraced its ultimate destiny to survey and colonize the universe. My own curiosity, and that of the people around me, would lead us personally into the most intense period of exploration since America reached its Manifest Destiny of consolidating the North American continent from coast to coast. When some historian in the fourth millennium writes of mankind's progress, I think the years from 1957 to 1972 will shine brightly through his words.

Our part was to run a race with Russia for dominance in space. We understood the goal. But we didn't understand the magnitude of the job ahead. Not at first, anyway. But neither did the U.S. government, nor the Russians. Certainly our own government hadn't thought carefully about the significance of a satellite passing over our country every ninety minutes. Could it photograph our most protected secrets? Could it eavesdrop

on the Pentagon? And more frightening of all, could it release an atomic
bomb on any pass, targeting every city in America including Washington,
D.C.? *Sputnik* raised incredible technical and political questions. But nobody had the answers at his fingertips. Getting them was going to be a bigger job than anybody thought.

Even the Russians were shocked at the world's reaction. Nikita
Khrushchev, the leader of the Soviet Union, went to bed after hearing the
Sputnik news from his scientists, apparently satisfied but not overly excited
by the event. Moscow's *Pravda* newspaper mentioned *Sputnik* the next
morning in a bland short story buried at the bottom of page one. Not until
the London and New York papers came out a few hours later, with huge
banner headlines, and long stories and analyses, did the Russians realize
that they had a technical marvel—and a huge propaganda coup—on their
hands.

The propaganda said that the Russians had a huge technical edge in
rocket technology and implied that this meant they also were ahead in producing and deploying intercontinental ballistic missiles. Congress reacted
with a hue and cry, demanding to know why America had fallen behind and
why our own space activities seemed to be stalled. Hearings were ordered
to pin the blame on somebody, and with Dwight Eisenhower's Republican
administration in power, that's where the Democrats, and some of Ike's
own party members, too, aimed their darts. An alarmist press fanned the
fires of fear and rage. Much of the sudden crisis came from a lack of technical understanding—in the press, in the politicians, and in the public at
large. It would be years before the truth, and then not all of it, was finally
known.

America could have beaten the Russians into orbit. It probably should
have. *Sputnik* came early in an eighteen-month scientific program called
the International Geophysical Year, a global effort to study Earth and its
upper atmosphere. Preparations for the IGY began in 1955. Both Russia
and the United States announced plans to put an instrumented satellite
into orbit during the IGY. The U.S. Army, which now employed Wernher
von Braun and much of his German rocket team, lobbied for the privilege.
But the Army's rocket was a military missile, and Eisenhower insisted that
Americans portray a peaceful image. The project went to the Naval Research Laboratory, which was developing a rocket with no purpose except
to send instruments into the upper atmosphere. With a little added development, NRL said, this Vanguard rocket could put a science satellite into
orbit. But Vanguard was nowhere as good as von Braun's Jupiter rocket.

When papers were declassified long after, historians learned that the Pentagon deliberately restricted the Jupiter's power in 1956 when it feared that von Braun would "accidentally" put the rocket's empty nose cone into orbit. Empty or not, it would have been the first man-made satellite in space.

Now the Russians were up there, Vanguard was still little more than a fizzle on the launch pad, and the press was roasting American technology. A cartoon in the Newport News newspaper struck all of us at NACA with bitter force. It showed a para-scientist who'd just tossed off a manhole cover and emerged from underground. There was a footprint on his fanny and he was looking toward the heavens in bewilderment. "Where the hell have you been?" read the caption. And it was the right question. *Space* was not a word found easily in the NACA Langley Library in 1957. We were *airplane* people and proud of it. That part of our pride stayed intact. Then for many of us, everything quickly began to change.

Exactly one month after *Sputnik,* Russia did it again. *Sputnik* 2, nicknamed *Muttnik* by the press because the new ship was considerably larger and carried a living dog as a passenger, was a stunning achievement. It proved that the first Russian success was no fluke; they could do it again, and even better. Within a month, the Naval Research Laboratory's Vanguard rocket, carrying a tiny U.S. satellite, exploded on the pad at Cape Canaveral. The world's press was watching and quickly accused the U.S. military of lagging dangerously behind the Russians. The popular phrase was *missile gap.* Another phrase showed up about that time, too: *space race.* Suddenly America and Russia were in a space race, and it was to last for the next eleven years.

I was on the way to a meeting at NACA headquarters in Washington when I heard the news of Vanguard's failure from my taxi driver and felt sick for my country. That night I listened to a television commentator berating Vanguard scientists and engineers for their lack of knowledge, compared to Russian rocket men, and felt even worse. The morale of the American people was low, but our morale as supposedly topflight engineers was even lower. Almost immediately, NACA began a series of courses for us, usually given in the evening at Hampton High School, on planetary astronomy, the fundamentals of space flight, and the many theories about the physiological and psychological effects of weightlessness on the human body. The California Institute of Technology and other major universities quickly put courses on film for us. I wasn't the only NACA engineer who was stunned at how much I didn't know and how much I had to learn.

Putting a satellite up was the number one American concern. There

were more problems with the Vanguard rocket into January 1958, and finally Eisenhower relented. A small satellite built by the Jet Propulsion Laboratory in Pasadena, California, was put atop one of the Army's Jupiter
rockets. It went into orbit on January 31, and America was finally in space.
At Langley, there was a sigh of relief. We'd answered the Russians with a
satellite of our own. A few weeks later, Vanguard finally worked and we had
two satellites up there. America was momentarily euphoric. But it wasn't
long before the larger questions about space exploration reappeared in the
newspapers and in the inner council chambers of the government. We were
certainly in a race with the Russians. What would it take to win?

While space policy and direction were being debated, NACA took action. Our chairman was Hugh L. Dryden, and he instinctively understood
that space flight was the next step up from airplane flight, and that the
people who knew aeronautics had to be the foundation for this new effort.
Plus NACA had the rocket division run by Robert R. Gilruth. Dryden asked
him to form a space exploration group within NACA, and Gilruth accepted. He took his key people with him: Maxime Faget, Guy Thibedeaux,
Aleck Bond, and Caldwell Johnson. Faget and Thibedeaux were Louisiana
boys who had aeronautical engineering degrees and had served in the war.
Aleck Bond was a Southern gentleman from Atlanta with a degree from
Georgia Tech and was an authority on thermal aerodynamics. And Caldwell Johnson was a local boy, a high school graduate with more intuitive engineering skills than most of the people around him. It wasn't long into the
spring of 1958 before we heard that Gilruth and his inner circle were talking about putting men into space, not just satellites. Gilruth took the idea
to Dryden and got approval to work on it.

By the end of April, Faget, Johnson, and Chuck Mathews, who had
joined the group, had sketched out a man-carrying space capsule that could
be launched by an Air Force Atlas missile. The capsule had a broad, blunt
bottom covered with a heat shield to protect it from the intense atmospheric heat of reentry. Retrorockets would start it down from space, then
be jettisoned. A big canister at the upper end of the capsule held a parachute and radio beacons. The capsule would land in the ocean, which
would help cushion the impact. A man would ride in it, cramped and
crowded by instruments. He wouldn't have much control over things and
wouldn't even have a window.

I knew little of this at the time. I'd see Chuck Mathews now and then;
he was obviously excited about the work, but was reluctant to share the details. Max Faget was only a little more forthcoming. He talked a little about

the blunt-body shape they'd adopted from proposals out of NACA's Ames Laboratory in northern California. I wasn't thrilled with the blunt body, but the pilots in the Flight Research Division were downright opposed. The idea of coming down through the atmosphere as a passenger with no control went against everything they believed in. They called the blunt body a "popcorn popper" and said anybody inside was just a guinea pig along for the ride. I had my own engineers working on various types of winged reentry vehicles. We were trying to find the best trajectories to minimize heating and, at the same time, find out about stability and control on the way down from space. We were also negotiating with the Martin Aircraft Company to get one of its prototype transonic bombers delivered to Langley so we could try out our gust-alleviation system at high speeds. The rush of events caught up with me in the early autumn of 1958.

The jockeying for position and the war of words between the military services came to an end when the Eisenhower administration decided to create a civilian agency to explore space, with the military free to exploit space for its own purposes including surveillance of enemy territory. NACA got the civilian job and would be the prime agency for aeronautical and space research, and for putting a man into space. Its new name, when it was officially reborn on October 1, 1958, was the National Aeronautics and Space Administration.

Chuck Mathews came to me almost immediately after Eisenhower signed the enabling legislation in late July. Gilruth was expanding his program and wanted me to come aboard. The job involved flight operations and testing. Nobody knew how to do these things yet, Chuck said, but it was obvious that they had to be done. Then Chuck gave me the downside.

"You're a Phoebus boy," he said, "and it's almost certain that we'll be moving this whole operation to somewhere else in the country. Bob will understand if you don't want to go, but why don't you take a couple of days to think it over."

I was flattered that Gilruth wanted me, and I'd already been considering job options that required moving away from my Tidewater roots. I don't remember the discussion with Betty Anne that night. But I do remember that I'd made up my mind to take the job before I ever got home. And I remember Betty Anne's final words before we went to bed.

"Of course you have to take it, Chris." She sealed it with a kiss, and I knew that deep down, she was relieved that I'd be moving on to a new challenge that might put my headaches and my ulcer far into the background. She was right, too. I'd already figured out that a lot of my stress came from the growing complexities of advanced airplane design, and from the need to understand higher and higher degrees of mathematics. I was a damned good engineer, but I was no mathematician. With each new assignment, I forced myself to learn the math and to work the formulas. It was an agony that I didn't look forward to, and the thought that this would be my life forever made my stomach ache and my head throb. Now I had the chance to join a new group, where I'd be an engineer working on flight operations problems while somebody else worried about the math. It just felt right.

Chuck Mathews got a big grin when I told him the next morning and said he was surprised at my speed. For a government group, I was surprised at theirs, too. A week later, I was sitting in meetings in temporary offices in a nearby wind tunnel building. For the first time in my adult life, I didn't belong to the Flight Research Division. Now I was part of something called the Space Task Group (STG) and I was off and running to a place no one knew anything about.

Thirty-five of us were on the roster when the Space Task Group was officially born on November 5, 1958. By then, we'd already done a lot of important work. Most of us, including Gilruth, Faget, Johnson, and Mathews, were engineers. A few were administrative and businesspeople, and the rest were the most important of all: the secretaries. We had another fifteen engineers from the Lewis Laboratory in Cleveland on temporary assignment to help with the early work in manned spacecraft design and operations.

We were a young crowd. Bob Gilruth was the old man, at forty-five. He was mostly bald, with a gray fringe, and he had a quiet way about him. But he knew what he wanted and he had our respect. His deputy was Charles Donlan, forty-two. Max Faget was chief engineer, and Chuck Mathews was chief of flight operations. Both were thirty-seven. I was thirty-four. Because Caldwell Johnson didn't have a college degree, he couldn't have a senior management position. Gilruth made him an engineering branch chief, though, with degreed engineers reporting to him. Caldwell was thirty-five.

Some of the other men in the Space Task Group were even younger, and I got to know them well. Glynn Lunney, John Mayer, Bill Bland, Aleck Bond, and a few others were in almost from the beginning and became important players in the space race we were about to run. This was Bob Gilruth's core team. Our history lay in airplanes and NACA. Now our country was telling us to get on with it, to get American men into space. But we had no roots to enlighten us on procedures, expectations, organization, or even on what we had to learn. We were a giant puzzle and it was Bob

Gilruth's job to put us together to form a picture. The job ahead had never been done before. We would learn by doing. So we got started.

The first step, a big one, was totally alien to our Langley way of thinking, and because there was no time to waste, we took it before STG became an official organization. We considered ourselves the do-it-yourself experts of aeronautics. When we needed something—an instrument, a radio—we requisitioned it, bought it, or built it ourselves. Except for airplanes, of course, which were generally sent to us by the company that made them, with the request that we test one and find its flaws. We didn't depend on the airplane industry; it depended on us. Now suddenly our roles were reversed. We were faced with writing an RFP.

An RFP, a Request for Proposal, is the government's way of defining its need for something, usually in excruciating detail, and then asking private industry to submit proposals for how they'll do the job, and how much it will cost. To give everybody a fair chance in the bidding, the RFP has to contain everything possible about the technical, financial, and business management parts of the job. We'd never written an RFP in our lives. We didn't know a damn thing about putting a man into space. We had no idea how much it should or would cost. And at best, we were engineers trained to do, not business experts trained to manage.

We did know a few things. We knew that, come hell or high water, we were going to put an American into orbit around the world. And we knew that more than national pride rested on beating the Russians to the punch. This space race wasn't a game. It was deadly damn serious, and the future of our American way of life was at stake. If anybody looked at later space efforts and considered them a crash program, they should have seen us in early November 1958. With not much more than a cocky self-confidence, tinged with occasional moments of total bewilderment, we embarked on the crash program that made the rest of them look like a walk in the park.

It was called the Mercury Program Request for Proposal. Charlie Donlan was one of the RFP leaders, pulling together the details with Bob Gilruth always looking over his shoulder. They called in Charles Zimmerman, an aeronautical engineer at NACA who also had airplane industry experience, to manage the business aspects, such as they were.

Max Faget and some of his key people, such as Caldwell Johnson, worked on the detailed specifications for the manned capsule, everything from its blunt-body shape to its electrical innards, pressurized atmosphere

so the man inside could breathe, and its retrorockets, heat shield, parachutes, and escape tower. And that list didn't begin to cover the total package of details. Max was a little guy, out of Louisiana Cajun country, and
he'd served on submarines in World War II. His energy put most of us to
shame, and we couldn't begin to aspire to the heights his design genius
took him. The only person who could keep up with Max was Caldwell, and
they made a helluva team. Before it was over, the two of them held key
patents on our spacecraft and were big-time players in all of America's
manned space flight programs.

(People wonder about Wernher von Braun's role in all of this. He didn't
have one. While the Space Task Group was putting a manned program together, von Braun was still working for the Army at Redstone Arsenal in Alabama. At first, he wanted nothing to do with NASA. He changed his mind
later, when he saw what we were doing.)

Chuck Mathews was in charge of the operations section of the RFP. "Our
part's simple," he told us. "Chris, you come up with a basic mission plan. You
know, the bottom-line stuff on how we fly a man from a launch pad into
space and back again. It would be good if you kept him alive." That was
Chuck, a typical automatic controls engineer who loved his work. He was a
sleepy-eyed blond who drove a 1948 Hudson Terraplane, a high-tech car with
an electric shift. It was the Edsel of its day, but Chuck thought it was a great
machine. Who'd argue with one of NACA's brightest engineers? His hair was
always a mess. Sometimes a shock of blond hair visible over stacks of reports
and engineering papers was the only way we knew he was at his desk. "Be innovative," he'd tell us. "Don't be afraid to try something new." We listened.

I had my orders: write the basic flight plan, or as it would be called in
the RFP, the design reference mission. I had Howard Kyle from Langley's
Instrumentation Division to help me figure out what our space man would
do and how he would do it. Mathews put some of his men, loads experts
such as John Mayer and Ted Skopinski, and a few others, to work on figuring out the path our space capsule would follow. In their experience, when
you put a man into a flying machine, he rolled down a runway, took off,
flew around awhile or maybe went from point A to point B, then landed
again on a runway. Now they were faced with no runway, a circular flight
path that went all the way around the world, and no runway again. Their
man would take off like a rocket, literally, and sooner or later land some-
place in some ocean. Pick one. The job of determining the important characteristics of launch, orbital, and entry trajectories wasn't going to be easy.

I had the added task of figuring out the recovery aspects of this whole thing. Once the astronaut and capsule were bobbing around in the ocean, how do you rescue him? The end game of a space mission was going to take planning and teamwork.

Howard Kyle and I sat down to begin sorting things out. He was an expert in radar, telemetry, and communications. I was an operations guy, accustomed to sending men up in dangerous airplanes after giving them specific instructions on what to do, how to do it, and to let me know if something went wrong. We listed the questions most immediate to our task:

How do we measure the trajectory during launch so that we know when to abort if the rocket goes haywire?

How do we know the capsule reached orbit? It'll be to hell and gone high out over the Atlantic. And if we know it's in orbit, how do we figure out what path it's on?

If it's not in orbit, where's it going to come down? And what do we do about it? Can we use the retrorockets to change the impact point, get closer to the recovery ships? (Note here: Ask Max Faget for details on the retros.)

What kind of radar beacons do we need? How powerful? What do we need on the ground to track them? (Ask Max about the capsule's radio antennas, too.)

The guy in the capsule (*astronaut* wasn't used yet) should have some instruments and dials and gauges. What does he need to know? How much should we tell him?

How do we track systems' health and performance? Consumables such as oxygen, hydrogen peroxide for the attitude-control thrusters, electrical power use, interior oxygen pressure and temperature? Do we want a caution and warning system that tells the space man and us on the ground that something is about to fail? Or just did? We need data on the ground, too. This list went on for pages. We supplied most of our own answers.

How about medical information on the man? Do the doctors need it? And if they do, how do we get it for them? High-quality remote sensing of human bodily functions doesn't exist. Yet. (We don't have any flight surgeons. Better get some, and fast.)

What kind of electronic aids do we need to find the capsule after it's on the water? (Ask the Navy for advice.)

What voice communications do we need during launch, orbit, and reentry? What radio frequencies are good for such operating conditions? (Ask everybody.)

By the end of the week, it was obvious that many of the answers to our questions would influence the capsule's design and the actual recovery operations. Any company that wanted to take on the challenge of building our hardware had to know it all. Chuck Mathews started sitting in with us, and we laid out a tentative schedule of activities to reach the day when we'd put a man on a rocket and light it off.

"Can we really do all of this?" was the question we asked as the details piled up to mountainous proportions.

The answer was always the same: "We have to." There was no other choice.

Suddenly, going to work was a joy again. An important goal, especially one involving the nation's future, is a mighty motivator, and there were moments when I felt euphoric. I couldn't wait to get to work each day, I stayed late each night, and many weekends were just more opportunities to do the multiple assignments that, like everyone else in the tiny STG organization, I had stacked on my desk.

Life at home changed. I was too excited about work to have headaches or stomach problems, but Betty Anne and the kids paid a price. We still tried to find an occasional weekend day for the beach and I seldom missed church, but my appearances at Little League grew further and further apart. The busier I got, the more Betty Anne had to become both mother and father to Gordon and Kristi-Anne. It affected the kids, though I didn't realize how much until years later when Betty Anne told me about Gordon's day at school. When the teacher asked what his father did for a living, he answered, "I don't know. He's not home long enough for me to find out." It was a smart-aleck answer, but there was a message in it, too. I just wasn't ready to fully understand it.

We did try to simplify at STG. While we were still writing down questions, we split the upcoming activities into two pieces. One was the early tests. We could use mock, or boilerplate, capsules to test the escape system, heat shield, parachutes, and other pieces of the puzzle. We could do a lot of that right here at Langley, or in the Atlantic Ocean next door, or at facilities provided by whatever company won the contract. We'd be in control of the tests, and the results would feed right back into the ongoing design. That part of the future felt good. Testing pieces of a space capsule wouldn't be that much different from testing airplanes, and in that field, we were the experts. But in the second leg of our test plan, we were still in kindergarten.

That part required real rockets, and we'd have to launch them with unmanned capsules before we could risk putting a man on board. We'd need support from the Atlantic Missile Range, run by the Air Force out of

Florida, if we were going to get a rocket launched. We'd need active help on the rockets themselves from the Army Ballistic Missile Agency at Huntsville, Alabama, and from the Air Force Ballistic Missile Division in California. And that meant getting official support from the Department of Defense, without letting the military actually take over. We'd have to seek out the right people, cultivate them, and learn from them. It wasn't exactly like going hat in hand. But there were going to be moments in those early days when we'd have to talk little and listen a lot.

A trio of smart fellows from the Lewis Laboratory in Cleveland stood out in these early days. G. Merritt Preston, Scott Simpkinson, and Andrè Meyer came down each Monday, stayed until Friday, and offered keen input into both Faget's capsule design work and our growing operations plan. They'd all end up handling operations that no one could conceive of at the moment. One of our own guys, Jerry Hammack from the Flight Research Division at Langley, quickly schooled himself in parachutes and recovery operations. Everybody's way of thinking was shifting from airplanes to outer space, and we were absorbing information, concepts, and even some basic physics faster than we ever had in our college days.

Gilruth pushed, prodded, and advised on every step. And then we were ready. In barely a month, the man-in-space RFP was finished, reproduced, and mailed to forty big aviation companies. Gilruth gave them two weeks to look it over, then show up for a bidders' conference at Langley on November 7. Thirty-eight of them did. They didn't see much of Bob Gilruth. He was a dynamic and innovative engineer and manager. But he wasn't much to look at—pushing middle age and already grandfatherly—and he wasn't a great public speaker. On that day, he established a pattern that would last for the rest of his career: He welcomed the crowd, said a few words about the importance of what was to come, then sat down. He sent his key people out front to deliver the briefings, engage in the debates, and in some cases become the faces and the personalities of American space exploration. He sat to one side, listened, absorbed, offered a quiet comment or a whispered observation, and not many even noticed that he was there. Bob Gilruth would always be a shadow compared to the names that became famous. But his guidance and his management were anything but shadowy; he ran America's man-in-space programs from their moment of creation, and all that flowed from that day forward came because of his vision and his leadership.

At the bidders' conference, Max Faget, Aleck Bond, Andrè Meyer, and some others were onstage. By the end of the day, they'd delivered the heart of the new program to the industry, answered a barrage of questions, and listened to scores of expert comments and suggestions. Some of those were incorporated into the final product of a detailed package, "Specifications for a Manned Space Capsule." It was shipped to the nineteen companies that were still interested on November 14. Gilruth gave them just four weeks to absorb it all. He had all of us on a fast track in this space race. Bids were due back on December 11, and he told us to plan on working through the holidays as we evaluated the various proposals.

"If they can write a proposal and submit a bid in four weeks," he told us, "then we can do our part and recommend a winner in four weeks, too."

I look back now and shake my head. Today the government may spend a year thinking about, then preparing, a Request for Proposal. There's a cover-your-ass mentality that has to dot every *i*, cross every *t*, and get some lawyer's initial on every page. It may be six months before contractors have to respond. Then the evaluations stretch toward infinity. By the time a major contract is awarded, the original idea may be several years old, technology has passed it by, politics has intruded and changed the mission or the rules, and the inevitable redesigns consume more years. The International Space Station is a perfect example of a bad plan followed by more than a decade of muddled modifications in both design and mission. If we'd had a 2001 bureaucrat running the Space Task Group in the late fifties, the Russians would have clobbered us in space, and the last forty years of the twentieth century may have played out far differently than they did.

On December 11, 1958, the tables were piled high with proposals from eleven hopeful contractors. We were ready. Charlie Donlan had divided us into teams, to evaluate certain parts of each proposal and rank them according to criteria and scoring methods that we'd set up in the past month. I was on several teams, looking at manual and automatic control systems, cockpit design and displays, caution and warning systems, pilot training, and flight test plans.

Christmas and New Year's Eve are blurs in my memory. I divided my time between family and work, with the family getting the short end. We delivered our scores to Donlan and Gilruth after three weeks of nonstop study and evaluation. My teams ranked Grumman at or near the top of our technical scoring. The McDonnell Aircraft Corporation wasn't number one on any list. At NASA headquarters in Washington, other teams had been looking at the financial and management sections of the proposals. Now they put them all together. We heard later that Grumman and McDonnell had tied. The final decision was up to T. Keith Glennan, NASA's administrator.

Grumman was overloaded with contracts for Navy airplanes, he said, and a new contract might disrupt national defense. On January 12, 1959, he awarded the contract, now called Project Mercury, to McDonnell.

A few years earlier, James McDonnell, president of the company, gave a speech in which he said that America could put a man in space by the end of the century. Now he had the job of building the capsule.

The deadline for getting it done was less than three years away. We didn't intend to fail.

The McDonnell Aircraft Company signed the Mercury contract in early February 1959. Company engineers hadn't waited. They'd been peppering us with questions from the day after the announcement was made. A signed piece of paper was only a technicality to them. They were as anxious as we were to get this thing rolling. We had some of the answers for them, but in the grand scheme of things, we were all still students with so very much to learn.

We didn't sit on our hands. Months before, while we were waiting for the companies to send us their proposals, we filled the time by learning everything we could about the military's missile operations. Bob Gilruth organized trips to the Air Force's Atlantic Missile Range at Cape Canaveral and to its Ballistic Missile Division in Los Angeles. We'd be working with the Air Force during launch and recovery on our early test flights, and getting to know the people and the organizations was a high priority. Gilruth put Chuck Mathews and me on the travel list, along with a few others in his immediate circle.

The trip was an eye-opener. The Air Force resented being shoved aside in the manned space business by this new National Aeronautics and Space Administration and wasn't going to make things easy for us. Individual Air Force officers took it down to a personal level and made it clear that it wasn't just NASA they didn't like. They didn't like Bob Gilruth, Chuck Mathews, Chris Kraft, or any of the rest. So the first battle we fought in the space race wasn't with the Russians but with our own military. We had to walk carefully, talk quietly, and find a way to win them over.

At the Cape, Gilruth arranged for briefings by both the Air Force and by the Pan American Corporation, its civilian support contractor. The Pan Am people were eager and helpful. No doubt they pegged us as a major customer for their services, and they were right. Air Force officers weren't so friendly. We were something worse than just a competitor: We were novices. The recovery guys at Patrick Air Force Base were particularly haughty. We were outsiders coming hat in hand to hear their gospel. And even though we knew that they'd had problems recovering payloads from the downrange Atlantic, we played the game because we needed them. It paid off when the smell of disdain in the way they treated us gradually faded and they started to answer our questions, volunteering information and anecdotes from their own experience in launching payloads and then looking for them in the vast ocean downrange.

For the first time, we heard about SOFAR bombs and SARAH beacons. We were instructed in the varieties and vagaries of dye markers. SOFAR is the acronym for "sounding, fixing, and ranging." A SOFAR bomb sinks to a preset depth in the ocean before exploding. Navy sonar receivers triangulate on the sound and pinpoint its location. SARAH beacons, for "search and rescue and homing," are ultra-high-frequency radios that send out a signal that allows search aircraft to home in on it. We quickly saw that we'd need to incorporate both devices into our Mercury capsule, as well a dye-marker package that would stain the seawater bright green or blue so it could be spotted from airplanes. We got some fast insight into the dangers of treading on Air Force turf when one colonel let us know, in no uncertain terms, that he, and he alone, was in charge of recovery operations at the Atlantic Missile Range. We did out best to stay on his good side.

It was even harder with the range safety people. Their job was to protect Florida's coastline and citizens from wayward rockets, and if it's possible to take such a grave responsibility too seriously, they managed to do it. Now we were adding a man on board to the equation, and I had to negotiate a whole new set of arrangements. Range safety's rules, regulations, procedures, and paperwork tormented me from the first briefing until that day far into the future when we reached a semirational compromise.

The questions were important: Who controlled the abort button, the Air Force or NASA? I wanted to control a capsule abort, but did that give me the right to blow up a rocket? Or if an Air Force range-safety officer needed to blow up a rocket, could he give me enough warning to save the Mercury capsule? What were the geographic lines that couldn't be crossed by a wayward rocket? What failures in the rocket or the capsule demanded an instant abort, and what could we live with while we studied a problem? More than anything, the overriding question was this: Who's in charge?

Even the damned paperwork just to get permission to use the range was mounds of gobbledygook and forced us to hire extra clerks. Everything had to be written down, and the Air Force had a set of acronyms and hieroglyphics that seemed like a foreign language. RSO was the range safety officer, but SRO was the supervisor of range operations. TM was telemetry. CMD was shorthand for command, and *verlort* was "very long-range tracking," a special kind of radar system. There were hundreds of these and we had to learn them all.

In volume after volume of paper, we had to tell them exactly what support we wanted, then provide detailed specifications on everything down to systems on the capsule itself. They wanted everything spelled out, from the size and shape of our payloads to the day and hour that we'd ship hardware to the Cape. I was just glad they didn't ask for the minute and second.

We came away from the Cape with our heads buzzing, but still with an exhilaration of spirit. We'd breached the enemy's defenses, learned more in a few short days about launch and recovery operations than we'd known existed, and even made a few friends among the Air Force and Pan Am people. We didn't know most of the answers yet. But for the first time, we knew a lot of the questions. And if unofficial Air Force policy said that we NASA types weren't very likable, the official line was that we'd get their help in the fullest measure. The real enemy was the Russians, and damned if we weren't going to beat them.

After a few days back at Langley to sort out what we'd learned at the Cape, Gilruth, one of his assistants, Paul Purser, Mathews, and I caught a flight out of Baltimore to another branch of the Air Force waiting for us in Los Angeles. This was the Ballistic Missile Division, commanded by General Bernard Schriever, the man who ramrodded development of America's first intercontinental ballistic missile. We wanted some Atlas rockets, and the only store in town was BMD.

We quickly discovered that ordering an Atlas rocket was not like ordering a new altimeter from a scientific catalog. The Air Force had procedures we'd have to follow. *Procedures* is another word for *paperwork*. When we mentioned that we'd need some modifications to our rockets—interfaces and fittings for the Mercury capsule, maybe some additional instruments and radiotelemetry gear, and a few other minor items—we ran into another set of procedures. It became clear that "our" Atlas rockets would always belong to the Air Force. The positive side was that both the Air Force and the Space Technology Laboratory felt responsible for the rocket's success. There is no such thing as a money-back guarantee in the missile business. But we left Los Angeles after days of briefings, meetings, and negotiations with the next best thing. If the rocket blew up, it was their fault. If it didn't, everybody's attention would be on that Mercury capsule up there in space.

These early meetings were my first exposure to the excitement a manned space program was generating in the aviation industry. The Convair people invited us to make an early trip to San Diego, and we knew without doubt that they'd go the extra mile to build us the best Atlas possible. That can-do, will-do attitude preceded us everywhere we went. American industry wanted to play a big role in putting American men into space, and in beating the Russians on this new high frontier. A few companies, very few, wanted in because there was money to be made. Even then, they were determined to deliver high-quality goods, and to do it on time or sooner. It wasn't just an American life that was at stake. It was America's future.

With NASA only a few months old, American industry and American technology were embarking on a voyage that would be the high point of the twentieth century. We still couldn't see that far ahead. Just seeing the day after tomorrow was hard enough.

Somehow during the last months of 1958 and the first quarter of 1959, I stopped being a flight research engineer and became an engineering manager. I could hardly remember getting up in the morning to go to work on gust alleviation. I was healthy, I was motivated, and I loved my jobs—even if my multiple assignments meant sleeping less and losing weekend time with Betty Anne and the kids. Except for the skirmishes I'd had over the F8U with the Navy Bureau of Aeronautics and the Chance Vought Company, this was a different world, and it held some alien beings. I'd spent my career working on problems with people I liked and respected. We could disagree, but we'd still make good decisions. Now I was confronted with some people whose motivations seemed sometimes strange and onerous. Management seemed to boil down to win/win, whether it involved negotiations, communications, or conflict resolution. I was facing people whose attitudes were different from mine. We weren't friends in a room looking for a solution. We were combatants looking for the upper hand. It didn't taste right, but it was exactly the challenge I needed. I once heard a sermon about the righteous man's dilemma of ending up in heaven with people he didn't like. In my new space assignments, I was in heaven all right. And I didn't like some of the people I had to deal with.

Some of the turmoil was internal. Any new organization generates instant politics. The biggest question inside NASA was, Who will own the Space Task Group? We couldn't stay at Langley. We'd overwhelm its relatively small facilities. A new center was already being established at Greenbelt, Maryland, to be called the Goddard Space Flight Center. Its director, the grapevine said, would be Harry Goett, and he was lobbying to have Gilruth's STG assigned to him.

I couldn't see Gilruth working for Goett. I knew Harry from his flight test days at the Ames Research Center in northern California, and his reputation as a manager wasn't good. More important, Bob Gilruth's stature far surpassed Goett's, both inside and outside of NASA. They'd be incompatible from the start. Gilruth talked about the possibility of moving STG to Maryland. "I don't know," he told me several times, "it just doesn't seem like a good idea." But then he'd brush it off in the press of other business.

Bob Gilruth did have friends in Congress, and he had the ear of Keith Glennan, the NASA administrator. Gradually the idea of making the Space Task Group a subordinate piece of the new Goddard Space Flight Center faded. Support for the manned space program in Congress and in the press was blossoming. We didn't know where we'd go, but it wouldn't be to Maryland.

5

Inventing Mission Control

ach day brought more lessons about the enormous differences between flying airplanes and flying into space. Airplanes are built at airports. The newest, most radical airplane rolls off the assembly line, goes through some ground testing, then taxis over to a runway and takes off on its short, first flight. When we tested airplanes at Langley, a pilot flew the thing to us over terrain that was familiar and well mapped. Nobody packed up an airplane and shipped it to us by rail, truck, air freight, or barge.

Going into space was something different entirely. Our rockets would be built in San Diego or Huntsville. The Mercury capsule would be built in St. Louis. Everything would be shipped to Cape Canaveral, tested, assembled, and tested some more before launch. The logistics of pulling it all together sometimes drove us to our knees.

But before any of that could happen, we had to provide a "design reference mission." The fine points of the mission changed as we learned more, but the core requirements were locked in place. We'd put them together while working on the Request for Proposal, partly using our airplane experience and partly taking our best guess about what we'd need. The water landing versus land landing question was settled early. Communications was another question.

We'd always had full-time communications with our airplane test pilots. They were never that far away. But the Mercury capsule would be over the horizon and gone within minutes after leaving the launch pad. And we needed more than a voice link with the astronaut. We needed to track the Atlas to orbit, then follow the Mercury capsule around the world. And we needed a flow of data about critical capsule systems.

The question of what to call our future space men had been settled by now. It came down to *astronaut* or *cosmonaut*, and *astronaut* won. When the Russians began calling their space men cosmonauts, we couldn't have been happier. The nomenclature was a simple way for the world to know who was who.

Our trip to the Cape gave us a new concern. The Air Force radars were good, but an Atlas on its way to orbit would be so far out over the Atlantic that it would be almost at the horizon when viewed from the Cape. We needed about thirty seconds of high-quality radar data to do a "short arc" calculation of whether we were going to reach orbit, and if so, what the orbital parameters were. In 1959, the short arc solution was still in the realm of theoretical mathematics. Nobody had done it for real. With the Cape radars staring at the far horizon, we weren't even sure that we'd get our thirty seconds. We needed our own radar site out there somewhere, along with a high-speed data link back to the U.S. mainland. We take these things for granted in the twenty-first century. In 1959, they didn't exist. We'd have to invent them or get somebody else to invent them. And quickly.

Bermuda was the only logical choice for our first radar site. The Air Force had a radar station there, but it would need extensive modifications to give us the tracking, voice, and other capabilities we needed to stick with our astronaut until we knew he was safely in orbit. Or otherwise. Bermuda is British territory, so now we entered the world of international politics. We were confident that the British government would approve the site modifications with no problem. But if we needed a site in Bermuda, wouldn't we need sites in other countries around the world? And would they all welcome us with open arms?

We'd arbitrarily decided that we needed to be in contact with the Mercury capsule at least every fifteen minutes. To get the basic information we wanted about space flight itself, we'd decided that the first orbital mission would go around the world three times, a journey that would take about four and a half hours. That was as much time in zero gravity as we thought a man should have at first, and three revolutions (orbits) would give us enough radar data to accurately control the landing point. But because the world was turning below, Mercury's track over the ground changed with each revolution. And should we also have a ground site in contact with Mercury at retrofire to help the astronaut with that tricky maneuver? Yes, that was mandatory. Finally, we wanted to track every second of Mercury's descent from orbit until it splashed down in the Atlantic.

We'd started developing the design reference mission to give the space capsule contractor—now the McDonnell Aircraft Corporation—basic guidelines on what the capsule had to do, and to give all of us some insight into recovery operations. Along the way, we discovered something else: We needed a worldwide network of tracking stations. That would soon turn into a major project of its own, involving friendly and unfriendly foreign governments, the U.S. State Department, the Pentagon, civilian contractors, and a lot more.

We were experts from the airplane world, but we were novices and naive about this new world of space travel. This whole business of putting a man into space was getting complicated. And we were too focused on getting the day-to-day assignments completed to see that it was only going to get worse.

It did, and quickly. April 1959 might not have been the cruelest month in NASA's short history, but it had its share of turmoil. We had more than two hundred people on the payroll now, and we still couldn't find a moment to relax. We traveled as much as possible in the evenings and on weekends to keep the work day open for work. I even bought a second car to get to work, or to leave at the airport, when being a one-car family put too much stress on Betty Anne in her sudden new role as both mother and father to a pair of active children.

That automatic abort system we wanted on the Atlas had been made a requirement for the Redstone rocket, too, and neither was going well. Bob Gilruth must have figured that I didn't have enough to do because he assigned me to oversee the work at both the Redstone Arsenal in Huntsville and at Convair in San Diego. I was eating more meals at my desk or on airplanes or in conference rooms than I was at home, but I was thriving on it and that old ulcer never did come back.

We were soaking up rocket technology daily. It quickly became obvious that if something went wrong during a launch, only an automatic system could save the astronaut and his capsule. But if an abort system was to be truly automatic, it had to sense that bad things were happening inside the rocket.

"What systems can cause a catastrophic failure," we asked, "and what measurements can forecast that failure? What if the rocket fails to stage?"

From the first reactions by the rocket experts at Convair and the Space Technology Laboratory (STL), we might have been asking about their private lives. They were particularly incensed at the staging question. They'd never had an Atlas fail to drop off its two booster engines and then continue with the single remaining sustainer engine. I made a mental note during that discussion. *Be prepared to deal with a rocket that doesn't stage.*

In meeting after meeting, the Convair engineers didn't want to admit anything about forecasting failures. In private conversations, some of them over drinks at night, we found out why.

"You're asking us to identify weak links," an engineer told us. "So we'd be admitting that weak links even exist. Do you think the Air Force is going to like hearing that?"

"If we figure out how to monitor and predict a failure," another asked, "does that mean we're responsible for saving an astronaut's life? I don't think we want that on our backs."

"You could be asking us to shut down a rocket that might conceivably survive," still another told us. "And you're asking for a whole new set of electronics that doesn't exist for any missile. That system has to be more reliable than the rocket itself if we're going to avoid shutting down a good rocket."

It sounded as if they leaned toward saving the rocket over saving an astronaut's life. The more the subject came up, and it came up a lot in the first months, the madder I got. "Damn it!" I told Bob Gilruth and Chuck Mathews. "We've got to have an automatic abort system."

They backed me completely. We told Convair and STL to design and implement an abort system for the Atlas rocket. There was grumbling and groaning, but they did it. They also extracted a trade-off. NASA had to take responsibility for Atlas changes and system additions, because neither the Air Force nor STL would jeopardize their reputations by approving something that they thought might increase the risk of failure, even though the benefits made the risk worthwhile. Time showed that Gilruth and his engineers could challenge their expertise at any level, and when it came to the final argument, it was our vehicle to do with as we chose.

The first abort system oversight meeting—now it was called the ASIS, for Automatic Abort Sensing and Implementation System—in San Diego set a pattern. No more than a dozen of us were actually involved, but the conference room was filled to overflowing with hangers-on eager to be part of the act. That was the day I met Phil Culbertson. He was a lead project engineer on Atlas, and we'd eventually have a long and successful professional relationship. But on that day, he was strictly the guy at Convair responsible for the ASIS.

I wanted everything I could get that might tell me that an Atlas was about to blow up, go off course, be underpowered, you name it. Culbertson and his engineers wanted the simplest possible system, with a minimum of additions to their Atlas. In the meeting, the remnants of the stonewall attitude still held on. Engineers talked at length about how their system worked. But when we asked how it failed, we usually got a blank look. The Convair engineers would scratch their heads or go into a huddle to figure out how to tell me what *might* go wrong. After the first few times, it got easier. They were good engineers, and I'd handed them a challenge that they eventually couldn't resist. So I got most of what I wanted until we moved into a heated discussion about the reliability demands on the system. Then Culbertson asked a tough question.

"How much reliability do you want?" he demanded. "What kid of numbers are you looking for?"

In a perfect world, I wanted perfect reliability. But Culbertson's question took me back for a moment. I knew what he was really asking: *What will you settle for?*

Later on, we established a NASA policy and the answer was easy: "Three nines." On critical systems, we wanted 99.9 percent reliability—one chance in a thousand of failure. But three nines didn't exist at the moment, and if Culbertson had me stumped, I wasn't going to admit it.

"We want the best and most reliable system we can get," I said.

Culbertson wasn't happy with that. "C'mon, Chris," he blustered. "That's no answer."

"It's what you're going to get," I snapped, proving that I could be as obtuse and stubborn as the most seasoned bureaucrat. "You figure it out and give me the best you can do." That made it a different kind of challenge, and I could see his reaction in the way he clamped his mouth shut and almost broke a pencil. He'd give us the best, and we damned well better like it!

That trip and a few others convinced me that we needed a full-time

NASA engineer on the West Coast to work with the Air Force, Space Technology Laboratory, and Convair. We talked it over and transferred Robert Harrington from the Cape to be my man in Los Angeles. He was a bulldog in keeping everybody on the ball and took a big load off my shoulders.

That left the Redstone rocket in Alabama. Jerry Hammack and I made the much shorter trip to Huntsville. Hammack was our Redstone lead engineer and worked with Richard Smith, his counterpart in von Braun's organization. The Redstone ASIS was a much simpler system, on a simpler rocket. It didn't take long for us to be comfortable with the work. Smith was digging out everything we'd learned on the Atlas and was translating it to the Redstone. That trip was my introduction to the Army Ballistic Missile Agency. I met a lot of engineers and managers. But I never laid eyes on Wernher von Braun. An antipathy between von Braun and Gilruth, and thus between von Braun and all of the rest of us, was simmering just beneath the surface.

Earlier in 1959, NASA made overtures to the Army about transferring von Braun and his full contingent of German and American engineers to us. The answer was a quick no. A lot of us guessed that von Braun was hoping for leadership of his own space program and that working for NASA wasn't in his plans. The man was brilliant, maybe the archetype for the term *rocket scientist*. But he had an ego and could be stubbornly opinionated. I'd find that out for myself before the year was over.

Bob Gilruth had a better feel for von Braun's ambition. One fine spring day, I was at lunch in the cafeteria with Gilruth, Max Faget and Chuck Mathews. We were discussing the Redstone rocket and our relationship with the people at Huntsville.

"It seems strange to be working with the same people we hated during the war," I said. "And I didn't get a chance to meet von Braun. What's he like?"

Gilruth looked up from his salad and gave me one of those looks that said *This isn't a good subject*. But after a moment, he found the words to describe everything he felt about the German rocketmeister in one short sentence:

"Von Braun doesn't care what flag he fights for."

We were already running into difficulties from an unexpected quarter. Most of us had been around airplanes and test pilots long enough to have little doubt that medical problems would be minor. A lot of flight surgeons agreed with us, but then the "medical community"—an elusive term for an even more elusive bunch of anonymous people—got involved.

Our troubles started in the White House when President Eisenhower appointed a group of science advisers. This President's Scientific Advisory Committee (PSAC) appointed its own medical advisory teams. Almost immediately, that new subset of advisers began telling PSAC horror stories about what could happen in space. We heard about it secondhand, and our first reaction was to laugh and shrug it off. Then we found out they were serious. The list of space dysfunctions being passed around the White House was enough to frighten a bureaucrat out of his starched white shirt. The concerns included loss of orientation, blindness, inability to eat or swallow, nausea and vomiting, panic, heart attack and brain damage, difficulty in breathing while enduring high g forces, rapid heart rates leading to unconsciousness, complete loss of bodily control, sudden and undetected ulcers, and a lot more. It was ludicrous. But important people were listening, and they started asking us questions that we couldn't answer. A major objective of Project Mercury was to determine man's ability to perform in space, particularly in zero gravity. But these guys wanted us to have the answer first, then fly later.

The Space Task Group was made up of engineers, not physicians, so Gilruth asked for help from the Department of Defense. Three doctors were assigned to help us sort things out: Dr. Stanley White from the Air Force, an experienced flight surgeon; Dr. William Augerson from the Army; and Dr. Robert Voas, a Navy psychologist who was allegedly an expert in flight training. At least in part, we were letting doubters into our camp, and we paid the price. Voas wasn't nearly as good as his advance notices, and I quickly learned not to trust much of what he said. He spent too much time trying to match wits with us and delighted in twisting our tail with strange theories or stranger demands for tests. Augerson was a good doctor, and even though I liked him, I thought he was a purist who often demanded more than the system could deliver. Only Stan White understood what we were up against and helped us to fight the "medical community" battle.

Then Keith Glennan, the NASA administrator, compounded our problems by asking Dr. Randolph Lovelace to chair a committee overseeing the medical aspects of Mercury. Lovelace operated the Lovelace Clinic in Albuquerque, New Mexico, to test and observe humans under high stress. All of the astronaut candidates would go through hell at his clinic. It was survival of the fittest, and only the best physical specimens became astronauts. Randy Lovelace was clearly one of the world's leading experts in how humans respond to physical stress. But his committee created sensational-ism over the question of man's abilities in space and, instead of helping NASA, only intensified the doubts. I knew any number of flight surgeons who thought that man would be just fine in space. But Lovelace's reputa-tion and the notoriety made it difficult to challenge him.

The only way out was to put both humans and a bunch of other pri-mates, mostly chimpanzees, through a series of onerous and costly tests. It was just one more item being added to our to-do list.

Lovelace caused us more problems, too, when he allowed some female volunteers to go through his tests without our concurrence. One of them was the wife of a U.S. senator, and before long we were being dragged through the "Why no women astronauts?" controversy. Nobody seemed to understand that President Eisenhower had ordered us to pick astronauts from active-duty military test pilots, and none of them were women. We had our orders, and the subject of including women never came up until it was raised by outsiders.

We got two infusions of people in April 1959. The first was just seven men. Gilruth, Charlie Donlan, and a couple of others had been sorting through files, reports, and recommendations to find just the right men to become astronauts. Hundreds of candidates had been whittled down to thirty-one, who then went through a hellish series of medical and psycho-logical tests. Gilruth wanted to bring six of them on board. One by one, they picked five. But two guys were a toss-up for the last slot. Gilruth looked at the files and gave the order: "Take them both."

I never knew who the two toss-ups were. But I can guess about one of them, and he gave me enough trouble one day in space that I made sure he was grounded forever from going back into orbit.

The seven astronauts—Scott Carpenter, Gordo Cooper, John Glenn, Gus Grissom, Alan Shepard, Wally Schirra, and Deke Slayton—were intro-duced at a press conference in Washington, and NASA was never the same again. To everyone's surprise, including the astronauts themselves, they were instant heroes. The press fawned and fell all over themselves trying to get photos, interviews, and comments. The astronauts' homes and families were put under siege by reporters. Friends, relatives, and grade school teachers were interviewed and quoted. It was a feeding frenzy.

By the time the astronauts reported to work at Langley, the media cir-cus had shrunk from three rings to one. At Langley, we'd been mostly ignored by the press, but now reporters and photographers were the norm. Our initial reflex was to duck our heads and run for cover. But before long we were immune to the attention lavished on the astronauts, and that spilled over onto us now and then. If giving an interview or a press briefing was to be part of our job, then so be it. We'd fit it in somehow, and most of us felt that we had an obligation to tell America what we were doing. It turned out to be more of a challenge than we expected. The aviation and science press knew even less about space than we did. Many of them would learn, even become pretty smart, but with a steady flow of new reporters as our space programs grew, it seemed that every one of them had to be educated.

Gilruth wanted the astronauts to be involved in the nuts and bolts of Mercury, and he told them so. They were all engineers and supposedly experienced test pilots. That wasn't quite true. Carpenter had virtually no flight-test hours in his log book. Glenn had more, but nothing compared to the others. Maybe that was why Carpenter gravitated quickly to Glenn's side; the two became close friends.

Each of the astronauts was assigned to a specific part of the program where Gilruth expected him to become a near-expert and to be a vocal proponent of the "astronaut point of view." At the same time, all seven were supposed to become knowledgeable about everything. Donald K. Slayton—"Deke" to everyone—drew the Atlas and Redstone, with a secondary assignment of getting familiar with flight operations. That put him squarely in my domain, and we quickly became friends.

Like all of the early astronauts, Deke wasn't a big man. Nobody over six feet tall could fit into the Mercury capsule, and the original seven ranged from Gus Grissom at five-seven to Al Shepard at five-eleven. Deke wore his hair close-cropped, and his craggy face made him look as tough as he really was. He was a test pilot through and through. The only time I saw even a slight indication that he was afraid of anything in the air or on the ground was when he had to be in a press conference or give a speech. But he grew into that duty, too, and eventually handled his persona like a pro. Gruff, but a pro.

Deke brought the others into flight operations, to meet us, get acquainted, and learn the ropes. Away from the spotlight, they were just regular guys, and my operations people warmed to most of them quickly. But again, it was Glenn and Carpenter who didn't show a lot of interest in our end of the business. As I watched him over the months, I decided that John Glenn's goal was to cozy up to our top management and thus improve his chances of getting one of the early Mercury flights. Scott Carpenter tagged along obediently. I had the old F8U confrontations in the back of my mind, remembering Glenn from his days on the Navy Fighter Desk. Now his attitude toward flight operations only confirmed my private opinion that Glenn had his "head up and locked"—a reference to a pilot missing the obvious by landing with his gear up and locked. But it didn't bother me. I knew that when the time came, flight operations was going to be the most important thing in his professional life. If he didn't come around now, he would later.

Our second set of new arrivals came from a most unexpected place. A group of good aeronautical engineers from England had been recruited by the Canadian company A. V. Roe to work on the Arrow, an advanced jet fighter. They were nicely settled in Toronto when the Canadian government suddenly canceled the program. The news came over the public address system and left the engineers instantly jobless.

Charlie Donlan heard about it and immediately headed north with some of his people. They met with James Chamberlin, one of the group's leaders, started matching résumés with NASA's needs, and then made a series of job offers. Chamberlin and twenty-four others jumped at the chance to work on the American space program. In one bunch, we got engineers who would make major contributions to getting us into space. Chamberlin immediately took over the Space Task Group's Engineering and Contract Administration Division. The others were salted throughout our technical divisions.

I got five of them, all experienced in flight testing and operations. One was a Welshman named Tecwyn Roberts. He quickly became one of my key people, and a friend, too. Along with another transplant, John Hodge, he was a major player in helping me design the room where we'd all work while an astronaut was in space. Like everything else, its official name eventually came with an acronym—MCC, for Mercury Control Center.

Before we could invent mission control, we had to know what we were controlling. So we got on with the job of inventing the Mercury capsule and overseeing the McDonnell people who would build it. After looking at the heat-resistant shingles they wanted to use on the capsule to dissipate reentry heat, and walls of drawings, I sat down with a trio of McDonnell engineers to talk about the Mercury cockpit. John Yardley was the project engineer, and the big question in his mind was what information and instruments to give to the astronaut. We started with the rocket, but still thinking in airplane terms—instruments to show rocket thrust, fuel consumption, that sort of thing. But it quickly became obvious from Air Force input that things happened so fast in rocketry that an astronaut couldn't do anything anyway. Our abort system would have to be automatic, firing the escape tower rockets and pulling the capsule away from a failing, and maybe exploding, rocket.

Still, pilots were pilots. They'd want to know something. Bob Foster was McDonnell's expert on caution and warning systems. "Let's give them indicator lights," he suggested. "One for each rocket engine. It's green for a good engine and red for a failed engine."

I thought about it for a few moments and decided that it was a perfect solution. A green light would give the astronaut a feeling of security. A red light would warn him to brace for possible abort. And the whole thing weighed next to nothing, compared to bigger and more elaborate instruments.

That led us to think about the overall system. We'd thought about three lights for each system being monitored: green for a good system, yellow if the system was starting to fail, and red when it failed.

"What can the astronaut do if the light turns yellow?" I asked. "Can he fix something?"

The McDonnell guys looked at each other. "No," Yardley finally said. "He's crammed into the capsule too tight to do much. And all the systems are sealed away or behind panels anyway."

The more we talked, the more obvious it became that it was going to be up to us on the ground to handle most problems up there in space. It was a sobering thought. We'd need to become experts on everything packed into a Mercury capsule, and we'd need to decide beforehand when a system still warranted a green light and when it should turn red.

"Forget the yellow light," I said. "We don't need it and it'll only give the astronaut something to worry about. We'll do the worrying for him down here."

By now, I was considered to be one of Gilruth's senior staff, and there wasn't much in Project Mercury that I didn't have a hand in, or at least a finger. But the program was growing so rapidly that we seemed to be bringing new people in almost daily, and then dumping this issue or that issue in their laps. We "senior staff" still had the management responsibility, but the details had to be handled by people who could work on the task full-time. And usually, full-time meant working a sixty-hour week or more.

Our tracking network was a typical case. The more Howard Kyle and I wrestled with the network requirements, the more we realized that we had a tiger by the tail. Even a simple look at a globe showed us that a Mercury capsule in orbit would pass over a lot of water and a lot of foreign countries. We'd need tracking sites, with radar, voice, and telemetry capabilities, somewhere in the South Atlantic, in Africa, on down to Australia, then up across the South Pacific.

Hawaii was already geographically perfect; Mercury would pass close to overhead and we'd need a site there. After that, the capsule would come up over Mexico on its first pass, then cross the California coast on its second and third orbits. The Air Force tracking site at Point Arguello, part of its Western Test Range, would be right below for retrofire. By adding some sites across the southern and southwestern United States, we'd have full-time coverage during the critical reentry from space.

Solving all the problems—diplomatic, logistic, and financial—was simply more than Howard and I could dream of doing. We explained the complex enormity of the tracking network to Chuck Mathews and Bob Gilruth. Within weeks, Abe Silverstein, Gilruth's nominal boss and assistant administrator for manned space flight at NASA headquarters, had listened to Gilruth's plea and created a new organization called the Tracking and Ground Instrumentation Unit. It was based in Langley's Instrument Research Division, and its acronym, TAGIU, was completely unpronounceable. We called it Tag-Eee-ououo.

Howard and I briefed the TAGIU people, most of whom we knew, on what we needed and handed them stacks of notes and papers. What they did next still boggles my mind. It was a majestic example of what a group of bright, motivated people can accomplish when time is short and the crunch is on. Almost immediately, they had specific sites targeted around the world. Seven of them were in countries that required the State Department to open serious diplomatic discussion, and to do it quickly. Two of the sites were in midocean, requiring negotiations with the U.S. Navy about modifying ships to handle the NASA tracking hardware.

Within days of getting their basic plan approved, TAGIU engineers wrote a detailed Request for Proposal to find a contractor who could help design, then build and maintain, the sites at far-flung locations, some of them so remote that the word *primitive* was the accurate description. And barely three months later, they'd awarded the contract to a team headed by the Western Electric Company. The Bendix Company would supply the tracking and communications gear. Bell Laboratories would do overall design. IBM would provide the computers and software. And the Burns and Rowe Construction Company would do the brick-and-mortar work.

The TAGIU guys would put tracking sites on the Canary Islands off the west coast of Africa; at Kano, Nigeria, in Central Africa; in Zanzibar on Africa's east coast; at Muchea, near Perth, and at Tidbinbilla, in Australia; and on Guam and Canton Island in the Pacific.

Then they'd modify a U.S. Navy tracking site on Kauai, Hawaii. They'd modify an Air Force site at Point Arguello, California, which was already part of the Air Force Western Test Range near Vandenberg AFB. New sites would be built at Guaymas, Mexico, and at Corpus Christi, Texas. Finally, they'd modify two of the Atlantic Missile Range's ships, the *Rose Knot Victor* and the *Coastal Sentry Quebec*. Those ships could be stationed as needed in the Atlantic, Pacific, or Indian Ocean.

Of course, there was the major site in Bermuda where they'd modify an existing Air Force installation and where we'd eventually build a backup control center.

Howard Kyle kept track of the project, and his reports left me in awe. We needed the network up and running in a year, during the last half of 1960, and TAGIU was going to deliver. It was thankfully out of my hands, except for developing the operational requirements at the sites and eventually providing the people to work there during missions. More than forty years later, I look at a world map and still consider it one of the major accomplishments of America's space program. We literally wired the world for communications. Nothing like our network had been done before.

The Space Task Group was growing faster through 1959 than any of us could track. One of the newcomers was a colorful little guy sent over by the Air Force at Gilruth's request to handle STG's relations with the press, and to be a buffer between the astronauts and reporters. Lieutenant Colonel John A. "Shorty" Powers was maybe five foot six, but his voice and his brash manner were at least six-eight. Shorty taught us about dealing with the press. He'd done it on defense programs and knew many of the key national reporters on a first-name basis. But even Shorty was a novice at dealing with the high-level attention the astronauts were getting. We made mistakes along the way, in how astronauts were made available to the press and in the events we sometimes staged just to get as many reporters as possible satisfied that they'd gotten to see something or to talk to an astronaut. Mostly we were trading off time. If we could make twenty or forty reporters happy in a two-hour briefing, it was better than submitting to individual half-hour interviews.

Sometimes that led to charges that we were orchestrating events to exploit the astronauts for political and budgetary purposes. Nothing was further from the truth. We weren't that devious, and definitely not that sophisticated in manipulating the news media. We were research engineers and test pilots by profession, and even with Shorty Powers guiding us through the minefields of publicity, we were nothing but novices in dealing with the press.

Eventually NASA headquarters stepped in and negotiated a "personal story" contract between the astronauts and *Life* magazine. I was surprised when it happened and never quite comfortable that *Life* reporters and photographers had exclusive access to the private lives of the astronauts and their families. But some of the arguments had merit. The families were being hounded by the press and needed a way to say no. There were increased expenses for the families—travel and new clothes, for instance—that weren't reimbursed by the government. And beyond the $10,000 government policy, the astronauts discovered that they couldn't buy life insurance at any price on the open market. They were facing some of the most dangerous moments in their careers, and if something happened, their families would be left with few resources.

The *Life* contract included both insurance and a hefty annual payment to each astronaut. I don't know of anybody in the Space Task Group who, after seeing the intrusions into the astronauts' lives, begrudged them the added pay. But there were other fallouts. NASA was sharply criticized for allowing the contract to be written, and other reporters developed an instant resentment of *Life* people. The contract was legal and the criticism gradually faded. And over the years, *Life* went out of its way to provide photos to NASA that were released to the press as "NASA photos." So the contract had some uses. But eventually, it faded away and none of us were sorry to see it go.

Deke Slayton and I went to Los Angeles in October 1959 at the invitation of the Society of Experimental Test Pilots. SETP was only a few years old, but its gatherings were already famous in the industry for two things: the quality of papers presented, which frequently triggered national news stories, and the amount of liquor consumed, which seldom got any publicity at all. That may have been because most of the reporters covering SETP were old-line journalists who could be found, after their stories were filed, bending elbows and swapping stories with test pilots in the Beverly Hilton lounge.

Deke was an SETP veteran, but it was my first time at the society's convention. I wanted to talk about the worldwide network we were building and about my idea of a flight control team on the ground. I knew test pilots only too well. These were rugged individualists, the modern reincarnation of the Old West cowboy. Some of them, such as Chuck Yeager, had refused to apply for the astronaut selection and were now vocal in saying that a real test pilot should have control of his craft. A Mercury astronaut, they said, was nothing but "Spam in a can." This was my chance to tell our side of the story and maybe change a few minds. Deke was even more determined. He didn't like the Spam joke at all. "If those guys think we're just along for the ride," he told me on the flight to L.A., "I'm here to tell them different."

The astronauts had only been on board for six months, but it was long enough for them to be deeply involved in the design details of the Mercury capsule. Deke was quickly becoming an expert on the Atlas and Redstone rockets and understood the escape tower system almost as well as the engineers who designed it. I used him as a sounding board for some of my ideas about flight control, and his feedback influenced me when the final design was done.

The whole concept was slowly forming in my head. I'd seen it over and over during airplane flight tests. We'd plan a test in great detail, both the pilot and the engineer on the ground knowing exactly what should happen, and when. We also knew in advance that a flight test seldom followed the plan. Something would change—the weather, fuel consumption, the need to repeat a maneuver—or the plane would react unexpectedly. The pilot was in control, but he usually accepted suggestions radioed from the flight test engineer on the ground.

Now I was moving beyond the "suggestion" phase. I saw a team of highly skilled engineers, each one an expert on a different piece of the Mercury capsule. We'd have a flow of accurate telemetry data so the experts could monitor their systems, see and even predict problems, and pass along instructions to the astronaut. We were asking McDonnell engineers to predict how their systems might fail, just as we were asking Convair engineers about their Atlas rocket. But I had a continual nagging fear: *What about the unknown unknowns? It's the surprise event that kills the pilot.*

The more I thought about unknown unknowns, the more convinced I was that we needed a team to work on those problems when they came up. And they would come up.

I got a polite reception from the test pilots. One look at a map and they understood my description of the worldwide network. A Mercury flight wouldn't be confined to the airspace over the Mojave Desert or Langley Field. There were some frowns when I explained the basic concept of a Mercury Control Center, but some smiles and nods when I added that we were going to build a procedures trainer for the astronauts—the first of ever-more-complex simulators—and that it would be tied in with the center so that astronauts and controllers could train together.

These guys were used to having their hands on the throttle of whatever they were flying, and to making it turn or climb or dive according to their input. So I went into some detail about what happens during a rocket launch and in orbit. Most of it was automatic, I had to admit. An astronaut was strictly a passenger when that rocket ignited and he began the ride into space. He'd feel some strong g forces, but even if something went wrong, there was nothing for him to do but ride it out. The abort system would sense a problem and fire the escape tower rockets faster than he could. It had to be fast. Human reflexes weren't quick enough to get away from the fireball of an exploding rocket.

In orbit, the astronaut could do more. He could fire the attitude control thrusters to point the Mercury capsule in any direction he wanted, including straight up or straight down. But his line of flight was fixed by the laws of physics and orbital mechanics. Even if he was pointed backward, he'd continue to follow the same line around the earth. It took more rocket power than Mercury carried to move sideways in orbit. When I described how the capsule and astronaut got back down, I was able to give the pilots a better feel for things. I explained how the capsule would be maneuvered to exactly the right position—blunt end forward so that the heat shield would absorb the high temperatures and pressures of reentry—before the retrorockets were fired. Again, the plan was to do it automatically.

"But all of the functions can, of course, be performed manually," I added. That brought some whistles and applause. None of us had the foresight to know how many times manual control would come into play, or just how important the astronaut would turn out to be. From the first Mercury flight to the last, he was never "Spam in a can." But without that control team on the ground, one or two might have been stranded in space.

Deke took over, and now the pilots were listening to one of their own. He gave it to them straight, and even funny. He challenged the doctors who were worrying over "the bloody astronauts," and the military planners who thought space was for a "college-trained chimpanzee or the village idiot."

With just a few lines, Deke had them on his side. I particularly liked one line in his prepared speech. It came from Deke himself, not from any speechwriter.

"If you eliminate the astronaut," he said in stern tones, "you concede that man has no place in space."

As far as Deke was concerned, that ended the debate. He talked for another half hour. Man belonged in space. He made the case over and over that day, slamming the medical community, the naysayers, even his own test pilot friends who had been so derisive. He didn't convert everybody, but he had most of them in his pocket when he was done talking.

SETP was paying our expenses and told us to bring our wives. It was Betty Anne's first trip west of the Mississippi. That night and for several more, we toured Hollywood, drove along Sunset Strip, and ate at some fine (and expensive) restaurants. Marge Slayton lost track of Deke that night after his talk. All seven of the astronauts were there, and they were quickly establishing a reputation for practical jokes. They got Deke passed-out drunk in the Beverly Hilton, then carried him on a lounge chair outside onto the Hilton marquee. The weather was warm and he didn't wake up until the sun was in his eyes.

A few weeks later, I got a call from the TAGIU about the worldwide network. The Bell Labs people they had under contract were designing flight control consoles and doing concepts for both the remote sites and for the control center to be built at the Cape. Would I like to bring some people to New Jersey for a look at their work? With a sinking sensation in my stomach, I quickly said yes.

I took a full team to Whippany, New Jersey, the next week—Howard Kyle, Tec Roberts, John Hodge, and a half dozen others whom I intended to be the core of my flight control team. Bell Labs was staffed with experts. They knew radar, telemetry, and both Teletype and voice communications. Their designs for the network's remote sites were well along, and they had a strong beginning on the control center at the Cape. That was the part I didn't like.

As strong as they were in the various electronics fields, the Bell engineers had no understanding—none—of what we needed for systems analysis, and for command and control of a capsule in space. One of their senior people, an obstreperous character named Bill Lee, was a psychologist. I never knew whether he was serious or being a smart aleck when he presented his vision of a control center.

"It's a desk with three telephones on it," he said, and then defended his position by wondering why we needed anything else.

"Is this guy stupid?" I muttered. I answered my own question by launching a series of discussions with Lee and his superiors about what we really needed in a control center. By the end of the week, we'd all reached the same conclusion. We already had the Philco Aerospace Division, out in Palo Alto, California, working closely with us on preliminary designs for consoles and sites. They'd been picking our brains and putting our ideas into sketches and schematics. It only made sense, both my team and the TAGIU engineers agreed, to have Philco take over and do the whole job.

It was the right move. We needed everything operational in the next year, and now we had some real experts working side by side with us to make it happen.

6

Rockets

We learned about failure early.

While I was still struggling to get a coherent control-center philosophy organized in my head, another part of the Space Task Group was firing rockets to test pieces of our hardware. Max Faget and his people were in charge of the Mercury capsule's overall design and manufacture. Max had been Bob Gilruth's right-hand man in NACA's Pilotless Aircraft Research Division, using leftover World War II rockets to power airplane models to supersonic speeds. Along with Caldwell Johnson and others, those PARD engineers got a ten-year-long education in how rockets work.

Even before NASA came officially into being in late 1958, Faget had Gilruth's approval to build a hurry-up rocket they could use to test parts of the Mercury capsule. They called the rocket Little Joe. A squat, one-stage thing, it could send a capsule one hundred miles high. It wouldn't go fast enough to reach orbit, but it was good enough to test the escape system and the heat shield. That is, it was good enough if it actually left the launch pad.

Gilruth and his PARD team did most of their rocket experiments from Wallops Island, just off the coast of Virginia. So that was where they put the

Little Joe launch pad. The first launch, in July 1959, was supposed to test the Mercury escape tower. Faget got a boilerplate Mercury capsule from Mc-Donnell and put a live escape system on top. If something went wrong during a launch, the escape tower's tractor rockets were supposed to fire. We'd agreed that the system would activate automatically if the ASIS detected a problem. But we also gave the astronaut an abort switch next to his left hand, put another in the blockhouse firing room with the launch conductor, and another on the flight director's console in the control center.

No matter where the initial command originated, the results were supposed to be the same. Explosive charges would cut the clamp that held the capsule to the main rocket. The escape rockets would pull the capsule far enough and high enough away to be clear of an exploding Redstone or Atlas rocket. Then the capsule's parachute would deploy and it would land safely—preferably in the water.

That first test of Little Joe and the escape system didn't go as planned. They were about thirty-five minutes from launch when an electrical component overheated and fired the escape tower. With the Little Joe still sitting on the pad, the Mercury capsule was pulled away and sent high out over the surf. Then its parachute opened and it splashed down safely. We heard the news a few hours later, both good and bad. The test was a failure. But though it wasn't part of the plan, we knew that the escape system could handle an abort straight from the pad.

On another day, one of Faget's test engineers was making a final check of the electrical connections at the base of a Little Joe. Cameras were already running to record the launch. He heard a loud click that momentarily froze him in place. Then he started running. He was a big guy, and clumsy, but the film showed him sprinting like an Olympic runner. He'd recognized the click as an ignition relay being activated. He was just far enough away to avoid injury when the Little Joe lit up and took off. The film was funny, but that didn't change the fact that the test was another failure.

Then came Big Joe. Big Joe wasn't a rocket, but the code name for the first critical test of the Atlas rocket–Mercury capsule. The two most dangerous times in a space mission are during launch, when the rocket might blow up, and during reentry, when the friction of slamming down through the atmosphere generates heat up to three thousand degrees Fahrenheit. That's hot enough to melt the metal that the capsule was made from, and if even part of that heat got through to the cockpit, our astronaut would be baked alive. How to protect the hardware and the man was literally a life-and-death question.

Engineers at McDonnell came up with several solutions. One of them was a beryllium heat shield that had the right heat properties, but was hard to machine into the right shape. There were also health problems from beryllium dust. The solution that Max Faget liked best was to cover the blunt end of the Mercury capsule with a special kind of fiberglass that didn't burst into flames. Instead it absorbed heat and slowly turned into a blackened char that flaked away. The process was called ablation. In lab tests, the more it charred, the more it resisted heat. But it was another iffy proposition, and some smart people had big doubts that it would survive the reentry forces. Yet if it did, the material would save a lot of weight. So we needed a flight test. That's where Big Joe came in.

We'd dropped the Jupiter rocket from our plans and bought our first Atlas from the Air Force for the heat shield test. It was a standard ICBM, except for an adapter at the top to hold a Mercury capsule and a simple mechanical interface. The Air Force, with less than gracious acceptance, allowed Aleck Bond, Scotty Simpkinson, and a few others from STG who were assigned to launch operations to use a small corner of one of the hangars along a row of industrial facilities. Hangar S became their home through the summer of '59 while they struggled to get the capsule ready. At the same time, Bob Thompson, my college classmate who joined us from NACA to head up the recovery effort, and his people worked with the Navy on what would be the first recovery of a Mercury capsule returning from the edge of space.

It was also our first exposure to Air Force launch operations and to the details of using the Eastern Test Range. I was back home on this one, but afterward was usually in the firing room or central control for Mercury launches. It lit off in the darkness at 3:00 A.M. The timing would give the recovery team the maximum amount of daylight downrange to find the capsule. But then things went wrong.

When I heard the story later that day, I remembered the meeting in San Diego when the Convair engineers assured me that no Atlas had ever failed to stage. I'd made a note to include nonstaging in the mission rules we were starting to write and to include it in some of our future training exercises. Now before we even got started on our own rule book, this one failed. The two booster engines didn't drop away, and the single sustainer engine couldn't keep the mission on its planned trajectory. A rule of my own took shape, too: "If somebody says that something never happens, be prepared because it probably will!"

At some point, the Mercury capsule's instruments recognized that something was wrong and began firing its attitude control thrusters. Those little jets could do nothing against the weight and power of an Atlas, and before long the capsule's fuel was gone. Then somehow it separated from the Atlas and got off on its own. It was coming down through the atmosphere when the radar and telemetry trackers lost sight of it.

We had Navy recovery ships stationed every five hundred miles along the flight path. One of them saw the fireball pass overhead. The next ship saw nothing. If it survived to make splashdown, it had to be somewhere in between. At just about dawn, a Navy search plane picked up signals from the SARAH beacon. An hour later, they spotted the green dye marker in the ocean. Mercury was bobbing peacefully on low swells, with its red and white parachute floating nearby. By 10:00 A.M., a fast-moving destroyer was alongside and hauling it aboard.

The next day we got our capsule back and the STG engineers at the Cape figured out from its onboard tapes what happened. The genius of the capsule's design had been proven. When the aerodynamic forces built up, the capsule flipped itself around to put the heat shield facing down. It was still oscillating back and forth because all the attitude-control fuel was gone. But it was stable enough to come down safely, pop out its parachute, and splash down intact.

It didn't go as high or come down as fast as we'd wanted. But the ablative heat shield proved that it could protect the capsule and the astronaut during reentry. Some small problems included a latch at the small end that welded itself shut from the heat it absorbed during the oscillations, and a few warped shingles. Max Faget put his people to work on the thermal properties of a cone-shaped body. What they found took nothing more than a minor change in shape to eliminate.

The Air Force was embarrassed by it all, and the mission did nothing to improve our relations with them. But we got a psychological lift. Mercury got a tough test that wasn't on our list, and it came through better than anyone could have hoped. There was a lift for Bob Gilruth, too. When they sealed the capsule, Aleck Bond and some of the technicians had left a note inside addressed to Gilruth. "The people who have worked on the project hereby send you greetings and congratulations," it said in part. Gilruth kept that slip of paper as one of his prized possessions.

The Space Task Group and Project Mercury were somehow pulling things together. Thousands of details—tens of thousands—had to be managed, but we were doing it. After the trips to California and New Jersey, I had a stronger vision of what our control center was going to be, and the Philco people were getting the design down on paper. Maybe if we knew exactly what we were facing up there, it would have been easier. But one thought kept driving me to ask for more and more capability: "What about the unknown unknowns?"

We had our plan for getting the data about both the man and the machine. The question was what to do with it. "Surveillance isn't enough," I said whenever we discussed what should actually happen in the control center. "We have to be more than onlookers. We have to be ready to jump in and help out."

The idea of active response took hold and we started a process that became a core principle of every space control center then and now. We started asking "What if?" What if this happens? What if that breaks? What if this happens and that breaks at the same time? It didn't take long before our what-if lists ran into the hundreds of items. Some of them were logical. What if the oxygen system springs a leak? Some were ludicrous. What if the astronaut decides to land in Russia? And some were frightening and sobering. What if the astronaut has a heart attack?

Zero gravity and a near-perfect vacuum were the enemy. We thought we knew how to deal with them, but we hadn't actually done it. Some of our job was obvious and not even very hard. We'd set the rules and control the consumption of things like fuel and oxygen. We'd also do the plan and then determine and control the orbital mechanics—how high the capsule would be, its ground track, when and where it would do retrofire and then come home to a safe splashdown. The first orbital flight would go around the world three times and last about four and a half hours. If something went wrong—when something went wrong—I wanted us to be ready to identify the problem and then either correct it or control the outcome.

Once that vision was clear in my head, I started on the real-world problem of who would do the work. "Who" meant function as well as person. Before Philco could lock in a final control center design, we had to know what consoles we'd need and what each would do. I sat down with my growing operations team for a series of brainstorming sessions and paper simulations.

Most of the capsule systems could have been watched over by one engineer. But the environment inside, we decided, was locked to the astronaut's survival and needed close monitoring. So we dedicated one console to environmental systems, with its displays and activities closely linked to the flight surgeon.

The surgeon's position was mandatory. He'd monitor the astronaut's EKG and chest expansions, giving him real-time data on both heart and breath rate. Conditions in the cabin were important to the surgeon, and he'd have the advice and counsel of an engineer whenever he needed it.

The capsule's muscle—its various control systems that let the astronaut point the capsule in any direction he chose—had to be watched. We marked down a separate systems console with its own engineer. He'd also monitor electrical power distribution and the batteries.

Communications was settled by deciding that only one person, probably another astronaut, would talk to the capsule. Any messages, commands, or questions would go through the capsule communicator (capcom) console.

The most complex functions in the control center were launch guidance, coordination with the range safety officer, trajectory calculations, and retrofire. We'd need both console displays and a number of graphs on big plotboards that everyone could see. It was too much for one person to handle. We split off control over firing the retrorockets and commanding the retrofire clock on board the capsule and made it a separate position.

Someone had to keep track of the worldwide network's remote sites and coordinate with the local range authorities. That would fill up another console. Since we were using the Air Force Atlantic Missile Range, we decided to ask the Department of Defense office at Patrick Air Force Base to assign an officer to that job.

Another console and set of recorders was needed to monitor the rocket for each flight.

My flight director's console would have access to every communications circuit; we called them loops. The primary communications circuit among controllers, we decided, would be the "flight director's loop."

Just to my left, we put a console we called Procedures. In effect, that flight controller was our hall monitor. We would write down every procedure for every console, and he'd be responsible for making sure that everybody followed them. The detail was enormous: what comm loops went where, what data came in from remote sites, who had access to what instruments and displays, and much more. When we started running missions, Procedures became my alter ego, almost an assistant flight director. He kept us on track and it was no small job.

Once we settled on the primary consoles to be manned by engineers directly involved in the flight, we looked at what else we might need. Mercury was a government program, and we all understood the bureaucracy. So across the back of the control center, we added three more consoles. One was for the operations director, who had overall responsibility for each mission. Walt Williams, a long-time NACA operations man, had just been called back from Edwards AFB, California, to fill that newly created job. As we completed the control center design, he was to be my new boss.

A second back-row console went to the Department of Defense. We wanted a senior officer, preferably a general, who could immediately call on any and all military assistance we might need—for added recovery forces, emergencies of any kind, countdown disruptions by boats or airplanes that encroached our area, and anything else that might come up. It was an active console and we needed it many times in the years ahead.

The third back-row console was almost an afterthought. We'd need an information officer to let the press, and thus the entire outside world, know what was happening during a flight. Having him at his own console seemed to be the most convenient for all of us. None of us understood how important and significant that public affairs officer would be, or that the PAO position would become one of the most notorious in the control center.

Finally, there were contingencies—back to that big worry over "unknown unknowns." So we made room for a number of generic consoles that would let us add monitoring functions when experience dictated. And it did. One of those consoles would be occupied by a control center supervisor, somebody from Pan American, the range support contractor. He'd be the one making sure that the control center hardware kept working and that we had the local range support when we needed it.

So we had our control center design. The engineer at a console, with the blessing of the flight director, would become a decision maker. Each console position would be backed up by a team of experts who knew their own systems down to the finest detail, so he'd have plenty of help.

The one console position that needed more support and would need it instantly at the end of each mission was for recovery of the capsule and astronaut. We added a separate room next to the main control center where our recovery experts could have their own consoles, plotboards, and communications lines to coordinate the location of recovery forces and to provide ship and airplane commanders with information both on the ongoing mission and on the predicted landing point.

While we were getting our control center design finished, NASA was continuing to expand. Wernher von Braun decided that he did want to be part of this new space program, and the Army reluctantly agreed to transfer him and his entire team to us. The rocket part of the old Redstone Arsenal at Huntsville, Alabama, became NASA's new Marshall Space Flight Center, with von Braun as its director. By now, there was quiet talk inside NASA of going beyond mere orbital flight and one day sending space ships around the moon. Von Braun's primary assignment was to supply us with Redstone rockets for our suborbital Mercury test flights. But he was already thinking about bigger rockets that could reach the moon.

I was invited to a Dallas meeting of the American Institute of Aeronautics and Astronautics to talk about Mercury operations and our flight control ideas. That evening, I finally met von Braun at a cocktail party, hosted by Chance Vought, that included a number of now-senior people I knew from my F8U days. Von Braun's notoriety and fame made him the only person in the room whom everybody knew or recognized. He was accustomed to being the center of attention. It didn't take him long to tell me that our control center concept was all wrong. His English was perfect, with that slight and penetrating German accent, and he seemed a bit surprised when I didn't back down and agree with him.

Wernher was an experienced pilot, going back to his early days in the Nazi rocket program. Flying a Mercury capsule, he said, wasn't any different from piloting any other flying machine.

"No, that's not right," I responded. "We've thought about this for a long time and it's very different."

Now neither of us would back down. Wernher had a Teutonic arrogance that he'd honed to a fine edge. He saw himself as the number one expert in the world on rockets and space travel and had polished that self-image with magazine articles, books, lectures, and technical papers. He was famous. He was a NASA center director, equal to Bob Gilruth, and probably trying to figure out ways to move Gilruth aside. As our argument grew louder and more strident, the rest of the room went quiet and we gathered an audience.

This may be the famous Wernher von Braun, I thought angrily, *but he doesn't know what the hell he's talking about.* I had a Scotch in one hand and felt the other one start clenching into a fist. If he'd said, just one more time, that ground control of a space flight was a dumb idea, I might have punched him.

Before it got that far, a lovely lady stepped up to Wernher's side and whispered something in his ear. His expression changed instantly and I saw the rigid stiffness in his posture relax. "Thank you," he said, whether to her or to me I wasn't sure, and let her lead him away.

"That's his wife," someone said quietly at my elbow. My own tension disappeared; I nodded and went to the bar to refresh my drink. She'd saved us both from a situation that was quickly becoming embarrassing. Then I realized that I'd be working with this man, perhaps even closely. I hoped that I'd at least started him thinking about a new role for engineers on the ground.

The medics kept giving us fits. They insisted on flying monkeys aboard a couple of early Little Joe hardware-test flights. The monkeys survived and that's all they learned.

We had other arguments with the doctors, too. They'd stuck their noses in the worldwide network design, insisting on full-time monitoring of the astronauts in orbit. If something happened—they brought up unconsciousness, swallowing his tongue, and sudden eruption of a twenty-four-hour ulcer—we should be able to bring him down and have a doctor at his side within two minutes. It took a briefing on geography, the laws of physics, and our budget to get them to back off on that one. If they wanted an instant deorbit, the astronaut still wouldn't come down for at least thirty minutes, and then he might land in the middle of Africa, somewhere in the Pacific, or maybe in downtown Phoenix.

The doctors did have one good request. If an astronaut needed medical attention on the launch pad, they also wanted to get to him in two minutes or less. The idea made sense to the pad experts, to the recovery people, and to the astronauts themselves. They expanded the concept to include hardware troubles and other emergencies, and the two-minute access rule went into our operations plan.

The question was how to make it happen. In one of our meetings, Dr. Bill Augerson came up with a screwy idea for rescuing an astronaut in an emergency. He wanted us to add a keyed latch to the clamp that held the capsule to the rocket. If something went wrong, he'd personally fly up on a helicopter, unlock the clamp, hook the capsule to the helicopter, then ride with it to the ground where he'd blow open the hatch and start treating the astronaut. It was a dangerous, complicated, and unworkable scheme. I listened with a straight face, admired Augerson's personal courage, then voted with everybody else in the meeting to forget that one.

But Augerson insisted that we needed something. If we didn't like this idea, what did we suggest? He had a point. We didn't have a quick answer, so we started looking. On one of the constant trips we were making to California, meeting with the Air Force in Los Angeles or with Convair in San Diego or both, we found it.

"How about a cherry picker?" one of the Air Force officers asked when we mentioned the problem. "We use them with Thor rockets up at Vandenberg. There's one on the pad right now."

The next day, Deke Slayton and I drove the 140 miles north of Los Angeles to Vandenberg AFB and the Western Test Range to check it out. You see cherry pickers used today by electrical-power-line or traffic-signal repairmen. It's a big basket at the end of a long, hydraulic arm, usually mounted on the back of a truck or something similar. It can be raised or lowered and maneuvered into place by an operator in the basket or on the ground. The Vandenberg launch pad people gave us a demonstration, then asked if we'd like to check it out in person.

Deke and I climbed aboard and they raised us right up next to the Thor's payload. I was satisfied until Deke said, "Just a minute." Then he started jumping up and down in the basket, rocking it back and forth, and scaring the hell out of me.

"What are you doing?" I yelled, grabbing for the railing and hanging on with all my strength.

"The rescue people will bang the crap out of this thing." He grinned and slammed back and forth to simulate their movements. "They're going to be in a hurry to get me out, right?"

The ground crew was all grins when they brought us down. "Hey, that was fun," Deke said. "Good hydraulics, and it'll take a beating. Right, Chris?"

I was seldom so happy to get my feet back on solid ground. I looked at Deke and nodded. We went back to Langley and recommended buying two cherry pickers, one for the Redstone pad and one for the Atlas pad. That made Bill Augerson happy, and they were at the ready when we started launching from the Cape.

Traveling with Deke was an adventure. We were quickly becoming good friends, and I discovered that he wasn't just a test-pilot barroom drinker. He was a connoisseur of both the art and the etiquette of bar drinks. We never went to Los Angeles without a recreational evening scouting the bars along La Cienega and the Sunset Strip, then finding one of the area's finer restaurants for dinner. My favorite for years was Scandia, and Deke agreed. "But first," he'd say, "we have to continue your education on the finer things of life."

So we'd hit one of the better bars—there were plenty to choose from—and he'd introduce me to a new cocktail. I'd always been a Scotch drinker. The first time out, he showed me that Scotch had other possibilities. "Scotch and Drambuie," he said. "It's called a rusty nail. You'll like it."

I did. We made a lot of trips to Los Angeles. He introduced me to gin gimlets, salty dogs, Galliano, green and yellow Chartreuse, brandies and cognacs of all kinds, and even eaux-de-vie. He knew them all, when they were appropriate and when they weren't, how to judge a bar by the quality and selection of its stock, and then how to measure a man by what he drank. Deke wasn't just a great test pilot and a fine engineer. He was a bon vivant with taste and charm. Not many people got to see that side of Deke Slayton. Particularly the charm.

I only had one occasion to question his taste. We flew to St. Louis for a meeting at McDonnell Aircraft. It was windy and about fifteen degrees when we stepped out to the airport sidewalk. Our transportation wasn't there yet. Then it started to sleet.

"Damn, it's cold," I complained, and hugged my thin overcoat tighter around me.

Deke popped open his briefcase, pulled out a bottle of clear liquid, and handed it to me. "Try this."

I unscrewed the cap, took a swig, and felt a volcano burning down my throat. "What the hell . . ." I couldn't say any more. The fire inside was too intense.

"White lightning," Deke said, taking a swig himself before putting the bottle away. I didn't ask for an explanation. By then, a warm glow was spreading through me and I forgot all about the cold.

We were falling behind. Bob Gilruth had approved an ambitious schedule in 1959. We thought we could fly our suborbital missions with astronauts in 1960 and have an American in orbit in 1961. That was when we were rank novices at the space business, with barely a glimmer of how complex and difficult it would be. Now in the dawn of 1960, we were semipros and still a long way from meeting those goals.

Everything was running late—the Mercury capsule, the Redstone rocket, and the Atlas. Part of it was our fault. The Space Task Group had turned into the most demanding, detail-oriented customer the industry had ever seen. Part of it was the fault of the various contractors, who uniformly underestimated how difficult it would be to design, develop, and test systems for space flight. And part of it was the bureaucracy, which could drag like cold, thick molasses.

My operations people insisted on knowing everything. I drove them without mercy in that direction because I understood from experience what it was like to have a pilot's life hanging on the next decision I made. If we didn't have the right information, we might make the wrong decision and kill an astronaut. Walt Williams understood, too. He'd put in more than ten years as NACA's operations boss in the California desert, where the nation's most exotic and tricky experimental airplanes are tested. He knew what it was like to go to a test pilot's wake.

Walt could be cranky. He worked well with the top brass, knew how to negotiate to get what he wanted, and had their respect. But then he'd turn around and run roughshod over the people working for him. His own people would shy away from him, just hoping that they'd done everything right. When they didn't, Walt could be sharp and abrasive. I'd seen it firsthand, working for him as a young kid fresh on the job at Langley. Gradually I'd learned to get along with him, and his caustic criticism faded away.

Now he was back, once again my boss, and I was glad to see it. He'd grown into a strong, experienced operations man in the desert, who knew how to deal with the military and the aerospace industry. He brought two sharp guys with him. Kenneth Kleinknecht had been a lead engineer on the X-15 rocket plane and knew what manned flight was all about. Martin Byrnes was the business and contracts expert who'd helped Walt set up a completely new flight test organization at Edwards. He had the smarts and the experience we needed to deal with both the industry and with other government agencies.

Walt understood exactly what I was trying to do with my people. When any complaint reached him about our nitpicking demands on hardware and the people who were building it, he brushed it off. In early 1960, to make the obvious official and to show just how strongly he backed me, Walt named me the Mercury flight director. It was a position of great authority. The amount of static I got from contractors subsided measurably after the announcement.

I understood the frustration we caused. Engineers are skilled at describing how their system works under normal conditions. In operations, we wanted to know a lot more. I'd run into resistance first when we'd asked how the Atlas rocket might fail. Now we were asking the same question and more about everything.

How does this system react when it's too hot or too cool? If this instrument fails, will the needle go off-scale high or off-scale low? Will it send false signals, or no signals at all? What happens to the oxygen system if pressure falls? Or suddenly increases? Will this box or that component fail, or will it taper off and give us some time to troubleshoot? We were bull-dogs because we needed to know.

It forced the aerospace industry into a new way of thinking. Most of our questions were aimed at keeping the system alive as long as possible. Engineers are problem solvers. Before long, the people at McDonnell, Convair, and elsewhere were intrigued and caught up in the process. Their managers complained because our demands added to their costs. But they did it. By making ourselves better, we made them better, too.

We asked for detailed drawings and schematics of everything. We'd tried doing them ourselves, but it was too much. The contractors gave them to us, and the books of drawings that we created became one of the bibles of flight control, always at our fingertips. The other bible was the book on "mission rules." But we weren't ready to tackle that one yet.

Designing the systems, then assembling thousands of pieces to make a Mercury space capsule, was a daunting process, and the estimated delivery dates fell further and further behind. Bob Gilruth's response to pressure from Congress, the press, and his own bosses at NASA headquarters was to simply say, "We'll only fly when we're ready." Gilruth was the man in charge, and almost every Mercury decision was made at his level or lower.

Even with the delays we encountered, Mercury was moving quickly. When Congress wonders why programs today take so much longer, they need only to look at where the final decisions are made. When the authority is taken by Washington bureaucrats, instead of being granted to the people doing the work, it adds years to the schedule and billions to the budgets.

I look back now and see how fast we moved. There were rumors that the Russians were getting ready for a manned flight, too. We were all determined to be first. But Gilruth's slow and steady approach kept us from panicking and doing something that we weren't really ready to do. He took the heat and we just kept working.

I suspected that Gilruth and others above him were being briefed by our intelligence people, but I wasn't invited to those sessions. Later, when I found out that we were indeed keeping close track of the Soviet space program, I turned down invitations to those briefings. By then, I was talking often to the press and I didn't want to lie to them. It was better to say honestly that I didn't know when reporters asked about Soviet space flights. So I learned about Russian space flights the same way almost everybody else did, by reading newspapers or watching television. The Russians never announced their plans in advance. By the time we heard anything, they had already launched. Aside from the emotional impact of seeing them in space, the Russian space program had little impact on our daily schedules.

On the Atlas side, Convair could deliver an Atlas rocket with the capsule adapter installed and a test version of the abort sensing system. The first two Atlases we bought had a "thin skin" upper body. The Atlas was almost a metal balloon when its propellant tanks were filled and pressurized. That skin would be fifty percent thicker on later models, but for now the Air Force wanted us to use existing rockets.

Our own control center was under construction in the Cape's old radar building, so on a wet July day, the Mercury people directly involved—Gilruth, Chuck Mathews, me, and several others—were at the range's control center about three miles from the pad. Walt Williams was at a console in the blockhouse, his first time as NASA's launch operations manager. But mostly, the mission we were calling Mercury-Atlas 1 was an Air Force operation.

I'd been invited to sit with the range safety officer (RSO), partly representing the Space Task Group inside the center's operations area, but mostly to watch the operation and learn everything I could. Outside, the clouds were low, and now it was raining so hard that we frequently lost sight of the launch pad. The RSO gave me a headset, and for the first time I listened to the multiple conversations going on between the various consoles, the launch director, and the recovery people. It was disconcerting, but after a few minutes my ear and my brain sorted it out and I could follow one or two discussions at the same time, forcing the rest into the background.

Walt Williams was being briefed, advised, and pressed for a go, no-go decision. This was far beyond his experience, or any of ours, in controlling airplane test flights, and he wavered at the mercy of the Air Force and Convair people who were overloading him with information. I listened to Air Force people telling him that the Atlas was a weapons system designed to handle the winds and rain outside. "War doesn't wait for the weather," I heard an unnamed voice advise him.

The range support engineers were less happy. Radar wasn't bothered by the clouds, but none of the launch cameras in the area would pick up more than a minute or so of the flight. Gilruth was skeptical as the weather got worse. I kept my mouth shut, but I'd never seen a storm at the Cape as bad as this one.

"It's your decision," Gilruth told Williams. A few minutes later, they picked up the countdown.

The first sixty seconds looked and sounded good to me. The radar plot boards showed a normal Atlas ascent and trajectory. Voice reports coming in to the RSO console were calm and matter-of-fact. Then it all changed.

"Tracking multiple targets." I heard the voice and assumed it was a radar operator. I looked up at the plot boards just in time to see the bright dot representing the Atlas fall off the trajectory line. Then the board went crazy with tracks that meant nothing to me.

"I didn't touch it!" the RSO said, looking me square in the eye. At first, I didn't understand. Then I knew he was telling me that he hadn't sent a destruct command to the Atlas. It exploded on its own. He shrugged and looked back at the boards. I looked at Bob Gilruth and Chuck Mathews. We had a big problem on our hands.

We put together a team and immediately sat down to review the telemetry data from the capsule and the Atlas. It took us three days, working through the weekend, to figure it out. It had to be structural failure. Every part of the Atlas and every part of the Mercury capsule was working perfectly until it all fell apart. But the clincher was missing. We didn't have film of the actual breakup; the Atlas was in the clouds when it happened.

We did have data from instruments that measured accelerations and motions. They showed sudden motions below the capsule. Then nothing. By Monday morning we were convinced that the upper section of the Atlas ruptured. The capsule adapter ring might have contributed, but we weren't sure. Gilruth was just about to pick up the phone to tell the Air Force what we'd found when a contingent of officers, along with engineers and managers from the Space Technology Laboratory and Convair, showed up at our door.

"We've got the answer," the lead colonel said. "The whole thing was caused by your capsule."

Gilruth was incredulous. We were in a conference room in Hangar S, the walls covered with data readouts and big sheets with our notes and drawings. Gilruth leaned back in his government-issue chair and said quietly, "I don't think so."

"It's obvious," the colonel said, and started into a song-and-dance routine that left all of us angry. "We think one of the thermal shingles ripped off the Mercury capsule and punched a hole in the Atlas."

We pressed for details. How did they reach that conclusion? What had we missed in the data that they'd found to support their theory? The argument was getting heated when they finally admitted that they'd all gone home for the weekend. They hadn't looked at the data, nor had they done any failure analysis. They'd had a meeting just before coming over, decided that since the Atlas was fine, the problem had to be the Mercury capsule.

It was a ruse and a fraud, designed to shift the blame to NASA and the facts be damned. Walt Williams was chagrined and embarrassed that he'd approved the launch without photography. He'd never again make that mistake. We all learned lessons that applied to future missions. One of them was to trust our own judgment, even when our supporting contractors and the Air Force said something different.

Gilruth had solid structures experience. Jim Chamberlin was even better and drew the job of sorting out the problems and recommending a fix. In the end, he reported that the probable cause was a resonance vibration between the capsule adapter and the upper part of the Atlas. Because the Atlas skin was so thin, it ruptured. The adapter between the rocket and capsule had to be strengthened. But even that wouldn't solve the problem of the thin Atlas skin.

Chamberlin, Gilruth, and Max Faget got together over an engineering drawing board and started sketching out solutions. Whichever one came up with the idea, it was always attached to Gilruth because he had to take it upstairs and fight for it.

"We need a belly band," he said in a briefing at NASA headquarters, "a metal strip around the upper section of the Atlas." A bright young headquarters engineer named George Low understood immediately and joined with Gilruth in getting approval from Abe Silverstein and Keith Glennan. With that in hand, and Space Technology Laboratory engineers on his side, Gilruth took it to the Air Force.

General Bennie Schriever's reaction was immediate. "Absolutely not!" he said when we presented it to him. The engineering effort was significant, and Air Force pride was at stake, too. We were still outsiders; anything we suggested was suspect. Schriever's engineering staff backed him up, and no matter how we presented the data, he wouldn't budge. Neither would we. We met alone to decide what to do. The Air Force was in a political box. If we were wrong and the next Atlas blew up, the stigma would all fall on Schriever and his Atlas program. If we were right, the Air Force might be branded incompetent because a bunch of NASA novices had solved a problem for them. And that could have major influence on Air Force budgets being argued in Congress.

"We're paying for these rockets and we're right on this one," Gilruth told us. "If Schriever can't help us, we have to go to the secretary."

Gilruth took us back into the meeting and laid it on the table. "Let the secretary of the Air Force decide," Gilruth said, "and I'll stake my job on the results." Schriever thought it over and agreed. A few weeks later, they had an audience with Secretary Dudley Sharp in the Pentagon. Gilruth got his belly band on the next Atlas, with the clear understanding that it was NASA's design and did not have Air Force approval.

Everyone understood, too, that if it failed, Bob Gilruth would be looking for a new job in Greenland or Alaska.

7

"We're Go, Flight"

W e had a control center. It wasn't quite finished yet, but we needed to get started on our operations training. McDonnell hired the Link Trainer Company, which was famous to pilots for its boxlike airplane trainers during World War II and had been improving the art of simulated flight ever since, to build two Mercury capsule simulators for astronaut training. One of these procedures trainers was installed at Langley, the other at the Cape. We quickly found additional uses for them.

Out first job was training ourselves to use this new control center and the worldwide tracking network that was also nearing completion. We'd planned a suborbital flight of the Mercury-Redstone combination, MR-1, for October 1960, so there wasn't much time.

First, we built some small plywood rooms in the procedures trainer building, one for each remote site around the world. Then we wrote a script of what kinds of data and voice transmission we thought each site would get during an orbital mission. We didn't have realistic equipment, so an intercom had to do. The controllers who'd be assigned to those sites went to their rooms, barely fancy enough to be called plywood boxes, and we read the script over the intercom. We timed it so that each site got the information we thought they'd get on a real mission, then sorted through it and passed the pertinent data back to the control center.

It was crude, but we learned lessons immediately. Discipline was the most important. We learned quickly to keep everyone informed, to update the entire remote network on mission status, problems, and timing. A controller who knew that a problem was coming his way in fifteen or thirty minutes wouldn't be surprised when it popped up over the horizon.

It was all done on the intercom, a better and more reliable communications system than we might have when equipment failed in the field. If a remote site controller gave us too much data, it started to clog even the intercom system. If he gave us too little, we were left wondering. To speed up communications—at some of the sites, the best available technology other than a high-frequency radio was a Teletype transmitting at twenty words a minute—we fell into the habit of using acronyms and jargon. Almost unconsciously, this new language became a way of life that still goes on today with the newest generation of controllers.

The language came naturally. ECS was the environmental control system, pronounced in the acronym fashion of one letter at a time. The flight dynamics officer was FIDO. That one was pronounced like the dog. I was the flight director and they called me Flight. The man called Flight forever after was the boss. His decisions during a space flight are the law, and even today he can be overruled on the ground only at risk of first firing him and handling the consequences later. And that has never happened.

It would take about ninety minutes for Mercury to go around the world. So we practiced one-orbit missions, over and over and over. In a week, we felt comfortable enough to add the Link procedures trainer to the system. About then, it was obvious that the Mercury-Redstone mission wouldn't be ready in October. We took the news with mixed emotions. The worldwide network was still being tied into the system, and all of us could use the time on training.

Our control center had direct communications with the sites in Bermuda, along the Atlantic Missile Range all the way to Ascension Island, and with Hawaii. Everything else from every site flowed into the computer and communications center at Goddard Space Flight Center, a thousand miles north in Maryland, then was relayed to us. Goddard had land lines coming in from the several U.S. sites and Mexico, submarine lines from Bermuda, Europe, Australia, and Hawaii, and high-frequency radio signals from Africa and across the Pacific. The Africa signals had to be relayed through London, and the Pacific signals went first to Hawaii.

The last stop was our control center at the Cape, and it all came over lines across the Banana River from mainland Florida. The lines weren't good. Until new and redundant lines were laid, we had a hell of a time keeping communications open during bad weather, or when the phone company was doing maintenance.

Recovery communications came from ships at sea and search-and-rescue aircraft, all piped through the Department of Defense communications web. Most of that came through separate systems and wasn't affected by the Banana River problems.

We flick on the television today and see live reports and events beamed by satellite from every corner of the globe. In 1960, no one had ever attempted to do a worldwide live network for any reason. We were about to control America's space program that way, and it was a mammoth undertaking. Space flight was changing all the rules.

Not all of the comm problems were external. We wore headsets with a boom microphone and a press-to-talk key attached to our belts. The Cape launch communications system was single-wire. When the key was depressed, it was talk only, no listen. In an early practice session, I couldn't get through to Scotty Simpkinson in the blockhouse. No matter what I asked, or how loud I talked, he wouldn't answer.

I had my fingers wrapped around that belt key and finally I exploded. "Damn it, Scotty, why won't you answer me?" I pulled my hand up and was about to pound on my console when I heard Scotty in my right ear.

"Well, Flight, if you keep your thumb off that mike key, I'll be happy to answer you."

The control center erupted in laughter. My guys had been watching and listening, just waiting for me to figure out that I had my mike key depressed. It wasn't the finest moment for the man called Flight.

This was our first experience with the real systems—capsule, rocket, communications, radar, all of it. Every system had its own idiosyncrasies and its own language. The communications lingo was so foreign to us that AT&T sent down a fellow to teach us the language.

Every hour brought a new problem, or five, and added to our education. Building a countdown, I discovered, was one of the most complicated things we ever did. Everybody had to come in at the right time with a go/no-go, and behind each decision was another long checklist or countdown. Everything might be perfect until the range discovered that a camera two hundred miles away was down, or that a radar site had blown a tube. A single no-go put the entire countdown on hold.

Hardware broke almost every day and had to be fixed. We'd transmit a command from the control center and it wouldn't arrive at its destination. Radios that worked suddenly didn't work. Then when the hardware worked, we'd bollix it somehow through our own inexperience. We were inundated with the newness of everything. Even words like *software* were new. What the hell did it mean? And our primary computers weren't at the Cape. They were a thousand miles away in Maryland. We'd run a training problem, and when we got one answer, the computer center would get another. It would turn out to be a spurious electrical signal or a fractional difference in timing. The littlest thing could screw us up.

Those were the details I never considered when we started thinking about how to launch a rocket. It was intriguing. It was exhilarating. But it gnawed at me all the time: *Have we got this thing right?*

The procedures trainer changed our work habits. It acted exactly like a capsule in orbit, with an astronaut or a substitute talking to us as the telemetry data flowed in. At the control consoles, it looked and sounded like an actual mission. We also generated realistic radar data and fed it to the computers, so that our plot boards and wall map followed the capsule as it went around the world. Finally we were training with real hardware, and our scripts were based on a real mission. It meant new faces, too.

Tec Roberts was my FIDO, and the core group of controllers included such young old-timers as Frank Simonski, Don Arabian, Carl Huss, and others. Glynn Lunney and John Hodge would be at the Bermuda site for this mission, and for training sessions they worked from their room near the procedures trainer. Either Stan White or Bill Augerson handled the surgeon console, and the military services detailed other flight surgeons for the remote sites. We had to train them, too. We brought down more and more people from Langley to become controllers, too. Dutch von Ehrenfried and Gene Kranz were in the early group. So was Arnie Aldrich. All of them would go on to long and illustrious careers in the nation's space programs.

We'd shifted to full-time training for the MR-1 flight, but it was the Florida monsoon season, and before we got a break, it had rained for thirty days straight. Cables and wires running from the control center to the launch pad were soaked. They'd short out and need repair. We didn't even have our own power supply. We'd be starting on a simulation or be halfway through when some fool of a contractor somewhere would dig into a power line. Even a good thunderstorm could black us out in the control center. I made notes every day. Fix this. Add that. Get a backup electrical system.

I don't know how many times we counted down in anger, our frustrations boiling over when something stupid and beyond our control had just happened.

The Redstone's powered flight would last only about 140 seconds, and from liftoff to splashdown it was fifteen- or sixteen-minute mission. These were real-time sessions. If a controller made a bad decision or a wrong input through his console, we lived with it and learned to do what we had to do to fix it. Later, the same thing would be true for astronauts in the procedures trainer.

My philosophy was simple: better to learn from our mistakes in training than from mistakes during a real space flight. We ran the full mission dozens of times a day, with variations mixed in to get us ready for failures of every kind. At least, that's what we thought. The one failure we weren't ready for was the unknown unknown.

Still, we were a ruthless bunch in the control center. After every simulation, I demanded that each controller critique his own performance, and the performance of his supporting staff. I'd add my own critique and so would the simulation supervisor. It was done in public, all of us together, and before long we'd all developed a level of trust and camaraderie that I'd never seen before. It was stimulating, infectious, and rewarding. It made us better than I ever thought we could be.

When we weren't training, we were talking. We tested each other on solving problems until it was obvious that we needed to start writing down the best answers. Before long we had a list of rules and observations for every system in the Mercury capsule, for the rocket, and for each flight control position. We noted a large number of what-ifs, too, along with what to do about them.

Then we printed the whole bunch in a booklet and called it our Mission Rules. It was the second bible of flight control.

By mid-November, we were wondering if we'd ever get MR-1 off the pad. Government per diem rules forced us to stay at the Cape through weekends. Our families were becoming distant memories and telephone friends. Betty Anne kept me up-to-date on things back home. The kids missed me and didn't understand why I was gone so much. She tried to explain it to them, she said, but I could tell that she was really trying to convince herself that the nation's needs came ahead of our family. Neither of us wanted a return to the old days of headaches and ulcers. But when we'd made the decision to become part of the space program, neither of us had any idea about the strains it would put on us. Somehow we'd just have to survive this.

Out on the pad, the technicians were having their own problems getting the capsule just right. Von Braun and his people had insisted that it first be shipped to Huntsville for interface tests with the Redstone. We'd assumed that would make the job on the pad easier. It didn't, and that was the last operational capsule that von Braun saw on his home turf. His only job in Mercury was supply to us with the few Redstone rockets we'd need early on. As the program matured, von Braun's small influence on Mercury disappeared completely.

Through it all, the press was hounding us. Almost every day, a new story appeared maligning us for not meeting our schedule. Bob Gilruth told us to ignore them. "We'll fly when we're ready," he said again and again.

The weather cleared and we started the countdown just after midnight on November 21. Redstone was von Braun's rocket, and his man, Kurt Debus, was assigned to the Cape as the Redstone test conductor. The countdown followed the German standard set during World War II; it included built-in holds that gave time to fix the little problems that would surely occur. This time, they were minor. We were all uptight. Our first launch as a control center team was actually counting down toward zero.

I heard Debus's accented voice in my headset counting the final seconds. In all our training sessions, he said, ". . . two, one, zero. *Fire!* Liftoff."

That last sentence told us that liftoff had occurred. I don't remember hearing any of that. My gut was sucked up tight and my eyes were focused on the television screen on my console. I'd never seen a Redstone launched before.

"Look at the acceleration on that son of a bitch!" I yelled. The launch pad was immediately obscured by a cloud of smoke. There was not a sound on any of our comm loops as the smoke started to clear. MR-1 was still sitting there. But its escape tower was gone.

Suddenly the entire capsule recovery system activated. The small drogue chute popped out. Then there was a new puff of smoke and an explosive canister deployed the main parachute. The SOFAR bomb discharged. The green dye marker package dispersed. And then the backup parachute discharged.

Our Mercury-Redstone sat there on the pad, with network television cameras and a few hundred reporters watching. The parachutes draped sadly over the rocket, billowing in the breeze.

"That reminds me of the time I took my kids to the circus and a crowd of clowns kept coming out of a Volkswagen," I muttered. The comment was picked up on the comm loop and I heard some coughs and throats being cleared. Then I focused on the issue. "What the hell are we going to do now?"

"Booster," I said in my best Flight voice, "what happened?"

Booster was another von Braun man, Joachim (Jack) Kuettner, and an expert on the Redstone. Jack was a little guy. When they were testing the V-1 at Peenemünde, they thought about making it pilot-operated and put a cockpit on one to try it out. Jack rode the rocket, bailed out, and they decided that it wasn't such a good idea. So he knew more about the Redstone than anybody else in my control center.

There was a moment of dumbfounded silence before he answered. "Just a minute, Flight. I'll ask the blockhouse."

That brought a wave of nervous laughter in the control center. If Booster was confused, what about the rest of us? The real problem was that we had a hot bird on the pad. Its propellant tanks were pressurized and full of liquid oxygen and kerosene. The capsule's retrorockets were armed and their battery was live. And the billowing parachutes were now threatening to topple the whole thing over. I listened to the discussion coming from the blockhouse about how to relieve tank pressure and get the lox and kerosene out of the Redstone. Then I heard a sentence fragment from Kurt Debus again, and my heart skipped a beat.

". . . a man with a gun," I thought he said. I concentrated on the conversation.

"What?" Walt Williams almost shouted. This was his second mission as launch director, his second mission gone to hell, and his chief Redstone man wanted a gun?

Debus explained. A rifleman could shoot holes in the Redstone to let the propellant pressures bleed off. In a few hours, they could send crews to the pad to finish up the job of making the Redstone safe. I was sorry that I couldn't see Walt's face when he told Debus what he could do with his rifleman.

The final decision was to do nothing. The sun's heat would boil off enough lox through the relief valves to solve the pressure problem. The wind was slaking, so it wasn't likely that our Redstone would fall over. We waited in the control center until the launch team moved the gantry back, then sent up a crew to cut the parachute shroud lines and remove the hatch bolts. Once it was open, Walter Burke, a McDonnell vice president, crawled into the capsule, disarmed the retrorockets, and turned off the rest of the Mercury systems. It was a brave act, and I held my breath while I watched the readouts in the control center go dark one by one.

The press had a field day. It wasn't just a funny scene on the pad. It was tragic, and America's space program took another beating in newspapers, on television, and in Congress.

It didn't take long to find the problem. One of the electrical umbilicals at the base of the Redstone was just a bit too short. The rocket actually lifted off about four inches. Then when the umbilical pulled out a fraction of a second too soon, the sequencer thought that the rocket engines had shut down, so it issued its own shutdown command. Redstone settled gently back into place.

The shutdown command triggered the escape tower to take off, just as it should have about fifty miles high. The barometric switch in the capsule recovery system detected sea-level pressure—it was, after all, still on the pad—thought the capsule had already passed through ten thousand feet, and started that sequence going. It was a comedy of errors caused by an electrical cord that pulled out twenty-one milliseconds too soon.

The Redstone people lengthened the umbilical; we put a new rocket on the pad and launched MR-1A on December 21, 1960. The old Redstone was used for spare parts. This time I heard everything, and when Kurt Debus said, "Liftoff," I saw the Redstone climb out of that cloud of pad smoke. It was quick, but it didn't have nearly the acceleration I thought I'd seen the last time.

I looked around our control center with a feeling of pride. Every man was concentrated on his job. Booster's call-outs were so normal that we could have been running another simulation. So were the advisories from Systems. Our capsule was handling its ride smoothly.

"Engine shutdown."

"Tower jettison."

"We're go, Flight."

With the escape tower gone, Mercury-Redstone was coasting up to the top of its arc nearly eighty miles high and well out over the Atlantic. At the high end of the trajectory, the capsule separated from the rocket. Its small thrusters fired automatically to turn it around, blunt end forward, and dumped the retro-rocket package.

"We're go, Flight."

"FIDO, give me an estimate." We were about to lose sight of the capsule as it went "over the hill"—dropped beyond the horizon. But there should have been enough radar tracking data at this point to make an estimate of the splashdown point.

"One sec, Flight." Our MCC computers were doing the calculations. "Looks like about thirty miles long, Flight."

"Let 'em know, Recovery."

"Rog, Flight." Recovery had been updating the Navy all along. Now he gave them our newest reading. I had the Recovery voice loop on my headset, and a few minutes later I heard the report come in. One of the Navy's search airplanes saw the capsule under its parachute and watched it hit the water. It was floating upright and looked good.

Forty minutes later, an aircraft carrier was alongside and lifting it aboard. The little remaining tension in the control center dissolved in a happy sigh of relief.

"Good job, team," I said. "We're on our way."

Our first mission was history. The Space Task Group had been formed less than twenty-six months ago. In what seemed like the blink of an eye, we'd put together a huge team of NASA and contractor people, designed a space-going system of rockets and capsule, created a worldwide tracking network, and built this control center. And this was just the start.

Our test plan called for two successful Redstone launches before we'd risk putting an astronaut on one. We'd be into 1961 before that could happen, and we knew the Russians were out there somewhere.

But I couldn't have felt better. I reached into my jacket, pulled out a cigar, and lit up. It was going to be a close race to be first into space. But I still thought we'd win.

I'm a lousy commuter. After writing our report on the control center and flight performance of MR-1A, we packed up, left the Ko Ko Motel, and took a charter flight home to Langley to spend Christmas with our families. The charter was supposed to be an improvement over commercial air travel or the train, but we still hated it.

A commercial flight home necessitated a rental car between Orlando and the Cape, usually shared by three or four people to keep costs down, then a flight to Washington, D.C., and a connection back down to Newport News, Virginia. On a good day, it was a ten-hour trip. The airlines were using Lockheed turboprops or the British Viscount, both of them with bad safety records. We lost some engineers when a Viscount crashed inbound to Huntsville, Alabama. The train was worse. It took twenty-four hours following a circuitous route through Georgia, the Carolinas, and Virginia.

Bob Gilruth approved the charter flights, run by Capital Airlines and flown by pilots off duty from their normal routes. The plane was an elderly Martin 202, also not one of the best models ever built, and all too often we got bounced around in the turbulence and thunderstorms so common along the East Coast. It didn't matter if we were going north or south. By the time we got to the destination, we saved five hours of travel—golden time that I could spend at home with Betty Anne and the kids. The beatings we took in that old airplane cost us all in bodily wear and tear, but it was worth it.

Still, this was the greatest time of my life. I remember the intensity and long hours, but I don't remember ever being overwhelmed. Nor did anyone else simply fold under the workload. We knew that we'd come together, all of us in the right time, at the right place, to be involved in mankind's ultimate journey. The events of our daily lives were to be among the century's greatest moments. We were inspired.

It only took a few news stories to deflate us again. Jack Kennedy had won the election, and while we were back at the Cape getting ready to launch MR-2 with a chimpanzee in the capsule, he appointed Jerome Wiesner as his science adviser. Almost immediately, Wiesner released a report that challenged the whole idea of putting men into space. By the next day, simply by calling people like Dr. James Van Allen for comment, the press had fanned it into a "manned vs. unmanned" controversy that is still acidic forty years later.

Van Allen's instruments on the first U.S. satellites discovered the radiation belts surrounding Earth. Now he became an overnight expert on space policy by telling reporters that automatic probes could do more for science, and do it cheaper, than men. Congressional critics jumped onto the Van Allen bandwagon, and a pro-and-con debate emerged from nowhere.

By then, I was spending more and more time with reporters. Walt Williams was too often unavailable or simply asked me to take an interview because he wasn't as familiar with the technical details. The press were growing in number, with more of them covering Project Mercury each week. I took the interviews and did the briefings, at least in part, to help educate reporters on what we were doing and how things worked.

Shorty Powers gave us some common-sense guidelines. "Stay away from controversy," he told us, "and stick with describing your jobs and the missions."

I developed a standard answer when reporters asked questions that moved into the political arena or were otherwise sensitive: "Those are good subjects, and I have no business discussing them." Since I was quickly acquiring a reputation for telling it like it is, I could make the answer stick. But now I was being asked about the manned-unmanned controversy and I gave a straight answer: "It's a ridiculous argument. In my opinion, we need them both."

I stood by that philosophy then and I stand by it now. Robot spacecraft can do only so much. They can't yet duplicate the human eye, the human ability to draw inferences, or the quickness in making emergency decisions. And no matter how good radar and other sensing devices may get, the best obstruction-avoidance device ever conceived is the human pilot.

The Wiesner Report was quickly disavowed by Kennedy himself. He supported Mercury and would continue to do so from the Oval Office. It wouldn't be long before he did much more.

Gilruth's "fly when ready" philosophy had a drawback. We'd have long periods of getting ready. The press and many of our critics in Congress, and now the science-academic world, seized on the dry spells and castigated us for delays. The unspoken subtext was that we were incompetent bumblers, while the Russians were doing everything right. And the criticism wasn't always unspoken. In 1959, the Russians had impacted a probe on the moon and gotten photos of the moon's hidden side with another. They lost some, too, when rockets blew up. Those weren't revealed for years. The tame reporters covering a Russian launch knew when to shut up.

I was frustrated and angry when critical stories appeared, and morale in the entire team would sink for a few days. Going into space was new and extraordinarily complicated. The devil was in the details, and if we didn't root him out, an astronaut might die. That responsibility faced us every day. The flip side of Gilruth's plan didn't occur to any of us until it was in our face. When we were ready to fly, we did. Then it was like a dam breaking. We'd finish one mission, start on the next, and be planning the one after that. Each mission did more than the last. It started that way with MR-2, just five weeks after we flew MR-1.

It was our turn for an animal flight. They'd flown those two rhesus monkeys up and down on Little Joes out of Wallops Island. Now we were putting a trained chimpanzee into a Mercury capsule and launching him into space. He wouldn't get into orbit. The velocity was too low. But he'd get more than six minutes of zero gravity, and the medics had rigged up a test mechanism. When a light came on, he had to flip a lever. Right hand. Left hand. It was random. If he did it right, he got a banana pellet. If he did it wrong, he got an electrical shock to the sole of his foot.

That was all there was to it. The medics figured that if Ham's performance wasn't harmed by the gravity forces of liftoff, followed by no gravity at all, followed by more high-g loads during reentry, then they'd clear an as-

tronaut for the next ride. Bob Gilruth had a private meeting with the astronauts and asked each of them to write a memo—who should fly first and
why. They couldn't pick themselves. What happened to those memos? I
think Gilruth destroyed them. He wouldn't discuss what the astronauts
said about each other. "It was interesting," he told me one day, and that was
all I got out of him.

In another private meeting, he told them the news. Al Shepard would
go first. Gus Grissom would go second, on an identical suborbital flight.
John Glenn was the backup for both flights. For now, all this was a secret. I
was sorry that Deke Slayton wasn't on the list. He would have been my
choice, ahead of Al Shepard. I asked Gilruth about it later. "Deke gets the
second orbital flight," he told me, and I went away happy. For now, our
focus was training with Shepard. He and my control center team had to become closer than brothers.

But first, we had to get MR-2 and Ham out of the way. Then in another
five or six weeks, Al Shepard would become the first man to ride a rocket
into space.

Somebody started calling it *mission* control. MCC had meant Mercury
Control Center to us, but *mission control* was okay, too. It had a nice ring
to it. My team of controllers had been driven to the edge of exhaustion and
the height of exhilaration to get MR-1 off the pad. They'd come together
as something almost more than a team; they were becoming a well-oiled
machine. I'd picked two of the best to alternate at the procedures console
and to be my alter ego. Gene Kranz and Dutch von Ehrenfried kept things
working smoothly during training and during flights.

January 31, 1961. It was exactly three years ago that America put its first
satellite into orbit. Now we were into the countdown that would send a
chimpanzee named Ham into space and back again, all in one fifteen-
minute mission. I had plenty to worry about, and the ASIS, the automatic
abort system, was near the top of my list. It was fully armed for the first
time.

We'd sweated over the details of ASIS. An abort system that triggered
prematurely, or without any cause at all, was almost as bad as one that
didn't trigger at all. It was a dangerous and touchy system. It was also absolutely mandatory. So we fine-tuned it to the best of everyone's ability.

One question was timing. When could we turn it off and jettison the
escape tower? We pressed von Braun's experts for an answer to that one. "It

depends on fuel consumption," they said. "You don't want to turn it off too soon." We decided on four seconds before the rocket used up its liquid oxygen. If everything was good to that point, it wasn't likely that something bad would happen in those four seconds. But Redstone experience showed that every rocket burned its fuel at a slightly different rate.

The Redstone engineers checked their records and came up with an answer. The hottest engine they'd ever tested—the biggest gas guzzler, in effect—ran dry of lox after 139 seconds. "Okay," I said, "we'll set tower jettison for 137 seconds."

Things looked good until just after 8:00 A.M. We'd been up all night, starting the countdown just after midnight and aiming for a 9:30 A.M. liftoff. Ham went aboard the Mercury capsule before dawn and seemed to be doing well. Then we started having problems. An electrical unit overheated. The weather was getting bad. There were rough seas in the recovery area. I ordered a hold and we waited. After three hours, Ham was still calm, but mission control was tense and fidgety. This was only our third launch attempt as a team, the area was overrun with NASA management from Washington, doctors from the animal program, and a press corps that was bigger than ever. All of them were waiting for me to do something.

"Surgeon, how's the chimp?" I asked. He'd been strapped into his couch for hours longer than the plan.

"Resting comfortably, Flight."

"Recovery, what's your status?"

"Rough seas, Flight, but they say that they're ready for us."

I looked at Walt Williams and he nodded. There was no point in waiting any longer. We wanted ample light in the recovery area and it was nearly noon. I pushed the mike button and gave the order.

"Okay, we're go. Launch Conductor, we're go."

I heard him respond immediately. "Picking up the count at T-2 minutes on my mark. Three, two, one . . . mark!"

The first 130 seconds after liftoff looked perfect, except for a call-out from FIDO (the flight dynamics officer, Tec Roberts) that the trajectory was slightly high. I should have realized that the Redstone was running hot. But everything else was perfect, even Surgeon's reports on Ham's performance with the levers.

"Five seconds to tower jett," FIDO reported. "Four, three, two, one . . ."

He never got to zero. Redstone shut down at 134.5 seconds. I watched and listened helplessly.

"Abort, Flight. We have an abort."

"Talk to me, FIDO."

"We've got capsule sep and the escape tower fired, Flight. Looks like a seventeen-g spike. We're calculating a new apogee."

"Stay on it, everybody. Surgeon, how's the chimp?"

"He got zapped when the g load hit, Flight. He's okay, back to working the levers."

The escape tower rocket added more velocity than we'd intended. FIDO told me that the capsule was heading up to 157 miles instead of 115. It would land well beyond the planned recovery point and probably endure more heat and a higher g load, too. My guys quickly refined the new landing point from two hundred miles long down to about one hundred thirty miles. Airplanes, helicopters, and ships all headed that way. Everything looked good when Mercury went over the hill and we lost contact. It didn't stay well for long.

Inside the capsule, Ham's equipment started to go haywire. He was getting intermittently zapped with a shock, no matter what he did. Then a relief valve opened and cabin pressure started to drop. Ham was sealed in his own unit, so he was safe. The onboard movie camera showed that he was desperately trying to do his job and stop getting that shock to the sole of his foot. It wasn't long before the parachute deployed. The capsule hit the water harder than it should have, and the cable from the lowered heat shield snapped back and punctured the cabin. Water began to leak into the capsule.

We didn't know any of this until later. It took nearly twenty minutes for the first recovery airplane to report seeing the green dye marker in the ocean. Then helicopter pilots radioed that the capsule was listing badly. One of the choppers managed to snare the recovery hook and was lifting the capsule from the water. An hour and forty-five minutes after splashdown, it was safely on the carrier deck.

Ham wasn't happy, but he was healthy, and when they evaluated his work tests, he got high marks. We looked at the entire flight and said the same. Even that open air valve was no big worry. If it happened with an astronaut on board, he'd be safe in his space suit and he could probably close the valve manually. Ham didn't have that option. The important operational items—countdown, launch, flight control, range support, and recovery—all got excellent grades. We were ready to put Al Shepard on the next one in early March.

At the postflight press conference, Gilruth broke his own rule and announced that Glenn, Grissom, and Shepard were training for our first manned flight. We knew that Shepard was the man, but Gilruth didn't want the press to disrupt his training or his mind-set by focusing exclusively on him. If the Russians were our only worry, Shepard would soon be the first man in space.

But they weren't. That all-knowing medical community immediately got in our way, and so did Wernher von Braun. Ham's performance hadn't been one hundred percent perfect. Almost none of it was his fault. To our eyes, he'd been a real trouper putting up with broken equipment. But suddenly those outsiders were demanding that a new series of extreme tests was needed before NASA could certify that it was safe for Shepard to fly. If forty-five or fifty chimps had to be tested, and some of them run at g loads that would kill them, that was a small price to pay.

The arguments found a sympathetic ear in Jerome Wiesner. He immediately ordered an independent study by outside scientists. *Independent* is a relative word. Wiesner picked the study chairman, Donald Hornig. While the medical outsiders were insisting that we kill some chimps on behalf of timid science, Al Shepard's place in history was settled by von Braun. The Germans were embarrassed by the Redstone's performance on MR-2, and by their failure to predict its fuel consumption. They also found some unexpected rocket vibrations during powered flight. We'd seen the vibrations, too, and didn't think much of them. They hadn't bothered Ham, and they had nothing to do with the fuel question. Von Braun brushed aside our comments, and over the strenuous objections of Bob Gilruth and Walt Williams, he ordered a series of engineering changes and demanded an additional unmanned launch.

We were furious. We had timid doctors harping at us from the outside world, and now we had a timid German fouling our plans from the inside. Gilruth took his argument to Washington, where the new administrator appointed by Jack Kennedy was James Webb. Webb was an old-line technocrat from the Truman days, with a good reputation, and both political parties supported him. Webb looked at the arguments and weighed the sound and the fury being generated in the press. The Hornig report was due April 12. There was less political risk to giving von Braun his delay than in letting Gilruth launch Shepard in March.

Jim Webb became a great NASA administrator. But his first big decision was terribly and historically wrong.

Gilruth's antagonism to von Braun was growing. When the Germans wanted to call their mission MR-2A, or MR-3, Gilruth refused. "Call it MR-BD, for 'booster development,'" he ordered. "That way nobody will confuse it with capsule development."

Before we could do von Braun's "BD" launch, we had a much more important mission. Only three weeks after Ham's flight—three weeks of intensive simulations for my mission control team—we were counting down to the launch of Mercury-Atlas 2 (MA-2). This was the "belly band" Atlas that the Air Force had fought against and lost. The Air Force had its own problems with the Atlas, and this mission should have gone up the previous November. But after another one exploded, the Atlas program was put under a microscope before more launches were scheduled.

Now Bob Gilruth's career was on the line, and so was the fate of Mercury. If this Atlas failed, the program might be washed away in a sea of criticism. We had a Mercury capsule up top. The flight plan called for it to separate after the Atlas engines cut off, turn around, and ram back down through the atmosphere to test the heat shield. It was essentially the same plan we'd had last summer when the Atlas blew apart in those stormy clouds. That wouldn't happen again. We'd added some stringent mission rules about weather conditions at launch time.

This was our fourth live countdown in three months and we were looking ahead to launches in March, April, and May. Beyond that, it was hard to predict. We should have been getting tired, but we weren't. We'd turned into veterans ready to take whatever they tossed at us.

". . . two, one, ignition . . . and liftoff!"

The words crackled in my headset. MA-2 was on the way, and I kept my eyes fixed on the flight dynamics radar boards. The last one blew just as the dynamic pressures from acceleration and the atmosphere reached their peak. If the belly band was holding the upper part of the Atlas tight, the rocket should breeze through the "max Q" moment with impunity. I held my breath and listened to FIDO calling out the moment.

". . . thirty thousand feet . . . nine hundred forty miles per hour . . . approaching max Q . . . Max Q! . . . It's through max Q, Flight . . . still accelerating."

I exhaled loudly, thankful that my mike button wasn't depressed. Tec Roberts's voice resumed its normal, low-key tone, and he made the callouts as first the booster engines shut down, then the sustainer engine a few minutes later.

After that, it was a piece of cake. The capsule performed perfectly, the heat shield did its job, and the recovery forces had it on board the *Donner* less than an hour after landing. Gilruth was completely vindicated. His job, and the Mercury program, were both secure. That night, we started a tradition. The entire Mercury team at the Cape, NASA engineers, contractors, and even some Air Force people got rip-roaring drunk at a bar in Cocoa Beach.

We called it a splashdown party.

Then it was back to work. We refused to waste a Mercury capsule on von Braun's unnecessary Redstone test. My mission control team was pressed into duty to support his launch. We considered ourselves professionals now, almost experts. We did the job for him, and we did it right. But we didn't smile.

We launched MR-BD on March 24. It worked, and whatever the Germans had done to it was enough for them to spend a lot of time patting themselves on the back. I was just glad to get rid of the damned thing. Its biggest impact wasn't a technological improvement, but a delay in getting America's first manned space flight off the ground. We announced April 25, 1961, as the day for MA-3, an unmanned orbital flight of the Mercury-Atlas combination, and said that MR-3 with an astronaut on board would go in early May.

That orbital flight meant that the worldwide network had to be ready and integrated into our mission control operations. We only had a month of eighteen-hour days to train for it. The capsule would do a single revolution around the world. We'd make all the calculations, and on my command a controller at the Guaymas, Mexico, station would transmit the signals for the reentry sequence and retrofire. A mechanical "man" was installed in the capsule to check out cockpit and environmental control systems.

At the same time, we were training for Al Shepard's historic suborbital flight. The date would be May 2, but that hadn't been announced to the press. We had our hands full, getting ready for two missions at once. Six of the seven Mercury astronauts were invaluable in helping us to prepare. Scott Carpenter was no help at all.

He came over to train on the capcom console and would be the single point of communications between mission control and Al Shepard. We got Carpenter, who'd been successfully avoiding us since becoming an astronaut. I sat him down with Howard Kyle at the capcom console for orientation.

Carpenter seemed eager enough, and we ran a series of malfunction simulations that forced controllers to think about the problems and decide whether the mission should continue or abort. At the end of the day, Scott came to me with a notepad in hand.

"I wrote down some questions," he said. "Would you mind helping me out?"

"Sit down and let's do it." I was impressed that he'd made the effort and was willing to stay late. A few minutes later, I could feel my distress growing. His questions were rudimentary, and he knew almost nothing about the mission.

Where the hell have you been? I thought. I'd never had technical contact with Carpenter, and his grasp of things was shallow. I've never been known for my patience with ineptitude, but I spent more than an hour giving him answers and trying to include a little instruction. I left work that night a worried man.

The next day we did several sims, with Kyle and Kranz and others showing Carpenter what to do. His most complicated task was to activate a set of sequence switches on the console to make the Redstone trajectory look like an aborted Atlas. When the switches were set, we'd receive data from the Goddard Space Center. After a few runs, I decided to let Carpenter do it alone.

The next sim was a disaster. A computer support technician at Goddard was immediately in my ear asking if we had noise on the line between the Cape and Goddard. We didn't. We ran the sim again and it was another disaster. This time, the Goddard technician reported that the capcom console switch settings seemed to be thrown at random. I ordered a new sim and told Kyle to look over Carpenter's shoulder.

"He's not only throwing random switches," Kyle reported, "but he's so confused that he's still flipping switches after the Redstone aborts."

I terminated the training session and sent everyone but Scott Carpenter on a coffee break. The more we talked, the more I knew that he couldn't do the job. A few hours later, I called Al Shepard.

"I want somebody else for the capcom job," I told him.

"Why?" Shepard asked. I didn't want to be unduly harsh, but he pressed for an answer.

"He can't do it," I said finally. "He just doesn't understand what's involved."

There was a pause before Shepard came back. "I understand your dilemma. I'll take care of it."

The next day, Deke Slayton was already sitting at the console being briefed by Howard Kyle when I walked in. Deke fit in with my team perfectly. Scott Carpenter was out of my hair temporarily. But I knew in the pit of my stomach that he'd be back.

I sent the first contingent of my remote site controllers off on April 11, two weeks before the MA-3 flight. For many of them, just getting to their sites in Africa or Australia would be a three- or four-day trek. That gave us a week, once they were in place and acclimated, to run training sessions with the worldwide network up and active.

At 4:00 A.M. the next day, a Wednesday, the phone rang in Shorty Powers's quarters. It was a reporter, and Shorty had been in a deep sleep.

"The Russians just put a man into space!" the reporter was yelling. "Do you have a comment?"

The words didn't register in Shorty's befogged mind, and he uttered the sentence that would be quoted around the world as NASA's official reaction. It was embarrassing for all of us, a sleepy slip of the tongue that we could only try to put behind us in the days ahead.

"We're all asleep down here," he mumbled, and hung up the phone.

8

Launch Day

We could have beaten them. We should have beaten them. Instead we took the brunt of criticism from Congress, the press, and others for our slow pace. Only part of the criticism was justified.

We were stopped by anonymous doctors in the civilian world who didn't know what they were talking about, by a bureaucrat in the White House who'd been stung when JFK shot down his position on manned space flight, and by our friend the German rocket scientist who got cold feet when he should have been bold. None of them got even a hint of blame. So it was Yuri Gagarin and not Al Shepard who was the first man in space. I didn't like it worth a damn, and Shepard was furious. But the only thing to do was to get back to work and do our jobs.

There was a positive side to Gagarin's flight. The Hornig Report on manned space flight was coincidentally to be delivered to the White House on the same day that Gagarin flew. A few days earlier, we heard that it would definitely recommend a halt to manned space flight plans until those fifty chimps went through severe testing on the centrifuge, including some chimps to be tested "to destruction."

The report was delivered, though we heard it was a few hours late. It didn't mention the chimp testing. Nor did it recommend that we slow down. Yuri Gagarin showed the doctors of the world that a man could do just fine in zero gravity, at least for the one-revolution, 108-minute flight he'd just taken. The headlines around the world showed the White House that slowing down was the last thing we should do.

We took out our frustrations by plunging into the final training for MA-3. It was our first orbital flight. We were going to put a Mercury capsule up there, follow it around the world, and bring it back down again. The Philco and Western Electric people who'd designed and built our control center activated a new display for this mission. It was a world map that filled most of the front wall and hadn't been needed until now. There was the path to be followed by the capsule, like a sine wave overprinted on the oceans and land masses. Each station in the worldwide network was clearly shown with a circle drawn around it.

It was a beautiful display. I understood what it was for, but I still thought it was superfluous. When I'd raised my objections in the planning meetings, I'd been argued into accepting it. *Now,* I thought, *let's see just how useless it really is.*

Our first training session with the map in play convinced me that I was wrong. The map was filled with vital information. The graphic format made it simple to grasp. A Mercury capsule symbol moved along the sine wave, or ground track. I knew instantly where it was.

The circles around the tracking sites were the radio and telemetry limits. When the moving symbol touched a line, it should be in communication with that site. I could see if it was going through the center of the circle and we'd have the maximum of seven or eight minutes of comm, or if it was cutting through the outer edge of the circle and we might have only a minute to get everything done.

The exact times for acquisition and loss of signal (AOS and LOS) at the upcoming sites were posted on a display to the left of the map. A glance told me most of what I needed to know to get ready for upcoming events. It was too bad that we didn't get to use it when MA-3 launched.

We'd gotten used to working all night on launch day, leading up to a morning liftoff. We'd also gotten used to the Russians in their "fishing trawler" hanging around just barely into international waters off the Florida coast. For a fishing boat, that trawler had a lot of antennas sticking out and generated a lot of radio and radar emissions. The Russians knew our launch schedule as well as we did. All they had to do was subscribe to *Aviation Week* magazine, ask the local KGB agent (we knew they were around) to pick up a newspaper, or even just listen to the area's commercial radio stations.

A few days before launch, the trawler would leave port in Cuba and take up station off Cocoa Beach. We didn't mind being watched. But there were trawler scenarios that scared hell out of us.

Could they disrupt a launch by sending spurious signals to the rocket, or even send commands that would change its course? It was possible. But the one we worried about most was the abort signal. It was a necessary evil that nobody liked. I could send the signal from mission control. Kurt Debus could send it from the launch blockhouse. The range safety officer could cause an abort by sending a cutoff signal to the rocket, which would be picked up by the automatic onboard system and trigger the abort. And on a manned flight, the astronaut could send it from the capsule. But the astronauts were never happy with the prospect of an abort after they entered the capsule. An off-the-pad abort was dangerous.

I wasn't comfortable with giving that power to Debus, either. We made sure that his switch was deactivated as soon as the rocket cleared the tower. I trusted the RSO and I trusted myself. That left the Russians, because if I could send an abort signal, so could they. Our radio frequencies were widely known, though we kept our command coding secret. We even devised a complex series of electronic signals for something as simple as resetting the onboard clock. But it wouldn't take a genius to start sending signals toward the rocket in hopes of making something happen.

The trawler was out there on April 25 when we hit the standard two-minute hold in the countdown and did a final check. Everything was go.

". . . ignition sequence . . . liftoff . . . cleared the tower." This was the first thick-skinned Atlas, and it was spectacular. The good feeling lasted only seconds and shattered when I heard the tone in FIDO's voice.

"Roll program . . . roll program not executed . . ." People think that a rocket goes straight up. It doesn't. After clearing the tower, the Atlas was supposed to roll to the proper heading that would carry it across the beach and over water. From there, it follows a slanted climb toward space. This Atlas didn't roll.

"It's off the track," FIDO reported. "Looks like it's going straight up."

"Flight, RSO. We've got an off-nominal track. I'm initiating shutdown and destruct commands." The RSO and I had agreed to inform the other before either of us did something permanent.

"Roger, RSO," I said quickly. "Flight concurs."

I watched the capsule sequence lights on my console flash through the abort steps. Howard Kyle was in my ear from the capcom console reading them off as they lit up. As an afterthought, I opened the guarded abort switch on my console and sent a backup command. If the Atlas ASIS signal didn't tell our capsule to get the hell out of there, mine would. In the postflight reviews, I discovered that my reaction time was slow and I'd sent the signal ten seconds after the fact. That was intolerable. I needed to work on my mental and physical reflexes.

The ASIS worked without my signal. The Mercury capsule cut free of the Atlas just before it blew apart. The escape tower rocket pulled the capsule to about twenty-five thousand feet, then dropped off. At ten thousand feet, the parachute opened and Mercury splashed down safely. The launch-abort recovery team was alongside within minutes, the helicopter pulled the capsule from the water, and in almost no time it was delivered to a medical operations site.

It was the best test we could possibly have. Nobody even considered blowing up an Atlas on purpose, just to test our beach abort procedures. But this one blew anyway. The Atlas abort sensing system worked. The Mercury escape system pulled the capsule away from an exploding rocket and saved it, and the mechanical man inside. The ground recovery teams performed flawlessly.

But while it was happening, the comm lines were close to melting from the cuss words, most of them from Air Force people on the loop. The only Atlas that had worked recently was the one with Bob Gilruth's belly band. Was this problem caused by the Russians? The search for answers started within the hour. It didn't take long for the Air Force to check its telemetry and see that the Atlas had not received extraneous or unauthorized signals. Later that day, they found one of the guidance system black boxes among the debris on the beach. It was shorted out.

Even worse, it was a single-point failure. The black box was the autopilot or programmer, and it had no backup. That was fixed, too. Every Atlas we bought thereafter had black-box redundancy built into the critical systems. It would take more than one short circuit before the RSO or Flight reached for the abort button.

We had a week to get ready for Al Shepard's Mercury/Redstone 3 suborbital space flight on May 2. We'd been practicing for it nearly two months, mixing and matching training sessions with the MR-BD and MR-3 missions. The final push would only put a sharper edge on our skills, and on Shepard's.

Shepard didn't need much now. He was one of the better pilots in the group, with quick reflexes and a quicker brain. But two months ago, he screwed up in our first simulation, missing some obvious systems failures and failing to take the required actions. We did our dog-eat-dog debriefing, criticizing each other sharply. It was our first time with an astronaut, too. Then Shepard didn't play. He joked about his poor performance and shrugged it off. "Let's go again," he said.

"No, it's coffee time," I said, and waved Al into the adjoining room.

"We bare our souls in these sessions," I said. "When we make a mistake or see a mistake, we call it out. It's how we learn."

He paused a moment and thought about it. I saw his expression change from defensiveness to understanding. "Like we do an after-action report in the Navy," he said. "If we're not honest, somebody gets hurt the next time out."

"That's it," I said. "And we're just doing baby runs today. Down the line, we make these sims as realistic as possible. Especially when we're working problems and emergencies. Gene Kranz calls them 'sweaty palms' sessions."

Al grinned. We went back to work. From that moment on, he was one of us in the debriefings. Today they call it team building. But it was just old-fashioned problem solving, with an occasional bite. It was respect and trust, too, growing until they were mutual and strong.

"Do you know the answers in advance?" he asked me after one particularly sweaty session.

"No. The Sim Sup [simulation supervisor] offered them to me once," I said. "He thought I'd be embarrassed if I screwed up. Do I look embarrassed?"

"Nope, and you screwed up on that last call."

"That's how I learn. I don't want it to happen when you're up there."

A few days later, Al brought in a request. He'd done his first flying at the end of World War II. It was customary for a pilot to give his airplane a name. Sometimes it became his radio call sign. Al wanted to name his capsule *Freedom 7*—*freedom* for the founding principles behind the space program, and *7* to represent the first astronauts. I liked it.

"Pass it up to Walt Williams," I told him. "If he says yes, we'll use it."

Williams said yes.

Launch day. At least, we hoped it would be launch day. We'd gone to work in mission control at 2:00 A.M. on May 2, started our countdown preliminaries, then scrubbed because of bad weather. We rescheduled for May 4, but the weather was so bad that I didn't even set my alarm for a 1:00 A.M. wake-up. The May 5 skies were supposed to clear.

Outside and up and down the Florida coast, it was like a Roman holiday. Every room in a hundred-mile radius was filled with reporters, government dignitaries, and aerospace industry engineers and executives. History was about to be made. I brought Betty Anne and the two children down for it. I didn't see them much, but they had a good time at the beach between rain squalls.

The usual problems cropped up. Up in Maryland, Goddard reported a balky computer. Our own comm checks with the pad had sporadic interference. All of it could be fixed. Outside, the clouds weren't clearing as fast as the weathermen said they would. We'd have to slip beyond the 7:20 A.M. launch time, but Al Shepherd started getting into his pressure suit anyway. They took him from Hangar S to a trailer at the pad to finish up.

It was dark when he came out. Searchlights lit up the Redstone gantry, and the white rocket with a black checkerboard pattern around its middle looked like some kind of a strange monument. I watched on a small black-and-white TV monitor on my console. Shepard looked up at the rocket, shielding his eyes with a gloved hand. The light reflected off his silver suit.

I felt a thrill pass through me. *He looks like a space man*, I thought. The moment was more exciting than I'd expected. I wondered what he was thinking. *Is his stomach tensed up like mine is?* Later, in the debriefings, I forgot to ask. We had more immediate things to discuss.

The TV camera followed him up the elevator, to the clean room at the capsule hatch. Pad leader Gunther Wendt greeted him, and they seemed to exchange a few words. Then Al climbed awkwardly and feetfirst into *Freedom 7*. Wendt knelt down and leaned inside, hooking up the suit hoses and the comm lines.

"*Freedom 7*, how do you read?" The voice in my ear was that of Deke Slayton, only a few feet away at the capcom console.

"Loud and clear, Capcom. How me?"

"Loud and clear. Okay, Al, we're proceeding with the suit checks."

It sounded so ordinary. We'd done it a hundred times in simulations. Now it was real, but the familiarity of the procedures and the words felt good. I could see a calm settle on my controllers, and some of my own tension faded. It was happening, and it was happening right. I watched Gunther Wendt seal *Freedom 7*'s hatch. All we needed was good weather.

That took a while. Dr. Stan White on the surgeon's console monitored

Shepard's heart rate and respiration. There was nothing unusual. Deke chatted with him, keeping him advised of our weather problem, assuring him that everything else was good. When an electrical inverter in the rocket failed and was replaced, Deke let him know.

I walked outside for some air about 8:30 A.M. It wasn't raining and the black clouds seemed to be breaking up. Ernie Ammans in the weather room was predicting good launch conditions between 9:30 and 10:00 A.M. We picked up the count, aiming for nine-thirty. If we needed more time, the traditional hold at T-2 minutes could be stretched out.

The biggest problem for Al was his bladder. He'd been locked in *Freedom 7* for nearly four hours, and I heard him ask Deke for permission to exit the capsule. After a pause, Gordo Cooper in the blockhouse got an answer. It was no. Later we found out Al solved the problem by urinating in his suit. The moisture was soaked up by heavy thermal underwear, and the constant flow of air through the suit dried it out. I had to respect a man who could make that kind of command decision and carry it out without complaint.

The countdown went on. I went through a go/no-go with the computer people at Goddard, the range, and each position in mission control. The answer from everyone was the same: "Go, Flight!"

I pressed my mike key to inform the blockhouse. "Test Conductor, Flight. We're go."

"Roger, Flight. Launch is go."

We had a brief hold for a minor capsule problem, then pressed on past the magic two-minute point. Suddenly I was nervous as hell. My palms were sweaty and I hoped my voice didn't start quivering. I knew I wasn't alone. I could hear the pitch change in people's voices at the blockhouse as they went through their final checks. This was no simulation.

It was time for a final poll of mission control. I called out each position.

"Go, Flight."

"Go, Flight."

"Go, Flight."

Stan White varied the report slightly. "Heart rate's touching two hundred, Flight, but Surgeon is go."

So Al Shepard was nervous, too.

We were go. "Test Conductor, Flight, we're go." I was sure that my voice was an octave too high when I made my last comm check with the range safety officer a few seconds later. That was the moment when I looked down and, in shock, almost turned to Gene Kranz to say that my microphone had fallen off. At least, it had disappeared. Then it quivered into view. I was shaking so hard that it was turning invisible.

I leaned my hands on my console and forced myself to settle down. It was tough. A man was sitting out there on top of a rocket. No matter how proven the Redstone was, the potential for disaster was never more than a moment away. Al Shepard's life and the technological expertise and prestige of the United States were out there, too. We'd been playing down the idea of a space race with Russia. But in our hearts, we knew it was a real race and the responsibility for winning was ours.

Out in the world, with the count almost over, the television networks went live. A hundred thousand people watched along the beaches, and 100 million watched on black-and-white TV as the count went to the final seconds. Shorty Powers was on the public affairs console and was giving a running commentary to the press that was also being picked up by the networks.

At zero, the Redstone lit up. Deke Slayton saw that all the charts and graphs looked good and keyed his mike: "You're on your way, Jose."

It was a reference to Jose Jimenez, the Reluctant Astronaut, a creation of nightclub comic Bill Dana, who'd become fast friends with the astronauts. Shepard's response was all business.

"Roger. Liftoff and the clock has started."

That call at 9:32 A.M. on May 5, 1961, dissipated my tension. The clock report was vital because it told us that the capsule sequencer system had started properly. If it hadn't, Al had a backup switch on board. He didn't need it. On the beaches, people were yelling, "Go! Go!" In mission control, we were saying the same thing in hoarse whispers.

I watched the plot boards and they were perfect. FIDO was in my ear, confirming that the Redstone was following its normal trajectory. Nothing at all was wrong. This time the rocket wasn't hot. The escape tower unlatched and fired to carry it away. Then the capsule separated without repeating the Ham incident. Out there over the Atlantic, Al Shepard was coasting up to 116 miles. He was in space. His reports were so normal that Shorty Powers reached for a new way to tell the press what was happening.

"The astronaut reports that he is A-OK," Powers said, and a new word entered the American language. Shepard had said no such thing, but it was such a perfect way to describe the events that Shorty used it over and over in the next fifteen minutes. We'd used it internally, picking it up from the communications engineers, but I'd never heard it uttered outside of our small and private realm.

I could only grin and shake my head. "A-OK. Okay, Shorty, if that's what you want." I kept my attention on *Freedom 7*'s systems. After separation, the flight plan called for Shepard to take over from the automatic attitude-control system and point the capsule in various directions. It was a test of both the hardware and the man. I could see that Shepard was handling his capsule perfectly. His maneuvers were sharp and precise, using minimum fuel in the attitude control jets.

I tried to imagine what it was like for him at that moment, weightless for a long six minutes, looking through the periscope to get a view of Earth far below, flicking the hand controller and feeling *Freedom 7* respond in an instant. It had to be a magnificent experience, and I felt a passing touch of melancholy that I'd only know it from here, standing on Earth at my flight director console.

Shepard reactivated the automatic control system. We saw it in mission control as quickly as he did it. *Freedom 7* turned around; its retrorockets fired, then separated from the capsule. On board, the sequence light stayed dark. Slayton assured him that the pack had separated.

A burned-out bulb was literally the worst of our problems. I listened to Shepard and Slayton matter-of-factly calling out events. Reentry was right from our training sessions. Almost as soon as the last station downrange lost radio contact, one of the waiting aircraft picked it up and relayed it back to us. I heard Al say that he'd hit 11 g's and felt fine. The drogue chute came out late, but the main parachute deployed on plan at ten thousand feet. "It's beautiful," he said when he saw it billowing above, a sentiment that would be shared by every astronaut who followed. He was moments from splashdown.

Then Shorty Powers told the world that *Freedom 7* was on the water and everything was A-OK. Within minutes, helicopters had the capsule secured. Al opened the hatch and slipped his upper body into a horse-collar rig lowered from a chopper, and he was lifted aboard. It was a procedure we'd just adopted.

When I heard that he was heading for the recovery carrier, I let the last touch of stress fade away. I took out a cigar and lit up. Then I keyed the mike.

"Helluva day, team," I said. "Well done. Congratulations."

A sudden euphoria blanketed the nation and we weren't immune. I remember few times in my life when I felt better than on the night of May 5, 1961. "A helluva day" didn't touch the reality of what we'd done.

There were splashdown parties everywhere in Cocoa Beach. Betty Anne and I went to as many as we could. The reaction to the short and happy flight of *Freedom 7* stunned all of us. President Kennedy followed the flight from the Oval Office and with no warning at all asked us to patch him through to Al on the aircraft carrier. We saw some of the television coverage, including people all over America—all over the world, except for the Soviet Union—ecstatic at our man in space.

It wasn't just the mission. Reporters were applauding NASA for allowing live coverage. I heard the words "brilliant, risky move" over and over, particularly because it showed the stark difference between the United States and Russia. We took our chances in front the world's cameras and notepads. They took theirs in secret.

I'd been scared as hell when I first heard that we would launch Shepard with the world watching. But the more I thought about it, the more sense it made. If we failed, the world would know immediately anyway. So with our open system of government, we really had nothing to lose. What we didn't realize was how much we had to gain. On that day, we turned everything around in the eyes of the world. There would be future criticism, to be sure. But always it came because we were open to the press, not hiding behind a shroud of secrecy. And I believe to this day that the world reaction to Al Shepard's flight was a major factor in what President Kennedy did next.

Shepard was already at Grand Bahama Island being debriefed, and Shorty Powers was issuing almost hourly reports on his condition (A-OK), on how *Freedom 7* worked (A-OK), and how Shepard proved that man could live and work in space. (I would have given that last one a triple A-OK.)

Al told us later that zero g was no problem at all. He was so engrossed in following his flight plan activities, he said, that he had to remind himself to take note of how it felt. "I knew it would be the first question I was asked," he said, "so I better have an answer."

The answer for all of us was that Shepard gave us our first real data on a human who had worked in zero gravity, sandwiched between the high-g forces of launch and reentry. A few weeks later, we were rehearsing our presentation for NASA administrator Jim Webb. One of the research psychologists, Dr. Robert Voas, compared Shepard's performance on the procedures trainer and in space. There were a few tiny differences. He jumped to the conclusion that Shepard's performance had deteriorated in zero g and that was what he wanted to tell Webb.

I stepped up and disagreed. "My experience in looking at pilots' performance in aircraft maneuvers tells me that this is as precise as it gets," I said. "Those differences aren't significant." I saw Shepard smile, and before Voas could argue, both Bob Gilruth and Walt Williams had agreed with me. They dropped Voas from the agenda and told me to include the data, as I saw it, during my operations discussion.

Dr. Voas was unhappy at the rebuke. I didn't care. He was wrong. But he didn't forget and we locked horns time and again in the future.

The only medical data that raised eyebrows was Shepard's heart rate, spiking at 220 and holding above 150 for several minutes. Both numbers were above anything seen before in Shepard's medical history, even during stress tests and centrifuge runs. But the surgeons weren't that worried. They'd recently gotten a report on race car drivers showing heart rates even higher, and lasting for several hours. Even sucking in a bit of carbon monoxide didn't affect their high-speed driving. Based on that, Shepard's heart rate seemed well within the norm for a person whose adrenaline was flowing.

Something had happened behind the scenes after MR-3 that was about to change the world. Three days after splashdown, Kennedy invited Shepard to the White House to receive the NASA Distinguished Service Medal. The other six astronauts and their wives went along. So did Bob Gilruth, Jim Webb, and a few more. The news coverage didn't include some private talks that Jack Kennedy had with them.

The Kennedy administration had suffered in April when Yuri Gagarin flew in space, followed by the botched Bay of Pigs invasion of Cuba. Al Shepard's flight put some glow back on American prestige, but Kennedy wanted more. Long afterward, Bob Gilruth told me what happened.

Sometime in mid-1960, Gilruth ordered a study of how to get to the moon, fly around it, and come back home. This manned circumlunar flight could be a follow-up space program. I was vaguely aware of the study; some of my people worked out the lunar trajectory details for it. But it was background noise compared to getting ready for Mercury.

"The president was impressed with the world's reaction to the Shepard flight," Gilruth said, "and wanted to know more about what we were going to do."

So Gilruth and George Low from headquarters told the president what was coming in Mercury, then mentioned the circumlunar study.

Kennedy interrupted, "Why aren't you considering landing men on the moon? If we're going to beat the USSR, don't we need to do something more than just flying around the moon?"

Gilruth was taken aback. "But I didn't want to sound negative," he said, "so I told him that landing on the moon was probably an order-of-magnitude bigger challenge than a circumlunar flight. But he didn't let go."

Finally Gilruth told him that it was technically possible.

"What do you need?" Kennedy asked.

"Sufficient time, presidential support, and a congressional mandate," Gilruth said, never believing that it could happen.

"How much time?"

Gilruth and Low talked it over quickly and gave him an answer. "Ten years."

None of us knew that Kennedy was asking his vice president, Lyndon B. Johnson, to chair a quick study and make some spectacular recommendation about America's future. It didn't have to be space, but it was. Over the next weeks, NASA provided great detail about a lunar landing.

Gilruth passed the word down through Walt Williams for us to watch Kennedy's special report to a joint session of Congress on May 25. "There's going to be some kind of announcement on the future of space," Williams said. That was all he knew. Whether it was good or bad was left to my imagination.

I was in the control center writing a report and watching the television monitor on my console with a combination of fear and excitement. It was a good speech. But I was losing interest when Jack Kennedy got to the line that became forever famous. Then my head seemed to fill with fog and my heart almost stopped. Did he say what I thought I heard? He did:

"I believe this nation should commit itself to achieving the goal, before this decade is out, of landing a man on the moon and returning him safely to Earth."

For the minute, I was paralyzed with shock. My mind was going off in a hundred directions and I was sorting through the most amazing thoughts. *The moon . . . we've only put Shepard on a suborbital flight . . . an Atlas can't reach the moon . . . we have mountains of work just to do the three-orbit flight . . . the moon . . . we'll need real spacecraft, big ones and a lot better than Mercury . . . men on the moon, has he lost his mind? . . . Have I?*

I wasn't alone. No one on the Mercury team was immune. We were overwhelmed with emotion, with a sudden new sense of enormous adventure, with pride and awe that the president had dared to stand there and say such a thing. Then slowly and with a deep concern, our sanity returned.

I remembered the old saying about engineers: They'll overestimate what they can do in a year and grossly underestimate what they can do in ten.

I hope so, I thought, *because we've still got a lot to do on Mercury.*

Gus Grissom had the next flight, set for July. It was similar to Shepard's mission, but Grissom's *Liberty Bell 7* capsule was new and improved. A window was added to the hatch, giving him more visibility. Shepard had complained about the narrow field of view with the periscope, but the engineers had already figured that out and ordered the window. The attitude control system that worked well for Shepard was to be even better for Grissom, giving him more precision in setting up for observing Earth, positioning for retrofire, and other attitude changes.

The biggest change was in the hatch itself. The old design had a latching system that made it too heavy for orbital flight. The new hatch used explosive bolts for release. It was fast, lighter in weight, gave a tighter seal on the pressure vessel, and would let rescue crews get in much quicker.

We had two months to get Grissom ready. His backup was John Glenn, and Gus had trained with Al Shepard. I spent the first month back at Langley with the other flight controllers, devising and perfecting a new flight plan. This time up, Gus would have some added chores to solidify the proof that man could work in space. We also started work on preliminary flight plans for a couple of unmanned orbital flights, leading up to putting an astronaut in orbit for three trips around the world. All of it, we thought, could yet be done in 1961.

The office work had another benefit. We'd been at the Cape almost full-time for a year. Now we had a month at home, sleeping in our own beds, eating dinner with our families, taking some weekend time off for beachcombing, sunning, and relaxing. I even got in some golf. While I was there, the local paper wrote about the rumors that we'd be leaving. We knew it was true. Putting men on the moon would require engineering and test facilities far beyond anything at Langley. Astronaut training would be much more complex than our simple procedures trainer linked to mission control. And even the control center was inadequate. To manage and control missions to the moon, we'd need a new and bigger center, along with changes still unknown in the worldwide tracking network.

Leaving would be hard. Betty Anne, the children, and I were natives. Bob Gilruth had come from Minnesota, but he'd been at Langley since 1937 and wasn't eager to leave. A lot of us were in the same boat. What we didn't know was when and where. We were simply resigned to the fact that, sooner or later, it had to happen.

It was a Florida summer, hot and humid, when we went back to the Cape in late June. By now, Gus Grissom was up to speed on the changes in his capsule, had spent close to a hundred hours in the procedures trainer, and was ready to be linked up with mission control. The main changes in the flight plan called for him to make visual observations of Earth during his zero-g minutes, give the new attitude control system a good workout, and then use the new hatch to get out and be picked up by a chopper after splashdown.

Grissom was good. We'd heard that he was a pilot's pilot, a talented engineer, and easy to like. I gave him the same speech I'd given Al Shepard about our debriefing practices. "Al told me all about it," he said. "He said that if anybody was a f---up, I should just say it."

"We'll get along just fine," I laughed.

We planned to launch on July 17. Gus didn't need to get too rough with anybody after each sim, and we were increasingly happy with him. The weather was the problem, and we all bitched about it. July 17 became July 19. We counted down through the night. Gus went to the pad—it was just as inspiring and thrilling this time to see the lights glint off his silver space suit—and waited in the capsule until we scrubbed again. After Shepard's experience with a full bladder, the space suit people designed a unit—we called it a motorman's friend—for orbital flights. Grissom was wearing the new contraption and reported that it worked as advertised.

We finally got to T-2 minutes on July 21. The high tension of Shepard's flight was gone. We were toughening to the job and settling into a nervous level that was so acceptable we finally forgot that it was there at all. Grissom lifted off at 7:20 A.M. And it looked perfect all the way to space and back again. He handled the maneuvers to perfection, using the three systems of automatic, manual, and rate command, a combination of the two.

His Earth observations were cogent, and his call-outs during reentry were on time and worry-free. Then he splashed down.

The radio link between the low-flying recovery aircraft, the helicopters, and mission control was touchy. We only heard part of it. Gus was down and safe. Our own telemetry readouts were worry-free. We heard Gus tell the helo pilot that he'd be ready for pickup in a few minutes. Then next we heard excited voices, too garbled to understand clearly.

"Is he in the water?" somebody asked in my ear.

"Sounds like it," somebody else said. "What the hell . . ."

Conversation on my comm line was getting confused.

"Do they have the capsule hooked up? . . . Did he say it was full of water? . . . Oh, hell, I think it sank."

We didn't know what had happened to Gus. After a minute of silence, a second helicopter reported that they were "attempting to recover the astronaut."

"What does that mean?"

"Shut up!" I snapped. "Shut up and listen!"

I remember thinking, *I hope it's not a body they're recovering.* The next minute dragged on forever. Then we heard it. Gus was aboard the chopper. They were returning to the ship.

Grissom was safe, though soaking wet, shook up, and tired. Our day's work was done. But what the hell had happened out there? Our recovery chief, Bob Thompson, flew out to the ship to take charge and get the story. It wasn't pretty and it led to a controversy that had an impact on the program and future programs for years to come. Already the press was reporting that Grissom had blown the hatch off prematurely. Had he?

After an intensive and detailed investigation, the final report was inconclusive. Grissom said he followed normal procedures, including removing the protective cover and pin from the hatch detonator. He was waiting for the helicopter to get into position when the hatch blew off and *Liberty Bell 7* started taking on water. He scrambled out and into the sea—his space suit had a new neck dam that certainly helped to save him from drowning, but he forgot to cover his oxygen hose—while the helicopter tried to save the capsule.

But the capsule kept filling with water and the chopper's engine overheated. The crew chief cut the capsule free and it sank in seven thousand feet of water. At the same time, Grissom was struggling to stay afloat. Even with the new neck dam, water was seeping into his space suit. It may have been improperly sealed, and water was coming in through his oxygen hose, too. That pocketful of souvenir dimes didn't help, and he was sinking. Only then did the second helicopter crew see that he was in trouble. They dropped the horse collar to him and he barely made it. His comments on board the helicopter about the crew's lack of attention were feisty. But by the time they got to the carrier, he was damned happy to be alive and told them thanks and forget it. He wouldn't file a complaint.

Why did the hatch blow? We never found out. Other covers were tested extensively in the lab, and none could be made to blow by themselves. Wally Schirra spent hours in a capsule wriggling this way and that in an attempt to hit the detonator with his shoulder, his helmet, anything. He couldn't do it. The only way to make it blow was to hammer the detonator with a fist, and that always left a bruise. Gus swore that he hadn't hit it and showed a bruise-free hand to prove it.

That didn't stop the controversy. The press roasted him and laid the blame squarely on his shoulders. Inside NASA, we had two vehement camps. One side claimed that Grissom must have panicked or gotten confused, hitting the detonator either on purpose or by accident. The other side was sure that some unknown factor was to blame. Not a shred of proof supported either side.

I listened, talked to Gus, and read the reports. Then I chose to believe him. I trusted Grissom's professionalism, and it was best for the program, considering all the circumstances, for Flight to take his word and tell everybody to put it behind us. We had too much work to do, and *Liberty Bell*'s fate only changed two things. From now on, the detonator cover wouldn't be removed until it was time to blow the hatch. And Navy recovery crews would be told to get the astronaut first, then go for the capsule.

Liberty Bell 7 was salvaged in 1999 and went to an aerospace museum in Hutchinson, Kansas, for cleanup and display. There was no way to determine anything about the hatch incident. The primer cord explosive was gone and the hatch itself is still on the Atlantic bottom. The only unusual things found were dimes on the floor, probably souvenirs that technicians back at the Cape hoped to recover.

The sinking aside, only minor problems cropped up on Grissom's flight. My report to Walt Williams recommended that we go straight on to orbital Atlas flights. There was little left to learn by flying more Redstones. Nor did it seem useful to require astronauts to fly a Redstone before flying an Atlas. Walt passed the report on, but before Gilruth could make a decision, the Russians flew again.

This time Gherman Titov stayed up for a full day, orbiting Earth seventeen times. Man could certainly survive at least that long in zero gravity. There were some problems. Postflight reports, including intelligence reports, said that Titov vomited, hallucinated, and was at times irrational. He was not a well-trained pilot, and we understood that the Russian engineers and doctors were unhappy with his lack of self-control. We also understood the air, or space, sickness. But skilled test pilots were not likely to be affected the way Titov was. The important thing, everyone decided, was that he carried out his required tasks. His eyesight, coordination, and work abilities didn't seem to be affected.

The world didn't hear about the cosmonaut's problems. He was the new hero and NASA was under fire one more time. This time, those sharp criticisms had the opposite effect on our morale. We stiffened our resolve and got on with the complicated job of putting our own man into orbit.

Here I am, Christopher Columbus Kraft, Jr., age ten, standing with the family car in the town that doesn't exist anymore.

Burning the midnight oil at Virginia Polytechnic Institute with Betty Anne looking on, 1941.

Betty Anne Kraft and me one year after our marriage in September 1950.

This is my favorite photograph—astronaut Alan Shepard looks up at the Redstone rocket on the morning of May 5, 1961, when he became the first American in space. Even Al, the man with "the right stuff," was awed by what he was about to do. [NASA]

After the flight, Alan Shepard jokes with doctors while in flight between the U.S. Navy Carrier Champlain and the Grand Bahama Island. [NASA]

Another one of my favorite photographs: Astronaut D. K. (Deke) Slayton (*far left*), Col. Keith Lindell, and Virgil (Gus) Grissom (*far right*) greet Alan Shepard after his exhilarating first suborbital space flight. [NASA]

My crew and I discuss the flight plan for Gus Grissom's suborbital flight the day before launch (*left to right*): Alan Shepard, Gene Kranz, Howard Kyle, John Glenn, Dr. Stan White, and Stan Faber. [NASA]

As flight director for the first Mercury-Atlas unmanned orbital flight, I discuss a recovery operation with Navy representatives at Mission Control. [NASA]

My first encounter with John Glenn was not a positive one. We eventually worked out our differences, and he became a great American hero. Here we are discussing a flight plan in 1961. [NASA]

Mission Control Center with *(left to right)* John Glenn, Gordon Cooper, Shorty Powers, and Dr. Stan White. [NASA]

I respected President Kennedy a great deal. If it were not for him, we never would have made it to the moon before the Russians. Along with John Glenn and Alan Shepard, I briefed the president on Glenn's first manned orbital flight in February, 1962. [NASA]

President Kennedy and NASA Administrator James Webb *(far right)* presented me with the NASA Medal for Outstanding Leadership at the White House following Gordon Cooper's one-day flight in May 1963 *(left to right)*: Senator Everett Dirkson, Mrs. Cooper, Gordon Cooper, Deke Slayton, Kenny Kleinknecht, Vice President Lyndon B. Johnson, Senator Adlai Stevenson, Senator Hubert Humphrey, and Senator John Tower. [NASA]

Astronaut Edward White makes the first U.S. "space walk"—EVA—during the flight of Gemini IV. I had to encourage him to return to the spacecraft cockpit as he cavorted in space. [NASA]

Mercury-Redstone 3, or Freedom 7 as Alan Shepard named it, just after liftoff from the pad at Cape Canaveral, May 5, 1961—the first manned flight of the U.S. space program. This was the most thrilling moment of my life as Flight. I think it was Alan Shepard's also. It certainly was the start of something big. The reaction to this flight prompted President Kennedy to say, "We choose to go to the moon." [NASA]

November 1965: Hampton, Virginia, honored me with Chris Kraft Day. I celebrated with my wife and two children, Kristi-Anne and Gordon. [NASA]

I knew getting to the moon was a national obsession once *Time* magazine put me on their cover on August 27, 1965. [*Time* magazine]

A picture of Gemini 6 taken by astronauts in Gemini 7 as they traveled around the earth. [NASA]

Here I am with (*left to right*) Charles Mathews, Dr. Charles Berry, and Robert Gilruth trying to defend the flight of Gemini 7. [NASA]

My official NASA photograph holding the lunar module, 1966. [NASA]

Celebrating the successful completion of the first manned lunar landing, Apollo 11, are (*left to right*) George Mueller, Robert Gilruth, Eberhard Rees (deputy to von Braun), George Hage (chief engineer to NASA Headquarters Program Office), and George Low. [NASA]

Astronaut Gene Cernan and I greet Congressman Bob Casey following Gene's flight to the moon on Apollo 17. [NASA]

I showed President Reagan Mission Control in 1982. He was a great supporter of the space program. [NASA]

9

Around the World in Ninety Minutes

We had two missions to fly before we put John Glenn into orbit. Less than two months after Grissom's flight convinced us to drop the Redstone rocket from the program—a decision that also dropped Wernher von Braun from the program, though none of us thought of it at the time—another Mercury/Atlas combination was on the pad. It was the same capsule, refurbished and improved, that survived the last Atlas explosion. This time we had a near-perfect countdown, a near-perfect flight, and recovered the capsule after one orbital trip around the world. But now it was mid-September, and the Glenn flight, we were sure, would slip into 1962. (Glenn's selection was still a closely guarded secret.)

A few days after that one-orbit flight, we got the news we'd been expecting and dreading. Henceforth the Space Task Group would be called the Manned Spacecraft Center. And we'd be moving to a place none of us expected: Houston, Texas. As much as anything, it was a political decision. Houston, like several other cities, met the requirements NASA wanted. But it also was home to the chairman of the House Appropriations Committee, and Lyndon Johnson was vice president of the United States. It didn't hurt that an oil company donated a lot of land to Rice University, which then deeded much of it (but not all) to NASA free and clear.

Bob Gilruth went to Jim Webb and all but begged to keep his operation at Langley, Virginia. Webb made the politics absolutely clear with a single question: "What has [Senator] Harry Byrd done for the space program?" So we were about to become Texans. Betty Anne slipped out of bed that night and cried in the bathroom. I was sorry to be leaving. But it had to be and we talked through it until we vowed that we'd make the move with positive minds, expecting to like Texas as much as we liked our native Tidewater Virginia.

MA-5 was to be our last animal flight, set for late November. "Why another chimp?" reporters asked. "Why not a man?" They had a good point. Russia had put two men into orbit, one of them for a full day. We'd put a chimp and two men into suborbital flights and sent a man-simulator around the world in a Mercury capsule. That put four U.S. missions in the success column—each with nitpicking, but solvable hardware problems— and my operations team was now composed of veteran space-flight controllers. We talked it over and rationalized our decision.

"We've made a public commitment to the step-by-step approach," Gilruth pointed out.

"And the medical committee will give us hell if we ask out of the chimpanzee flight," Walt Williams added.

I thought, but didn't say, that it would feel good to have one more unmanned flight to make sure all those nitpicking problems had been solved. This time we weren't neck and neck with the Russians to put a man in orbit, and I'd rather lose a chimp than an astronaut. We fended off the press questions and stuck to the plan.

MA-5, on November 29, 1961, was a true dress rehearsal for manned flight. We sent off our remote crews and this time deployed four astronauts to key sites where they'd handle capcom duties as needed. Only John Glenn and his backup, Scott Carpenter, stayed behind, along with Deke Slayton, my primary capcom in mission control.

The Atlas was still a fickle rocket. I crossed my fingers on every launch. This one, with our Mercury capsule and a chimpanzee named Enos on top, was on the mark all the way to orbit. FIDO was quickly calculating the numbers.

"How's it look, FIDO?"

"Go for three orbits, Flight."

I couldn't ask for a better report. The capsule set itself into the right attitude and Enos looked good as it went over the hill. As it passed over Australia, the only problem worth watching came from the capsule's thrusters. They were firing too often to keep the capsule stable in the roll axis. We'd seen that on the last flight, too, thought the low temperature of space was turning the fuel to slush, and added some heaters. Now it was happening again. With a man on board, it wasn't a big problem. He'd simply switch the system out of automatic mode. Without one, fuel consumption had to be watched.

Enos had a series of tasks with levers and lights, more than the medics had asked of Ham. He was doing just fine, Surgeon told me.

I'd ordered go/no-go decisions for each orbit. While our capsule was over Hawaii, I polled the team. We were go for a second orbit, and when it passed the California coast, I made the call. Then things started to go bad.

"Roll disturbances are still high, Flight."

"Propellant consumption is above the line, Flight."

I listened without comment.

"Flight, ECS. Cabin and suit temps are going up. Looks like we have a hot inverter, too."

"What's your call, ECS?"

"We're go, Flight. But we need to watch it."

"Surgeon?"

"The chimp's okay for now, but his body temp is a bit elevated."

"Go or no-go?"

"He's go, Flight. For now."

The key words were *for now*. I heard warnings in every report. *This is what we trained for,* I thought. *This time, it's for real.* I activated my comm line to the remote sites.

"All sites, this is Cape Flight. Repeat, Cape Flight. We have several situations for immediate, repeat immediate, attention. One, roll disturbances and propellant usage. Two, cabin and suit loop temperatures. Three, inverter temperatures. Four, crew condition. His body temp is elevated.

"All sites, monitor and report immediately. Repeat, report immediately."

I thought about it for a moment, then clicked back onto the loop. "Hawaii and Cal [California] Capcoms, Cape Flight."

"Go, Flight . . ."

"We might have to bring him down on the next rev. Be ready to change the retrofire clock, or to fire the retrorockets on my mark."

"Rog, Flight."

Forty minutes later, the Australian sites reported that temps were decreasing and stabilizing. But the thruster problem was worse. One of them had failed. The capsule was automatically maintaining its attitude within our limits, and the jerky, back-and-forth roll maneuvers weren't affecting the chimp. It was propellant depletion that worried me. We needed enough to hold steady during retrofire. An astronaut on board could have switched off the automatic system that was malfunctioning. We couldn't do it from the ground.

"Systems, how's it look?"

"We're checking, Flight."

I wanted more than that, but I gave them time. By Hawaii, all the temps were just about normal, including the chimp's.

"Flight, Surgeon. He's go for a third orbit."

"Systems. I want a go/no-go before California."

"Rog, Flight."

I meant exactly that. Go/no-go decisions had to have an element of gut feel to go with the hard data. Enos was reportedly good for the third and final orbit. The surgeon didn't know that the chimp's test apparatus had developed a problem. Like Ham, Enos was getting an electrical shock no matter what he did. He was a strong little fighter and he was tearing the equipment apart. He'd also ripped out a urinary catheter and was beginning to bleed. We didn't have instruments to tell us that piece of bad news.

But my concern was having enough attitude propellant to set up for retrofire. Gordon Cooper was Capcom at the California site. I warned him to be ready to initiate the retrofire command.

At the systems console in mission control, they were still pondering the data when the capsule's signals were picked up in California. Arnie Aldrich on the systems console out there reported immediately. The errant roll maneuvers were still a factor, and propellant was getting low. I turned to Systems in my control center.

"Systems, what's your recommendation?"

I got equivocation. And I was getting a bit upset myself. I glanced at the mission clock and did a quick calculation.

"You've got twelve seconds, Systems!" I snapped. That did it.

"No-go, Flight. Bring him down."

"California, Flight. Retrofire on my mark." There were only a few seconds left to make it, but Aldrich was ready. "five, four, three, two, one, mark!"

The signal went up from the California station at Point Arguello. Arnie gave us a running commentary as a perfect retrofire sequence unfolded in space. We confirmed it on our own consoles at the Cape. Enos was on his way down.

We learned a few things that day. It was the first time I faced a life-and-death decision in my role as Flight. No matter that the life was a chimpanzee's. In those last minutes, all that mattered was getting him home safely. I don't remember thinking of him as a chimp. He was my responsibility as much as any astronaut would ever be. My team learned the value of making those decisions, too. The next time I asked for a recommendation or a decision on anything, I'd get it in a hurry. In the back of my mind, I made a note: It's better to make a conservative decision now and end the mission than to wait for the perfect decision and maybe lose everything along the way.

We'd missed the chance at that third orbit, but the rest of the problems were minor. The cabin and inverter temperatures had stabilized on their own. The thruster problem was trickier and taught us that the obvious answer isn't always the right one. On MA-4, we thought the problem was the cold, so we added heaters. Now we discovered that heat from the thruster firings was the culprit. It caused valves to warp, creating erratic fuel flow. This time we inserted a shunt, a piece of metal that carried heat away from the fuel lines and into the main capsule structure. The problem never came back.

Enos didn't get his third orbit in space. But until his test equipment failed, he'd done an admirable job of paying attention and working in zero gravity. Even when it got too warm for a while, he kept working. The medics had to be satisfied.

At the first postflight press conference, Bob Gilruth told the world that John Glenn would get the first manned orbital flight and that Deke Slayton would get the second. I was delighted for Deke. He'd be a joy to work with. Glenn would be okay, too, even if he'd shown a tendency to be overly picky while backing up Shepard and Grissom. We'd work it out.

Everything was coming at us at once. The moon program was to be called Apollo. A new two-man program called Gemini was inserted between Mercury and Apollo to keep us flying in space and to work out many of the technical problems of going to the moon. And we still hadn't put our first astronaut into orbit. Mercury, Gemini, and Apollo. It was a tall order for a team that was still testing airplanes three years ago.

We needed some time off and Walt Williams sent us home for Christmas. Even that generated critical stories in the press. How dare we? We dared, all right, and I spent a week with Betty Anne and the kids. It wasn't all rest. We started planning our move to Houston. And I started setting up my new job. Just before I left the Cape, Williams called me in. Chuck Mathews was going to Apollo, to help in spacecraft design.

"And you're the new chief of the Flight Control Division," Williams said. "Merry Christmas."

"Thanks a lot," I said. It was going to be a helluva increase in responsibility, with three man-in-space programs to support. I thought about how much we'd had to learn to get this far in Mercury. The depth of our ignorance on running long-duration missions with two men up there was heart-stopping. But Gemini would be simple compared to what we needed to learn before we could send men to the moon. *How the hell do we control a mission when the crew is a quarter million miles away?* I thought with a shudder. *It's hard enough when the capsule is only 120 miles overhead.*

It was daunting, but taking over the division was an exciting challenge. The first phone call I made when I got home was to my old friend Sig Sjoberg, who was special assistant to both Mathews and me. He jumped at the chance to be my deputy. We put in some of that Christmas holiday time establishing an organization that could support Mercury, Gemini, and Apollo at the same time, with each program in a different phase of work. One decision was easy. We needed to hire a lot of new engineers and technicians in the next months. They'd report to work in Houston. The rest of us would follow in mid-1962, after we'd finished the Glenn flight and the Slayton flight.

No matter the promotion and new responsibilities, I was still Flight and had a mission control center to run. Sjoberg and our administrative assistant, Chris Critzos, took on the mounds of paperwork that came with expanding the Flight Control Division and moving everything fifteen hundred miles away to the Gulf Coast of Texas.

The phone rang on my console in the control center while we were setting up simulations for Glenn's flight. We hoped to get him off the pad in January 1962, but so far the Atlantic coast weather wasn't cooperating. Walt Williams was on the line.

"Glenn and Voas were just here," he told me. "They want to change the flight plan."

It was late in the game for that kind of crap and I was testy when I asked, "What'd you tell them?"

"I told them to see you. It's your call." Before I could ask more, he hung up. *It better be good,* I thought, *or I know what my call will be.*

Dr. Robert Voas was a thorn in my side going back to the Shepard flight, and I didn't like that he was involved in astronaut training. He seemed to be getting a lot of the credit for the detail work being done by Ray Zedekar, George Guthrie, and others. If he understood pilots better, I might have felt differently.

He and Glenn met me in Williams's office. It was just the three of us and Glenn got right to it.

"We're concerned about doing retrofire under manual control," he said.

That surprised me. The flight plan called for retrofire to be done in automatic mode. Manual control was strictly a backup and Glenn had practiced it in the trainer under a variety of off-nominal and emergency conditions. He was pretty good at it, too.

"What's the problem?" I asked.

"Disorientation," Voas said. "If the capsule's spinning out of control, he'll have to stabilize it first. He hasn't had that kind of training."

I bit my tongue. *Of course not, you fool,* I wanted to say. *How do you simulate a zero-gravity condition like that here on Earth?* Instead, I asked the obvious question.

"What do you propose?"

"On the first orbit, turn off the automatic system and have John spin the capsule in all three axes," Voas said smugly. I'm sure that he was thinking that I should have thought of this myself. "Then let him regain control."

"So you want to create an actual emergency situation in space just so he can practice recovering from it, in case the same emergency happens by accident?" I tried to conceal the sarcasm in my voice.

I saw a light dawning in Glenn's eyes, but Voas was a bulldog.

"It's not an emergency if he spins it on purpose," he said.

"The capsule is doing exactly the same thing either way," I said with more heat. "What's the difference?"

Voas still didn't understand. Now we were arguing. "Glenn's going to have plenty to do up there without adding this dumb idea," I said.

"If he can't recover from a spinning capsule, you'll have his blood on your hands," Voas said, pointing to Glenn.

"And whose hands are bloody if he deliberately spins it and still can't recover?"

That ended the argument. Glenn and Voas got up and left. Whatever shred of respect I might have had for Voas was gone, and I hoped that none of his ideas rubbed off on Glenn in the future. We ran the mission without creating a "practice emergency." As it turned out, Glenn had to face problems of another kind anyway.

Close to a thousand reporters, both U.S. and foreign, were at the Cape for the John Glenn flight. The weather was so bad that he suited up five times before we let the countdown go to zero. The reporters' moods were as foul as the weather. They were stuck in Cocoa Beach, overrunning their expense accounts, and the only story they could write was about one more delay.

In one press conference, John Finney of the New York Times asked me about the odds of a successful launch the next day. I thought about it briefly and gave him an honest answer: "If I thought about the odds at all, we'd never go to the pad."

The crew of that Russian trawler had it just as bad, shuttling back and forth between Florida and Cuba over rough seas. Our coded signals to the Atlas and the Mercury capsule were kept secret. We used one set for simulations and another on launch day. One morning when we thought we'd get a weather break, the Bermuda site chief opened his safe to get the code and found that the metal bands on the security box were broken. That was one more problem we didn't need. Then he reported a second time. The box had been in the safe so long that the bands broke from stress corrosion. Nothing about John Glenn's flight was easy.

We finally did it on February 20. We'd had a few minor problems and holds, pushing liftoff into midmorning. Then John Hodge reported a problem with the Bermuda radar. I called a hold while he gave me a lengthy explanation.

"Go or no-go?" I asked. He gave me more explanation. "It's your decision, go or no-go?"

He wouldn't give me an answer. I looked at my watch and waited until the second hand was straight up.

"You've got two minutes. Go or no-go? Make up your mind!"

There was a long silence and the tension in mission control was thick enough to touch. I stared at my watch.

"We can live with it, Flight," Hodge finally said. "We're go."

"Goddard, Flight. Are you go?"

"We're go, Flight," the computer supervisor up north said instantly.

"Launch Director, Flight. We're go whenever you want to pick up the count."

Ten minutes later, the count went down to zero. Scott Carpenter was Capcom—we called the position Stony—in the launch blockhouse. His words at the T-10 second mark became famous: "Godspeed, John Glenn."

The man about to become America's most famous astronaut waited until he felt the thundering Atlas rocket begin to move. Then he gave the standard and now classic reply: "Liftoff and the clock is running."

Everything looked good for a while. Glenn was in orbit safely. After a few moments' disorientation, he'd adjusted quickly to zero g. He tried out the control system. It worked just as he'd expected. The capsule's nose pointed wherever he wanted, changing its position with a few quick blips of his thrusters. It was different from flying an airplane, but it wasn't hard. He crossed Africa, looked down on Australia and saw the lights of Perth, and sped toward sunrise near Hawaii. That was when he saw the fireflies. Sparkling in the sun's sudden light, his *Friendship* 7 capsule was surrounded by dancing bits of brightness. Nobody knew what they were. I heard his call and shook my head. This was something new.

We didn't have much time to think about it. Suddenly a lot was happening. It started with a phone alert from the White House. President Kennedy wanted to talk to Glenn. I wasn't happy with the interruption. But I told the backroom comm people to set it up. We'd patch him through when Glenn crossed over the Cape in about thirty minutes.

Then Glenn switched his control system back to automatic, and the capsule nose swung to the left. He used his hand controller to bring it back. Then it swung again. He brought it back. It wouldn't stay. My systems guy was Don Arabian. He was about to get very busy.

"Systems, what's happening?"

"Looks like a problem in the yaw thrusters, Flight."

"Recommendation?"

"Take 'em off-line, Flight."

"Tell him, Capcom."

Glenn dutifully pulled the switches to turn off the offending thrusters. He had others that would do the same job. Almost immediately, another problem jumped up to bite us.

"Flight, Systems, I've got a Segment 51."

"More details, Systems."

"Segment 51 is a live switch on the heat shield collar. It's saying that the landing bag deployed."

My gut told me instantly that it was a faulty signal. A quick schematic check showed three switches on the collar. Any one of them—or all of them—could be sending the signal. It could be serious. The landing bag was stowed between the heat shield and *Friendship 7*'s pressure vessel and would cushion splashdown. If it had actually deployed in orbit, that meant the heat shield was loose. That only raised more questions in my mind.

How could all of that happen without Glenn hearing something? Wouldn't there be loud noises when it happened and probably some clanging and banging from the heat shield every time the capsule's thrusters fired? And wouldn't Glenn instantly report if his own instrument panel showed deployment? I discussed it quickly with Al Shepard on the capcom console. He'd been through a splashdown and agreed. It was probably a spurious signal.

"Don't mention it to Glenn," I said. "Not yet, anyway. He's got enough to do without worrying about this."

"Rog, Flight."

At the same time, Gene Kranz on the procedures console sent Teletype alerts to the remote sites, asking them to watch for the Segment 51 signal. As Glenn crossed the Atlantic, then Africa for the second time, the reports came back. Segment 51 showed up in some telemetry signals, but not in others.

In the back room, President Kennedy's call to Glenn was canceled. By then, Shorty Powers was reporting the Segment 51 problem to the press and the world. I'm sure that Kennedy understood.

Top management now reared its head in mission control. Walt Williams was still my boss as operations director. He'd never interfered before. But now he got involved, setting up a backroom briefing with Max Faget, who'd designed much of the capsule, and with John Yardley, McDonnell's chief engineer. As Glenn approached Australia on the other side of the world, they appeared by my console. They were worried.

Arabian, Shepard, and I were convinced that nothing was wrong beyond a faulty signal. Gordo Cooper was the capcom in Australia. I told him to ask Glenn about noises. The answer came back in minutes: no noises. But now Glenn was alerted. Something was bothering us and he didn't know what it was.

"Tell him to cycle the switches on the landing bag system," I told Cooper. I hoped that turning the switches on and off would make the signal go away. Cooper passed the word and Glenn complied. Nothing changed. Segment 51 was still there in the Australian telemetry.

Now Walt Williams was getting antsy. Half the mission was still to go, and he was listening to Yardley and Faget suggest that we leave the retro-rockets in place after they'd fired. Their reasoning was simple: If the heat shield was loose, the retros might hold it in place during reentry, at least until the retro pack burned away.

I was aghast. If any of three retro rockets had solid fuel remaining, an explosion could rip everything apart. The fuel always burned completely, Faget said. That didn't make me feel any better about leaving that heavy and nonaerodynamic retro pack in place. About that time, one of the remote sites let Glenn know what was happening. Now he understood our worry. But again he reported no noises and no indications that the landing bag was anywhere except where it was supposed to be.

In mission control, our discussions were getting heated. None of us knew that the outside world was even more worried than we were. Shorty Powers's reports, consistently accurate, revealed that Glenn's heat shield might be loose. The press was smart enough to report that Glen's reentry might be short, fiery, and fatal. That Flight considered the signal to be spurious didn't change the tone of the reporting. John Glenn was in danger. Millions of people stopped to watch and listen.

As Glenn approached the United States on his third and final rev, we had to make a decision. He'd maneuvered *Friendship 7* into the correct position for retrofire. He'd checked his landing bag switches again, with still negative results. I looked at Walt Williams and Max Faget. None of us had changed our minds. It was my call—jettison the pack after retrofire or leave it on.

"Okay," I said, "you're the hardware experts." That wasn't quite true about Williams, but Max still insisted that leaving it on wouldn't put Glenn in even more danger. "I'll do it your way."

"California, Flight. Tell him we're thinking about leaving the retro pack in place after retrofire."

Wally Schirra did it and Glenn just as quickly wanted to know why. Schirra didn't have time for an explanation. The final word, he told Glenn, would come after retrofire. I wanted to know that those rockets had worked and that the chance of unburned fuel remaining was just about zero.

In mission control, everyone held his breath and watched the clock. We heard Glenn's report. One by one, those rockets that would bring him down fired on time. I asked for confirmation.

"Retro, you're sure they all fired?"

"Rog, Flight. We got three good ones."

"Okay. Texas, Flight. Give him the word. Keep the retro pack in place." We didn't have an astronaut at the Texas site. The local systems engineer had to pass along the critical instruction and he didn't know why.

Within moments Glenn had his instructions. Now he was both curious and concerned. But Texas couldn't tell him anything more. He'd have to wait a few minutes until we had him in sight from the Cape. Then Shepard used the few minutes before communications blackout to give Glenn a quick summary of our thinking.

Glenn's answer was short. He was starting to feel the forces of gravity build up. "Roger," he said. "Understand." In the next minute, he reported seeing chunks of burning material going by his window. A fiery strap wrapped around the window, then blew away. Things bumped and banged off the capsule, and the outside world was turning bright red. Then *Friendship* 7 went over the hill and we could only wait in silence with shoulders and guts tensed.

Everything worked. We heard his voice relayed through the recovery planes. The capsule had come through the reentry heat and the communications blackout, and now its parachute was out. Glenn splashed down forty miles long, but just five miles from a destroyer. When they had him on the deck of the *Noa*, I lit up my cigar. I heard later that America had gone crazy in that last hour. The celebrations from coast to coast were just short of delirious. John Glenn, on that day in 1962, and in that last hour before splashdown, became a true American hero.

While America rejoiced, I sat for a few moments and thought it over. We'd taken an action that I still considered dangerous and foolhardy in leaving that retro pack on and letting it burn away. We could have killed John Glenn just as surely as not. We were lucky. When Faget's engineers examined the capsule, they found just what I suspected. One switch was misrigged and was sending a bad signal. There had been nothing wrong up there at all, not until we left the retro pack on. That was wrong. It didn't kill Glenn, but it was still wrong.

We learned some hard lessons that day. One was that we needed telemetry from all three switches individually. If we'd had that on *Friendship* 7, we would have known that two of the three hadn't activated and that the heat shield hadn't come loose. Redundancy of that kind forces us into a new way of thinking and would soon make my flight controllers the best systems engineers in the world.

Before I went out to join the splashdown parties, I also jotted down a new mission rule: "Never depart from the norm unless it is absolutely required. Once you do, you enter a regime where events are unpredictable. Make a change on the fly and it might bite you and bite you hard."

My flight controllers and I were a lot closer to the systems and to events than anyone in top management. From now on, I swore, they'd play hell before they overruled any decision I made.

10

The Man Malfunctioned

The next time out we had real problems. The Mercury capsule worked, but the astronaut didn't. He damn near killed himself, and the effects on the space program could have been devastating.

We were still basking in the afterglow of John Glenn's flight. The capsule problems were generally minor. We were getting used to making fixes and doing some quick redesigns between missions. But none of us were used to the public attention that followed the splashdown of *Friendship* 7. President Kennedy didn't need the press reports to get excited about Mercury. But he knew a good thing when he saw it and came to the Cape to greet Glenn personally. Then he wanted to tour mission control.

He arrived in a crowd of NASA executives, Secret Service men, and a few members of Congress. Glenn was at his side, and they both walked up to me, hands extended. Suddenly I had a problem. Which hand should I shake first? I hesitated just long enough for Kennedy to understand and get a kick out of the situation. Neither of them backed down, so I grabbed John Glenn's hand and congratulated him on his outstanding performance. Then I shook Kennedy's hand. His grin told me that he didn't mind being second to America's first man in orbit.

I was the designated briefer, so with Al Shepard tagging along, I took the president around the room, explaining each console and introducing him to the controllers. Suddenly there was a loud commotion behind us. Reporters and photographers were wrestling with each other for position, and it was starting to sound angry. One of the *Life* photographers, Ralph Morse, popped up in front and took a picture, which remains one of my all-time favorites, of Kennedy and me at the FIDO console. Through it all, Kennedy seemed oblivious to the press. Maybe it was an everyday thing to him, but it was my first experience with rowdy journalists. The next time I saw reporters with elbows flying, I wouldn't be so surprised.

Within a few days, Kennedy hosted all seven astronauts at the White House and gave Glenn the NASA Distinguished Service Medical. Glenn rode down Pennsylvania Avenue with Lyndon Johnson and addressed a joint session of Congress. Then it was on to New York for a ticker-tape parade. The mayor of New York awarded medals to both Glenn and Bob Gilruth, who finally got some of the praise he deserved.

It was quite a turnaround. Only a few months earlier, NASA and Gilruth were being publicly criticized for going too slow in the space race with Russia. Now Glenn and Gilruth—mostly Glenn, whose status would never fade—were heroes, and through them, so were the rest of us. It was too good to last.

Deke Slayton was next up. If they'd let him fly, he would have been the next American hero. Instead he was dumped. Slayton was the victim of overly fretful doctors and a NASA hierarchy that turned timid when it should have been bold. What happened to Deke Slayton shouldn't have happened to a dog.

Deke had a heart murmur, more precisely an atrial fibrillation. Once in a while his heart zipped instead of thumped. It never lasted long. It didn't bother him and had no physical impact. He flew fast airplanes. He ran. He survived high-g centrifuge runs. He was experimenting with different foods to see if something he ate caused the problem. He'd been cleared by NASA and Air Force doctors, who weren't at all concerned.

Except for one. An unnamed internist in San Antonio, who'd examined Deke, secretly wrote a letter to NASA administrator Jim Webb recommending that Deke be disqualified for space flight. With the Glenn flight just over, Webb decided to look into Deke's selection. He appointed a trio of noted cardiovascular physicians, presumably to certify Deke for space flight. Instead they came back with a conservative report. Deke may not have a problem in orbit, but why take the risk? Jim Webb was a political animal. He took the easy way out and ordered Slayton to be replaced.

Deke was furious. So was I. He was treated shabbily and became a political pawn. I was even more furious when Walt Williams recommended that Scott Carpenter be given Deke's flight. My experiences with Carpenter convinced me that he was the least qualified of the seven astronauts. Unfortunately for all of us, I'd kept that opinion to myself. Now I couldn't hold back any longer.

I went to Williams and laid it on the table: "Carpenter will screw it up. He's the worst possible choice for the mission."

Williams was surprised and defensive. "I don't think so. Why are you bringing this up now?"

I told him about Carpenter's inability to understand the capcom job in mission control. I told him that Shepard hadn't even argued about replacing him. Carpenter had been Glenn's backup and, in training sessions, wasn't half as good at handling problems or emergencies. I told him that in my opinion Carpenter's engineering skills were substandard and that he might well jeopardize himself and the program during the mission.

None of it mattered. Williams had already given his recommendation to Gilruth and wasn't going to retract it. Even if Carpenter was as bad as I claimed, the mission was just another three-orbit test not much different from Glenn's.

Williams looked me in the eye when he said it: "Carpenter stays."

I bit my tongue and went back to work. I should have gone over his head, directly to Gilruth. But insubordination wasn't in my nature. Later I wished that it were.

It could have been our best mission yet. Carpenter's Atlas put his *Aurora 7* capsule into a near-perfect orbit. He did his turnaround maneuvers on time and to my surprise seemed to be doing everything just right. He casually reported a disagreement between the capsule's attitude and the instruments, but didn't mention it again for nearly two hours. He'd combined the comment with a reference to his gyro instruments that cloaked the real meaning of his report. Then he passed over Africa. From then on, almost every time he reported, he was behind on the flight plan. He had problems with his camera's film pack. He was fascinated by the view and spent too much time, and used too much fuel, pointing the capsule to look around. He got the camera working and lost himself in taking pictures.

By the time he reached Australia, I was beginning to worry about his fuel consumption. It was worse up over Canton and on to Hawaii. He'd spotted

John Glenn's fireflies and was maneuvering this way and that to track down their source. As he passed Guaymas, Mexico, Gordo Cooper gave him a first warning.

"Start to conserve your fuel a bit," Cooper radioed.

But Carpenter didn't. As he reached the Cape at the end of his first revolution, he told Gus Grissom that he was "a little ahead on fuel consumption." The subject was dropped while he tried to release a balloon for one of the scientific experiments. It failed to inflate. Then Grissom reminded him again to cut down on fuel use.

We still didn't know that Carpenter's instruments weren't giving an accurate reading of *Aurora 7*'s attitude, which way it was tilted, turned, or pointed. He wasn't reporting it, and our telemetry could only report instrument readings, not what he was seeing out the window. At this point, it didn't matter anyway. He was joyriding through space and using fuel at an alarming rate. The time when he'd need precise alignment and control would come at retrofire and reentry. As he reached Hawaii for the second time, I was jolted upright by his casual comment to the ground: "Fuel is forty-five and sixty-two percent."

That was low in both systems, considering how much fuel he'd need for a normal reentry. It wasn't dangerous yet. All he needed to do was follow our instructions about conserving fuel for the rest of the flight. But instead of being concerned, Carpenter reported that he was looking at fireflies, most of which now looked like snowflakes in space. We could see his fuel dropping as he maneuvered. After that, it only got worse.

Part of the problem was that Carpenter either didn't understand or was ignoring my instructions. His reports never alerted us to the difference between where his instruments said the horizon was and where he could see it out his window. With little more than one revolution left before he'd need to set up in a precise attitude for retrofire, and with his fuel situation approaching the danger level, I wanted to know more about that instrument problem. I told the Hawaii Capcom to get more details.

"*Aurora 7*. This is Capcom. Would like for you to return to gyros normal and see what kind of indication we have; whether or not your window view agrees with your gyros."

"Roger. Wait one." It took Carpenter thirteen seconds to give that answer, during which he traveled sixty-five miles beyond Hawaii and toward California. His next report came after another thirteen seconds and sixty-five more miles. It was about the fireflies. He was completely ignoring our request to check his instruments.

Two and a half minutes later, *Aurora 7* had moved beyond Hawaii's range and Carpenter still hadn't given me what I needed to evaluate whether he had a faulty instrument. I told Al Shepard on the California capcom console to find out what the hell was going on up there and left no doubt that I was getting frustrated with Carpenter. Shepard tried to stress my feelings by resorting to military terms.

"General Kraft is still somewhat concerned about auto fuel," he said. "Use as little auto—use no auto fuel unless you have to prior to retro sequence time."

Somehow Shepard missed that we'd been asking Carpenter questions about his ASCS problem. If something serious was going on up there, Carpenter's casual answers didn't address it. When he reached the Cape a few minutes later, fuel was down sharply again. Finally he agreed to keep hands off his thruster systems and just drift.

But over Africa, he was chasing those fireflies again, apparently determined to find out what they were. His fuel dropped another three percent before he went back into drift mode and stayed that way for a while. Over Australia, he finally solved the firefly riddle. When he rapped on the hatch or on the capsule walls, the "fireflies" appeared in droves. They were frost particles that accumulated on the capsule's exterior whenever moisture-laden air was vented.

But he should have been getting ready for reentry. *Aurora 7* was still pointing small-end forward. Carpenter hadn't bothered turning the capsule around into retrofire attitude.

By his last pass over Hawaii, his fuel was down near the critical levels. He was behind in stowing equipment. He hadn't gotten a good start on his retrofire procedures, and as I listened to him, he sounded confused. The Hawaii capcom tried to direct Carpenter through the retrofire procedures, telling him to get into the correct attitude, blunt-end forward, and to update the retro clock. Then came an electrifying report. After four hours of ignoring his instrument problem, Carpenter suddenly told Hawaii:

"I have an ASCS problem here. I think ASCS is not operating properly. Let me . . . emergency retro sequence is armed and retro manual is armed. I've got to evaluate this retro—this ASCS problem . . . before we go any further."

That mangled report about his thrusters only made his situation sound more confusing. I tried sorting it out in my mind. Was it an individual thruster problem? Was he having trouble maintaining a stable attitude? What the hell was going on up there? The one thing he didn't tell us was the most important: His instruments didn't agree with the position of the true horizon that he could see out the window.

I was ready to tell the Hawaii capcom that we needed clarification when Carpenter finally volunteered that his instruments were off. He was almost beyond range when Hawaii ordered him to go to retro attitude. The auto system would have gotten him close and he could have done manual corrections. Instead he tried to do it all manually. As he went over the hill beyond Hawaii, Carpenter made one more confusing report:

"I think we're in good shape. I'm not sure just what the status of the ASCS is at this time."

Was he in good shape or not? We didn't know. I told Shepard in California that he was the last hope to get Carpenter squared away. Then suddenly we lost communications with California. Our voice and data lines were dead. That was all I needed to completely ruin my day. But the Goddard comm supervisor was quickly in my ear, telling me that AT&T was rerouting the lines. Before Carpenter reached California, we had everything back. Later they told me that a backhoe operator in Phoenix had dug up the lines.

When Shepard got radio contact with *Aurora 7*, less than two minutes was remaining before retrofire, and Carpenter was still facing the wrong way. Shepard listened to Carpenter trying to position his capsule manually, then quickly talked him through the steps to get into position. Carpenter had switched to manual control, and that was a mistake. In the automatic mode, the capsule would have been close enough to the right position for him to use the scribe marks on his window to make the final corrections. Shepard understood perfectly that, in the next thirty seconds, Carpenter would come face-to-face with death. To make it worse, with the auto system locked out, he'd have to fire the retrorockets manually.

Shepard counted down, and Carpenter almost got it right. He fired the retrorockets three seconds late, one after the other. If any of the three rockets didn't burn exactly right, Scott Carpenter was about to be in deep, deep trouble. The Lord was with him and the rockets were perfect. But while *Aurora 7*'s pitch attitude (nose up or down) was close, the side-to-side yaw was off by a huge twenty-four degrees. He hadn't had time to line it up. His entry path would not follow the desired line.

Our initial radar data indicated that he'd land well beyond the planned recovery point. We'd get that yaw detail in our postflight evaluations. For now, it was obvious that *Aurora*'s decrease in velocity—the delta-V—was less than it should have been, and I knew immediately that he'd held his capsule's attitude wrong during retrofire. Whatever happened next was beyond my control, and all I could do was wait and worry. If I'd known just how far off his yaw position was, I might have been praying out loud.

"Flight, Retro. He's coming down about two hundred and fifty miles long."

"Get 'em moving."

"Rog, Flight." Within moments, John Llewellyn had passed the word to the Recovery Control Center. This was only our fourth orbital mission, but my controllers were already pros at deciphering the radar and calculating impact points.

Getting to splashdown was my real worry. Now *Aurora 7* was plunging through the atmosphere, with Carpenter using his thrusters to keep the capsule from wobbling around and perhaps tumbling end over end. Only a minute later, he ran out of fuel in his manual system. His capsule and his life now depended on the little fuel remaining in the automatic control system. Instead of conserving every ounce, he used some of the now precious auto fuel to turn him around and give him a view of the Earth. He didn't seem to know that it might be his last.

"I can make out very, very small farmland, pastureland below. I see individual fields, rivers, lakes, roads, I think," he radioed. He seemed awed by the scenery, but not by his own danger. "I'll get back to reentry attitude."

In mission control, I was gritting my teeth. Bob Thompson was in my ear from the Recovery Control Center, telling me that the recovery forces—ships and planes—were moving. At least things were working right on the ground.

We could see from our telemetry readings that the capsule was oscillating wildly as it passed over Cape Canaveral. It was supposed to be stable. And the surgeon told me that Carpenter's EKG looked as if he were having heart problems. That news somehow leaked to the press and caused a stir among reporters. We figured out later that he was pulling 12 g's instead of the normal 9 g's because of his shallow reentry angle, and he was trying to talk at the same time. The combination caused confusion in his medical instruments. But all I knew at the moment was that we had one more potential problem, an astronaut who might be having a heart attack.

Gus Grissom on the Cape capcom console reached Carpenter just before blackout. Through all of this, Carpenter still hadn't closed the face plate on his helmet, a fact obvious to my environmental systems guy because the oxygen flow rates were too high. Grissom was calm. "Do you have your face plate closed?" he asked. Carpenter caught on and closed it. Then *Aurora 7* went into blackout, when the fireball surrounding it was so intense that no radio signals could get through. We could only wait and hope that it came out the other side.

It was a wild ride, with severe oscillations. The last of the auto fuel ran out, and now the oscillations got worse. Finally Carpenter popped his drogue chute. He was late on that, too, and it didn't do much to stabilize the capsule. Then finally he deployed the main parachute just below ten thousand feet. It billowed open and the final descent was normal. It was just about the only normal thing that had happened in hours. A Navy search aircraft picked up the signal from *Aurora 7*'s radio beacon. The plane was about one hundred miles from the splash point and heading that way. We heard Carpenter's radio call a few moments later.

"Hello, Mercury recovery force. Does anyone read *Aurora 7*? Over." There was a quick cheer in mission control. Grissom managed to have a brief conversation before the signal was lost.

"*Aurora 7*, *Aurora 7*, Cape capcom. Be advised your landing point is long. We will jump air rescue people to you in about one hour."

"Roger." Carpenter replied. "Understand one hour."

But apparently he didn't. He was sitting in his life raft enjoying the calm weather when frogmen jumped from a plane behind him and startled him by swimming up. At least we knew he was safe. Shorty Powers had been keeping the world informed through the full reentry and now told the press that Carpenter was on his way to the recovery carrier. His rescue helicopter landed on the ship about four hours after splashdown. A reporter was there, representing the news media, and Carpenter's quote was flashed around the world: "I didn't know where I was, and they didn't either."

When I saw that in the next morning's paper, under a photo of Carpenter floating casually in his raft, I was furious all over again. We knew exactly where he was and had even told him that the frogmen would be there in an hour. They actually made it in less than forty minutes.

I'll give him credit for wising up fast. Before he got back to the Cape, Carpenter knew that he'd screwed up. His first words on the tarmac to NASA administrator Jim Webb were an apology for his performance and a promise to take all the blame in the upcoming press conference. Now Webb was aghast. The last thing NASA needed was a mea culpa astronaut telling the world that he was a foul ball. Webb grabbed Walt Williams and ordered, "Shut this guy up! He's a hero and that's what we're telling the world."

But Carpenter knew otherwise. Not until Williams explained the situation to Carpenter's wife, Rene, did he make any headway. Rene Carpenter understood all the implications. By the time Carpenter got to his press conference, he had his story straight. Yes, there had been some equipment problems. But between the ground and himself, they'd gotten him down safely. All's well that ends well.

That wasn't the end. We did our standard debriefing a few days later. I was anxious to hear why he hadn't given immediate attention to the different readings on his instruments and why he hadn't been forceful in reporting the problem to us. But Carpenter was still Carpenter. He started with the standard thanks, then launched into a discussion about the fireflies. I wanted to stand up and tell him that the damned fireflies didn't matter.

Finally he got to the important stuff: "On the first rev, I saw a difference between the horizon indicator and the true horizon. But I thought it would fix itself before retrofire. I realize now that it was wishful thinking."

Wishful thinking! I was so angry at his cavalier dismissal of a life-threatening problem—and a problem we could have fixed by giving him instructions from the ground—that I didn't hear much after that. I swore an oath that Scott Carpenter would never again fly in space. He didn't.

A few days later, I asked Slayton to set up a meeting with the astronauts so that we could review the mission's flight plan and rectify the situation we found ourselves in. We got together in the privacy of a beach house rented by *Life* magazine. Only John Glenn and Carpenter were missing. The others were embarrassed by Carpenter's performance, and Shepard said that Scott had put a veil over their reputations within NASA.

Most of the talking was done by Al Shepard, Wally Schirra, Gus Grissom, Deke Slayton, and Gordo Cooper.

"Here's what we're going to do," they said. "We're going to prove forever that astronauts can control their capsule and their emotions. That we can minimize fuel use and stay with the flight plan. And mostly, we want you to know that we promise you total cooperation with the ground, no matter what."

Wally Schirra was the next to fly. He was particularly adamant that he'd do whatever it took to make his flight perfect. I wanted to stand up and cheer. We shook hands all around.

"Thank you, gentlemen," I said. "That's exactly what we're going to do."

11

The End of Mercury

I n the first years of space flight, life for all of us was a juggling act. That included the families of everyone involved with the operations end of NASA. Between the Glenn and Carpenter flights, Betty Anne and I made a house-hunting trip to Houston. At the time, it was the sixth largest city in the U.S. in population and the biggest in area, at more than four hundred square miles.

Our Manned Spacecraft Center was to be built on sixteen hundred acres about twenty-eight miles south of downtown, close to the shore of Clear Lake, which opened into Galveston Bay. When we first drove by on FM 528, a narrow farm-to-market road in poor repair, the surveyors were still there and workers were driving stakes into the ground. It was hard to imagine that this flat, low-lying piece of Gulf Coast scrubland was going to sprout engineering and test buildings, a new mission control center, and office space that would be the focal point of the human race's first steps in off-planet exploration.

I turned our rental car around and headed for a small town about ten miles southwest. Friendswood was a Quaker village that Deke Slayton had found on an earlier trip. A local developer was preparing a new subdivision along Clear Creek, and Deke was seriously considering it. Betty Anne and I instantly saw why. Imperial Estates was lightly wooded with native oaks and an occasional pecan tree, and the setting was beautiful. The small-town atmosphere reminded us of home, and the people were friendly and open. The schools were good, too. We heard praise from the townspeople, and when we visited the school that Gordon and Kristi-Anne would attend, we liked what we saw.

Before we drove back to Hobby Airport for the trip home, we'd decided that Friendswood was for us. In the next month, I made a deal with a Houston building company to buy the land and provide architectural services to design our perfect house. With luck, we'd move in before school started in September. A big selling point was the $15 per square foot building cost in Houston. We were delighted with what we could get for the dollars we could afford. My exalted government salary at the time was around $17,000 a year. After several iterations with the architect, we had house plans that included a nice apartment for my mother, and I stuck my neck out for the loan. Maybe, we all decided, moving to Houston wasn't going to be as traumatic as we'd feared.

My mission control team finished our reports on the Carpenter flight in early June, about the time that construction crews were beginning to frame the house in Friendswood. It was a grueling period. We were flying men in space, putting together huge NASA-industry teams for Gemini and Apollo, designing two generations of advanced spacecraft beyond Mercury, designing the Manned Spacecraft Center, and getting ready to move household and family halfway across the country. I expected to be alone for a while, but Betty Anne sold our house in Hampton, Virginia, quickly. By mid-August, the whole family was crowded into a Houston apartment. Their Houston greeting was six consecutive days of temperatures over one hundred degrees. The extreme heat broke, and by Labor Day weekend, we were moving into our new house in Friendswood.

A few days later, I took my mother to Jim Baker's grocery store. She looked around and finally approached Jim.

"Where are your cigarettes?" she asked.

"We don't sell cigarettes or any other kind of tobacco product," this devout Quaker grocer told her.

She was indignant. "You don't sell cigarettes!" she said loudly. "Well . . . !"

We found another store a few miles away. She'd buy her cigarettes there, but always went back to Baker's Grocery, where she and Jim became great friends. Poor Jim eventually bowed to the will of so many "heathens" moving into Friendswood from Virginia and started stocking a few brands. By then, my mother had quit. She was afraid she might burn our new house down.

I was constantly involved in meetings, debates, and decisions about Gemini and Apollo. But my primary focus had to be Mercury and the upcoming missions. By now, Gilruth had named Deke Slayton head of the astronaut office. He couldn't fly in space, but he could and did exert major power and influence over who flew and what happened on each mission.

It was obvious that we needed more astronauts. Just before I left Langley, Charlie Donlan sat down with me to discuss it. We agreed that much more weight had to be given to the recommendations and comments by any candidate's associates and superiors. That was a defect in selecting the original seven astronauts. Scott Carpenter was our bad example. He had slipped through the process without a college degree and virtually no test pilot experience. Honest evaluations from his commanders and his peers might have told us more about him. We still wanted test pilots. Flying in space was dangerous, and test pilots had proven their mettle in handling life-threatening situations. Some good ones hadn't made it the first time around, and I told Charlie that I hoped they'd be given a second chance.

Finally, I told Charlie that we used the astronauts in ways that their pilot training hadn't anticipated. They suggested design changes to the hardware. They followed the manufacturing and assembly of the capsule. They worked as capcoms during missions. They were not just astronauts. They'd also become systems engineers. So I wanted to make sure that engineering training and skill was given due attention when astronauts were selected.

Beyond that, I had no input and was as curious as anybody when the second group was chosen. Two of them, Pete Conrad and Jim Lovell, had been in the last go-around. All of them were, indeed, skilled and experienced engineers, and their test pilot backgrounds were impeccable. To a man, they were highly recommended by both peers and supervisors. Of the nine, seven were either active-duty Air Force or Navy officers. Two of them were civilians. I'd met one of them before. Neil Armstrong was a NASA test pilot who'd flown the X-15 rocket ship at Edwards AFB. Elliott See flew for one of the big airplane companies. The others were Frank Borman, Jim McDivitt, Tom Stafford, Ed White, and John Young.

They reported to Houston in September 1962, and we met them in a round of formal and informal introductions. They were bright, eager, and anxious to get to work. I was impressed.

Wally Schirra's upcoming flight in October was to be a showcase for what an astronaut could accomplish in space. Deke and I pored over the flight plan, working out details that would let Schirra prove the case. Wally named his capsule *Sigma 7*, to emphasize that the flight was a summation, an integration of the whole of Project Mercury. Though we were now based in Houston, the astronauts and the mission control team spent weeks in Florida getting ready. Schirra's performance made him a real pleasure to work with during the grueling days of training and simulations. He didn't miss a beat and neither did we.

We'd forged a working team and I modestly thought it was brilliant. My flight controllers and the astronauts were ready to make the world, and the press, sit up and take notice.

But for a while, it was the damned Russians who grabbed the headlines. They'd launched Adrian Nikolayev one day, then shocked all of us by putting up Pavel Popovich the next day. The two ships passed about three miles apart on Popovich's first orbit. The press tried to make it into a rendezvous in space and credited the Soviet space team with an amazing feat.

It was a ruse. Nikolayev and Popovich were simple passengers in ships that had no maneuvering ability. Their crossing speed was more than five thousand miles an hour, so fast that their "rendezvous" was barely a high-speed hello-and-good-bye in orbit. They never got that close again, and the idea that they might have maneuvered to get close to each other, or even link up, was news media science fiction.

The real news, buried behind the rendezvous sham, was that the Russians could launch two missions only twenty-four hours apart, and that both cosmonauts stayed up for record-setting times in space. Nikolayev was in space for ninety-four hours, and Popovich for seventy-one. Our flight surgeons noted that they survived their exposure to zero gravity, though grapevine reports said that each suffered bouts of space sickness. Still, they were paraded through Red Square and accepted their medals with no obvious ill effects. It would be several years before we matched and beat their endurance record with our Gemini spacecraft.

President Kennedy countered the Russian success by bucking up our morale and giving us a pat on the back. He refocused attention on the American space program by visiting Cape Canaveral and Huntsville, touring launch facilities and looking at mock-ups of the big rockets Wernher von Braun's people were designing. Then Kennedy flew on to Houston, where I was to brief him. His trip gave me a chance to spend some time with John Mayer's mission planning people and to get up to speed on the latest thinking about going to the moon. I had to condense it down into a fifteen-minute presentation, complete with big three-by-four-foot charts.

Kennedy started his Houston visit with a speech at Rice Stadium. It was a typical hot, muggy day, but the stadium was packed. I listened to his speech on the radio and wanted to cheer when I heard his line about going into space. "We choose to go to the moon in this decade and to do the other things," he told the cheering crowd, "not because they are easy, *but because they are hard.*"

Yes! I thought. *He understands. And he is certainly right. These things are hard.*

We were set up for him in one of our rented buildings off Telephone Road. As soon as he got into the air-conditioning, he took off his coat and tie, which were wringing wet from riding in an open car from the stadium. One of the people with him was Senator Alexander Wiley of Wisconsin, who'd been in the Senate since 1938. Kennedy walked over to me with his hand out. This time it was the only hand extended, so I didn't have a problem deciding to shake it immediately.

"Good to see you again, Chris," he said. I was impressed. He either was well briefed or had a great memory. "Why don't you folks tell me about going to the moon?"

Max Faget was up first. He described the moon vehicles he and his team were designing. One was a combined command and service module (CSM), which would carry three astronauts to the moon and go into lunar orbit. The other was a lunar excursion module (LEM), docked to the CSM. Two of the astronauts would drive it to the moon's surface. It was new and exciting.

My job was to explain how it would all happen. Kennedy followed alertly as I told him about putting it all into Earth orbit on a big Saturn rocket, refiring the Saturn's third stage to head for the moon, and extracting the LEM from its stowed position. The more I talked, the more I realized just how much we still had to learn. I told him about going into lunar orbit, sending two men down to the surface for exploration, then launching off the moon to rendezvous with the CSM. Publicity about the Russians notwithstanding, nobody had ever done a rendezvous in Earth orbit. Yet here I was telling the president of the United States that we were going to do it at the moon.

Apparently I wasn't exciting enough for Senator Wiley. I noticed him napping through most of my talk. But near the end, he suddenly woke up and interrupted me in a loud voice: "Son, do you really think you can do all that?"

Kennedy burst out laughing. I managed to stammer out, "Yes, sir. I think we can." The president stood up, and this part of the tour was obviously over. Between chuckles, he thanked us.

"I'll be following your progress," he said, "and I'm looking forward to that lunar landing."

Less than a month later, we had Wally Schirra in space. There were only minimal differences between the *Sigma* 7 and *Aurora* 7 capsules, but we expected a major difference in mission results. Schirra was going to double Carpenter's space time by going around the world six times, then landing in the Pacific Ocean. That was necessary because of his orbital track. After nine hours and six times around the world, he wouldn't be over the right part of the Atlantic and it would be nearly nighttime. But at the Pacific splashdown point, it would be midday.

The big changes for this mission, officially called Mercury/Atlas-8, or MA-8, were in the quality of the astronaut and in the way we meticulously planned his activities in space. Schirra was involved in every step of the planning and was determined to get it right. He would conserve fuel in both the automatic and manual modes by doing slow and deliberate maneuvers, and by using manual control most of the time. The automatic system, we now knew, was a real guzzler.

Battery power had to be conserved, too. Many of Schirra's capsule systems would be turned off except when he needed them. Telemetry and radar transmitters, which had been left on full-time in earlier missions, would be turned on and off by ground command at each site.

The Pacific landing meant a lot of training for Navy forces who were getting their first exposure to Project Mercury. We assigned a large number of our own recovery engineers to the Pacific for several months, and it paid off. Not only was the Navy ready for Wally Schirra, but it would be ready and experienced for future missions, too.

Retrofire would occur over Australia instead of California. And I wanted as much coverage of Schirra's reentry as I could get, so we moved one of our tracking ships under the reentry track. Al Shepard went on board as the capcom. No astronaut had more experience with handling the ground portion of reentry than Shepard, and Schirra trusted him completely.

I had a big question about manual control of the capsule at retrofire. Scott Carpenter told us that he had great difficulty in determining his yaw—side-to-side—attitude by using the reference hash marks on the window. That kind of ability was going to be critical in Gemini and Apollo, not so much for reentry, but for positioning a spacecraft during rendezvous. If the computerized systems being developed for rendezvous failed, then the astronaut would have to do it manually. But if he couldn't determine his yaw with the window marks, he'd be out of luck. With Schirra's help, we set up a number of tests using the ground, stars, and sun as references, to see if we had an intrinsic problem.

Now Wally was up there. The Aerospace Corporation engineers had thrown us a last-minute monkey wrench when they insisted on off-loading one thousand pounds of fuel from the Atlas rocket. They were worried that the common bulkhead in the Atlas could be overloaded. I strongly argued against the change. We needed the fuel, I said, to guarantee that Schirra would reach orbit. I lost that argument, and the Atlas came perilously close to shutting down early. (On the next mission, the Aerospace engineers had a new set of reasons for keeping the fuel on board.)

We immediately had a problem with Schirra's suit temperature. It was running hot. On the surgeon console, Dr. Charles Berry was worried enough about dehydration that we discussed bringing Wally down after one rev. By acquisition at the Guaymas, Mexico, site, the situation seemed better.

"Surgeon?"

"He can take one more rev at these temps, Flight. But we need to watch it."

"Okay. Guaymas, Flight. Tell him he's go for another rev."

Schirra was working the settings on his suit circuit to regulate coolant flow through the capsule's heat exchanger, and he was doing it right. When he reached the Cape a few minutes later, his temps were decreasing. With a few more adjustments, he got it just right and the problem went away.

On the next rev, Schirra got to work on our detailed plan. He kept fuel consumption on the predicted curve and was delighted with his ability to make precise changes in the capsule's attitude. He also worked on the yaw reference problem. It turned out to be no problem at all. He used the hash marks on the window to determine his attitude in all three axes, not just yaw. Setting up manually for retrofire would be easy, he said. I extrapolated forward and felt better about manual maneuvers in Gemini and Apollo, too.

As the mission moved into its third, fourth, and fifth revs, Schirra let *Sigma 7* drift for long periods, conserving both fuel and electrical power. It was no problem at all, he said. We were already planning a much longer flight for MA-9, and that news was a green light to our extended-mission thinking.

With plenty of fuel remaining, we went with the planned automated retrofire. It was textbook. Wally Schirra landed within five miles of the waiting aircraft carrier *Kearsarge,* so close, he said, that he thought he might come down on the carrier's elevator. In mission control, I winked at Deke Slayton and lit up my now traditional cigar. Schirra was the perfect astronaut and he'd just carried out a perfect mission. The lingering taste of Carpenter's "Stigma 7" escapade had been erased by Wally Schirra and *Sigma 7.* I was a happy man.

We were ready to do more. The big questions were what and how. Our confidence was high and we could still learn a lot by flying several more Mercury missions. But Gemini and Apollo were taking priority. In January 1963, the Manned Spacecraft Center had twenty-five hundred employees spread around in rented buildings in Houston. Only five hundred of them worked on Mercury. The compromise decision was that we'd fly just one more Mercury mission and modify the capsule to keep Gordo Cooper in orbit for twenty-two revolutions, almost a day and a half.

The changes to the capsule—more oxygen and water, more fuel for the thrusters, and bigger batteries—took time to complete. To compensate for some of the added weight, we told McDonnell engineers to remove the periscope, a redundant thruster system, and a couple of radio transmitters. It would be the spring of 1963 before we could fly.

My position as Flight made me a natural target for the press. I'd made up my mind early on that it would be stupid to lie to reporters, or even to try to mislead them. Then there'd be hell to pay. So I answered questions honestly and gave as much detail as reporters could absorb. As I got to know reporters individually, I was blunt in suggesting that they do their homework before coming to an interview or a press conference. I had to do mine, and fair's fair.

So as Mercury neared its end, I had a reputation with the press as a straight shooter. Reporters like a source who won't lie, and I like reporters who take the effort to get the story right. Mercury was a hot story. I got to know a number of newsmen quite well. Roy Neal of NBC was a longtime associate of Walt Williams's, dating from the experimental flight program at Edwards AFB in California, and I was frequently invited to join them for a dinner or drinks. The *New York Times* had given Richard Witkin prime responsibility for covering Mercury. He was a dogged journalist with roots in the airplane world, and when he showed the same interest in space, I spent long hours explaining space flight to him. We became good friends, and he reciprocated by sharing his knowledge of Europe, and particularly Russian space politics, with me.

The *Washington Star*'s Bill Hines was a bulldog in press conferences. He challenged everything, but his stories were accurate. All of the major newspapers and magazines assigned people virtually full-time to Mercury. Howard Benedict of the Associated Press and Al Webb of United Press International had bureaus in Cocoa Beach. In our retirement years, Benedict became a close friend of the astronauts' and mine. Many of the names became well known to the American public—Richard Lewis from Chicago; Earl Ubell, also of the *New York Times*; George Alexander of *Aviation Week* magazine; and Martin Caidin, who wrote a shelfful of books about the space program and then wrote a novel that was turned into the TV series *The Six Million Dollar Man*.

Another was Jules Bergman of ABC, a boisterous and sometimes offensive character who liked to search for the sensational. But the more I tried to teach him about my world, the more I came to like him. He was a pilot and one weekend day asked if I'd like to fly up to the Daytona 500 with him. The roar of those stock cars when you're standing in the pits was almost the equal of a rocket's roar at T-zero.

The least savory character covering us at the Cape was a screwball named Doug Dederer, who seldom wore anything but shorts, a T-shirt, and beach clogs. He wrote for a local paper and did some stringing for the *New York Daily News*. He claimed to have sources in every NASA office, and no doubt he had a few. But some of his stories could have won awards for fiction. I did my best to stay away from him.

The group of journalists who were both insiders and outsiders came from *Life* magazine. Their contract with the astronauts gave them an inside track on personal information and sometimes got them access to the rest of us. I didn't feel guilty about using the *Life* beach house for meetings that were better held away from our own offices. And the photographer Ralph Morse became a friend to all of us, particularly the flight control engineers. He could be loud and intrusive, but he was meticulous in setting up pictures that told their own story about what we were doing.

The MA-9 flight was to last about thirty-four hours. With the countdown on the front end and the recovery operations on the back end, mission control would be up and running for at least fifty hours. I looked in the mirror and when I didn't see Superman staring back at me, I knew that it was time to create a second Flight. I called in my key men to work out a new plan.

My natural choice for the second flight director position was John Hodge. Since his arrival at NASA with the engineering contingent from Canada, I'd relied on him for one crucial assignment after another. He'd helped create the global tracking network. He'd helped in designing mission control and setting up our training procedures. And I'd sent him to Bermuda where he was flight director at our backup control center for each of the Mercury flights. Now he had to put together his own team of controllers, too.

We worked out a schedule for Gordo Cooper's flight. Hodge and his team would start the countdown. This usually meant showing up for work around midnight or 1:00 A.M. I would come in with my team just before launch and work the first ten hours of the mission. Then we'd hand it back to the Hodge team for the next twelve hours. In that time, I'd try to get at least six hours' sleep—and encouraged my own controllers to do the same.

Finally, we'd come back on duty for the last twelve hours and hold it through splashdown and recovery. It seemed reasonable. Cooper agreed that working with two different control teams and two different Flights was no problem, particularly since we'd all train together anyway. Before we locked the plan in stone, we ran a full-blown simulation of the entire flight. It worked. Mission control was growing up and growing into the new realities that long-duration flights in Gemini and Apollo would become routine.

Most of our remote sites didn't need extra people. The Earth's rotation meant that Cooper would only see them sporadically anyway. We did send two complete controller teams to Hawaii and added one extra man at several sites where the acquisition periods were long.

MA-9 also put strains on the Navy recovery forces. In thirty-four hours, Cooper would pass over every part of the world from about thirty degrees north to thirty degrees south latitude. We had a number of contingency landing sites, and we had to be prepared for an emergency landing almost anywhere. From our earliest Mercury flights where maybe a thousand people were involved, we'd come to this last mission with tens of thousands of people deployed around the world.

A few days before launch, I was going over things with Walt Williams in his office at the Cape. We were doing almost too well, with only minor problems, and both of us were relaxed. Suddenly a jet roared at low level directly overhead, went to full afterburner, and almost blew us out of our chairs. We rushed to the window and saw a NASA F-102 buzzing the launch pad, then going into a near vertical climb.

"It's that goddamn Cooper!" Williams blustered. "I'll have his ass on a plate . . ."

Then the phone started ringing. Everybody and his brother wanted to know what the hell was going on and what Williams was going to do about it. The airspace directly over the Cape was restricted, and a low-level pass was forbidden under any circumstances. Walt was madder than I had ever seen him and was threatening to pull Cooper from the mission and replace him with Al Shepard. I counseled patience, if not forgiveness. By the time a sheepish Gordon Cooper reported as ordered to Walt's office, the boss had calmed down a bit. But even a calm Walt Williams could be abrasive. He delivered a tongue-lashing that he later said was one of the best of his career. A subdued Cooper went back to the astronaut quarters, still assigned to his flight, but not anxious to cross paths with Williams anytime soon.

After a one-day delay to fix a radar problem in Bermuda, Cooper was on the pad early on May 15. He was a cool character, given to slow-talking comments and wild-ass highjinks. As he waited for the countdown to reach zero, his heart and respiration rates dropped so low that Doc Berry on the surgeon console thought he'd fallen asleep in the capsule he'd named *Faith* 7. He had. We woke him up several times in the last hour with communications checks. No man about to ride a rocket was ever less nervous.

He was that way in space, too. His biggest problem early on was the same cranky suit-temperature settings that had plagued Schirra. Once he found a stable setting, he went to work up there with inspired ease. He had new cameras to use, including a small TV camera, and a number of scientific experiments to carry out. The only one that failed was an instrumented balloon that failed to deploy.

He did create some controversy. Cooper looked down on the Himalayas, then on the oceans, and on vast continental stretches. He could see things, he said. He saw railroad tracks and the smoke from a train's engine. He saw runway patterns at airfields. He saw the wakes of oceangoing ships and followed them until he spotted the ships themselves. He saw highways and mountain trails. The doctors didn't believe him. Nobody's eyes were that good, they said. Some of the military intelligence people didn't believe him, either. How could he see such detail from more than one hundred miles out in space? Cooper insisted after the flight that he had, indeed, seen everything that he reported. It would take astronauts flying in Gemini spacecraft to settle the debate in Cooper's favor. On a clear day, all of that and more was visible to someone who knew what to look for and took the time to study the terrain.

After seven revs, I took my team off-shift and got some sleep. When we came back on, the mood in mission control was boredom. John Hodge and his controllers couldn't report a single significant problem that they'd had to handle. Cooper was asleep up there for a good part of their shift, and they had little to do but monitor systems and stay awake themselves. *That's great*, I thought. *We've established a multiple-shift control center, and everything is working.*

It didn't stay that way. After eighteen near perfect revs, the .05-g light on Cooper's instrument panel lit up. It wasn't supposed to come on until *Faith 7* was on its way back to Earth, and it was a crucial part of the automated reentry system. We talked Cooper through enough tests to show that the circuit had malfunctioned. He'd have to initiate some of the reentry sequences manually.

We'd just given him the new series of activities when he reported that his main inverter had failed. Then the backup inverter wouldn't come on-line. Without them, the entire automatic reentry system was dead. Now Cooper would have to do everything manually. He'd trained for this contingency and we weren't that worried.

But the gods of failure weren't done with him yet. Key elements of the environmental system went out, and carbon dioxide was beginning to build up in the cabin and in his space suit. Then one of his attitude thruster systems failed. We were scurrying in mission control to put together backup procedures for him and to get him down safely. Somehow Cooper was the calmest of us all.

"Things are starting to stack up a little," he radioed from over Africa on his twenty-second and final revolution. Then he listed a series of systems and hardware that weren't working. "Other than that," he added in his Oklahoma twang, "everything's fine."

But everything wasn't fine. One by one, the systems aboard *Faith 7* were failing. We didn't know if it was because of the long stay in space or if our designs were simply at fault. I stood at my console and watched Cooper finally reach Hawaii. John Glenn was out there as capcom and he counted Gordo down into the reentry sequence. It was perfect. Cooper had lined up his capsule using the hash marks on the window, and he fired each retro rocket exactly on time. I breathed a sigh of relief.

After that, it was in Gordo's hands and he could not have been calmer or more professional. Through reentry, blackout, and parachute deploy, he did everything right. When *Faith 7*'s parachute billowed, he was less than four miles from the carrier *Kearsarge*. He'd beaten Wally

Schirra's splashdown accuracy, and he'd set a new American record for time in space.

When we got the capsule back, our engineers tore into it to see why so much had failed. They found the same cause for everything. Cooper's urine collection system leaked. A fine mist of urine infiltrated into all of the electronics, and one by one, short circuits put systems out of order. We'd learned something new and important. Future spacecraft would have much better systems for handling both waste water and potable water. And the electronics themselves would be better sealed against the intrusion of moisture.

Gordo Cooper was cramped and tired after thirty hours inside a capsule that gave him less room to move around than the driver's seat of a Volkswagen. But he was healthy, happy, and ready to move on to Gemini.

So was I.

SECTION III

THE GEMINI MISSIONS

12

How Do You Get
a Man to the Moon?

The end of Mercury was a celebration. There were more ceremonies in the White House Rose Garden, and I was proud to have Betty Anne standing there with me when President Kennedy and Jim Webb gave me the NASA Outstanding Leadership Medal. Kenny Kleinknecht, who'd been Mercury program manager, got one, too, and Gordo Cooper received the NASA Distinguished Service Medal. None of us have many days in our lives like that one.

But it was just starting. We flew on to New York City and rode in police-escorted limousines past streaming fireboats on the river and on to the Waldorf-Astoria. There was a great dinner and a Broadway show that night. I almost choked when I saw which one: *Stop the World, I Want to Get Off*. The next day we rode through town in a ticker-tape parade to city hall, where all of us received keys to the city. Being in a ticker-tape parade is an event that I can never forget—people yelling praises and all excited because of something I'd had a hand in making happen. But I also remembered all the people who'd done so much work and were still back home. And the baseball fan in me kept thinking of the old Brooklyn cliché: "Yesterday's heroes are tomorrow's bums."

Finally there was another banquet, with a lot of speeches. Former president Herbert Hoover was there, sitting next to Walt Williams at the head
table. When Jim Webb got up to talk, I noticed Hoover whispering in
Williams's ear. I asked about it later.

"He asked who that was," Williams said. "When I told him Jim Webb,
he turned his hearing aid off and asked me to poke him when Webb was
finished."

Day three had us going back to Houston. All of us were paraded in open
cars through the streets of downtown Houston, and we had to listen to still
more speeches. That was the day Houston really realized that it was home
to the Manned Spacecraft Center and was going to be privy to momentous
events. They welcomed us like native sons. It was another warm, muggy
day, but the streets were lined for miles with cheering people standing six
and eight deep at the curbs. Many of them were waving American flags.
Betty Anne squeezed my hand. We were a long way from the Tidewater
country of Virginia. But we were home.

A police escort took us back to our house in Friendswood. We'd been
gone a little over seventy-two hours, and it occurred to me that we hadn't
stopped for a single red light the whole time.

The end of Mercury didn't mean the beginning of Gemini. My operations
team and I had been juggling Mercury, Gemini, and even some Apollo duties for more than a year, and tens of thousands of people at NASA and
working for contractors had been working full-time on the complexities of
our next—and rapidly approaching—manned space programs.

President Kennedy's commitment to a lunar landing had changed
much within NASA. Until then, our forward thinkers and futurists were
considering follow-on programs that included space stations and manned
exploration of Mars. Landing on the moon wasn't on anybody's priority
list. Wernher von Braun was a strong advocate for Mars. Max Faget
thought a space station should be next. NASA gave a number of contracts
to private industry to study those options and others, then make some recommendations.

Battle lines had also been drawn between the advocates of manned and
unmanned space exploration. James Van Allen and Jerome Wiesner went
so far as to call the manned programs a kind of hysteria, and the press, as
usual, actively fanned the controversy.

Nobody of importance asked my opinion. If they had, I would have sided with the space station faction. I'm still confident today that we had the kind of skilled engineers in the early sixties who could have built both the transport vehicles and a space station operating in Earth orbit. If we'd gone that route, we would have developed even better vehicles for exploring the moon, Mars, and probably much more. Certainly we could have done it with less risk, less money, and more productivity than the way things turned out. We would have had American men and women leaving their footprints on Mars in the 1990s. Instead we're still waiting for our first space station.

Was the moon a valid next step? Of course. Politics always plays a bigger role in major national decisions than does simple pragmatism. Kennedy's decision was politics-driven, and that does not mean it wasn't right. The reality is that landing men on the moon gave a proper and popular challenge to the technical community. The great strides in human progress always come in the face of a great challenge. Jack Kennedy put that challenge to us and we responded. In the longer historical view, it may have been a lesser challenge than we could have handled. But the beauty of Apollo is that the United States, and the world, profited from it anyway: economically, technologically, and spiritually. History will put the most value on the last of those.

We didn't know exactly how we would get to the moon, but Bob Gilruth had promised President Kennedy that we'd figure it out. Two concepts quickly emerged, and as one of NASA's few operational experts, I was asked to comment and evaluate them.

Gilruth initially favored "direct ascent." They'd build a giant rocket that would send a spacecraft straight toward the moon. Once there, it would turn around and use retrorockets to land. After a brief exploration, the craft would take off again and fly straight back to Earth. The more we looked at direct ascent, the harder it got. Even Wernher von Braun didn't have plans for the huge rocket it would take. And the shapes for direct ascent—basically looking like rockets with built-in crew cabins—were all wrong to make a safe lunar landing because their height made them all too likely to tip over.

The second plan was "Earth-orbit rendezvous" (EOR). Using smaller rockets, we'd launch several vehicles into orbit. Astronauts would have to rendezvous with the pieces, assemble them, then blast toward the moon for a straight-in landing. That still left the problem of getting back home. And assembling all those modules in orbit involved complexities that we could barely imagine. At the time, it was all we could do to get John Glenn up there all by himself.

Each of those ideas had their advocates, but each could be validly challenged on technical grounds. It took a third concept, one that seemed wild and harebrained, to break the deadlock.

The idea came from a group at Langley who started with the premise that neither direct ascent nor EOR were possible in the foreseeable future. Their payloads were too heavy. The answer had to lie in bringing payload weights down to the point where 1960s rocket technology could lift them. John Houbolt and Clint Brown asked for help from my old friend and resident genius Hewitt Phillips and some of his engineers. The wild ideas began to flow, and one of them, from William Michael, seemed almost logical. It was quickly refined into a plan.

Michael's idea called for dividing the lunar payload into separate pieces. One of them was a relatively small "bug" carried to the moon by a "mother ship." The "bug" would take astronauts down to the surface and back up again. Then it would be thrown away. All that was required was for the bug to rendezvous in lunar orbit with the mother ship. Nobody had done a rendezvous anywhere, but the Houbolt team assumed that this hurdle could be surmounted.

Houbolt became the archetypal madman with a mission. When this lunar orbit rendezvous (LOR) scheme was shown around NASA, it was universally met with disdain. But Houbolt wouldn't give up. In 1961, Bob Gilruth and Wernher von Braun had come to terms on building a big rocket for Apollo, to be called the Saturn 5. Several Saturn 5 launches could make the EOR plan work. An even bigger von Braun rocket, to be called Nova, could accomplish a direct-ascent mission. But Houbolt was certain that LOR could be done with just one Saturn 5. Gradually he started bringing people to his side. Max Faget decided that it could work, and his analysis convinced Bob Gilruth to stop thinking about direct ascent. Dr. Robert Seamans at NASA headquarters bought in. And in June 1962, in a historic meeting at Huntsville, so did von Braun.

I wasn't there, but Max and Bob came back with smiles. After listening to the pros and cons all day—the arguments against being made by his own people—von Braun didn't even bother to consult with his staff. In that abrupt, Teutonic way of his, he simply announced that he'd been swayed to favor LOR. So the moon plan was locked in. A command module with a tanklike service module carrying a midsize rocket and other systems— quickly called the CSM (command and service module)—and an LEM (lunar excursion module) were born. Just one of von Braun's Saturn 5 rockets could put the whole stack into Earth orbit, then send it on the way to the moon. Once free of Earth's gravity, an astronaut would pilot the CSM away from the rocket, turn around, and plug the CSM's nose into a docking mechanism on the LEM and pull it away, too. Only the CSM/LEM combination would go to the moon.

Once there, the service module's rocket would brake them into lunar orbit. Two astronauts would crawl into the two-piece LEM, back away, and use its descent engine to land. After exploring the surface, they'd blast off in the upper part of the LEM, the ascent stage, and rendezvous with the third astronaut waiting in the CSM. They'd dump the LEM stage and use the service module's rocket again to head for Earth. Finally, in the last hour, they'd dump the service module. Only the command module would make it all the way home, splashing down in the Pacific Ocean after a fourteen-day mission.

Looking back across forty years, it still seems amazing that we all bought into the idea. On that day in June 1962, we'd only flown John Glenn for three revs and had some hairy moments. Mercury was a capsule, not a spacecraft. It could point this way or that, but had no ability to maneuver in orbit. Rendezvous was a theory, not a fact. Our flight surgeons were leery about putting an astronaut into space for fourteen hours, much less fourteen days. We had difficulty communicating between continents. The moon was eighty times as far away as Africa. What the hell were we thinking about?

The answer was Gemini. It was first conceived as a sort of make-work program to keep everybody sharp until we could get to the real thing, Apollo. It may have been "interim," but Gemini quickly evolved into the most important and complex stepping-stone to the moon that we could imagine. Gemini would be a real spacecraft, carrying two men. It would have enough rocket power and fuel, plus an onboard computer, to change its orbit in space and thus to rendezvous with another ship.

It would come with front-opening doors so astronauts could get out of their seats, float free (albeit tethered) in space, and learn how to do work in zero gravity. It would be big enough to carry oxygen and supplies for a fourteen-day mission, which we'd reach in increments to measure the effects of space on the human body.

The only thing that Gemini wouldn't do was go to the moon. But rendezvous techniques were the same everywhere. If we could master rendezvous in Earth orbit, it would be possible in lunar orbit, too.

Except for designing and building a new mission control center capable of handling all that, building a complete Manned Spacecraft Center in Houston, and overseeing the various contractors handling those duties, plus the contractors building the rockets, the Gemini spacecraft, and the various pieces of the Apollo spacecraft, that was all we had to do. Bob Gilruth once said that if Jack Kennedy had been older and wiser, he would never have committed us to landing on the moon. The same was true for all of us working for Gilruth. If we'd been older and wiser, we would have known that we couldn't get it all done. But we weren't. So we did it.

While Mercury was still only halfway finished, I had to be closely involved with planning our new control center in Houston. We divided the work into two pieces, a state-of-the-art, real-time computer complex (RTCC), and then two separate, but identical mission control rooms. We assumed that we'd be flying missions so closely together that one room would be dedicated to the current mission and the other to the upcoming mission. The computer complex had to support both rooms, so we had let that contract first.

Only two companies made computers powerful enough for our needs, IBM and UNIVAC. IBM provided computers for our Mercury control center and we'd been happy with them. But a number of other companies were good at software and bid on the basis of buying IBM or UNIVAC computers. Finally we were ready to present it all to Jim Webb in Washington. The Source Evaluation Board was not allowed to make a recommendation. It merely presented facts and scores. But IBM was obviously best qualified for the job. When the presentations ended, Webb waved me into his office. Our conversation wasn't strictly by the rule book, but it showed that Webb was more interested in doing the right thing than in following the exact letter of procurement regulations.

"What do you think, Chris? It's going to be your control center. Who do you want?"

"IBM. They've got the best proposal."

"You've already got them on Mercury. Are they doing a good job?"

"Yes, sir, they are."

"So you want to keep them for Gemini and Apollo?"

"That would be my choice."

"Well, Chris, I don't see any reason why I shouldn't choose a company that's done good work for you already. But I think I need to pick the one that's best and meets all the requirements, don't you?"

"Yes, sir."

"Then don't worry about it. Have a good trip back to Houston."

The next day Jim Webb picked IBM to design and build our real-time computer complex.

John Hodge, Tec Roberts, and a few others led the design effort for the control rooms. A few years back, the idea of a control center was brand-new. Now we got at least ten proposals from companies actively involved in the new technology of command and control. The evaluations were tricky and complicated. Our internal estimate put the cost for both control rooms at about $35 million. The bids ranged from $15 million to $150 million. Once again, Webb asked me to his office when the presentations finished.

"Who do you want, Chris?"

"I like the Philco plan best. Their technology is good and so is their cost."

"Didn't they work on the Mercury control center?"

"Yes, sir. They did a good job."

Several days later, Webb picked the Philco Aerospace Company to build and equip our mission control and support rooms. I had exactly the team I wanted, IBM and Philco. I knew going in that putting it all together would be a massive and complicated job. We were dealing with technologies so new that some invention would be required. There were certain to be problems. But with IBM and Philco, I figured that at least half of the worry had been lifted from my shoulders. Between them, they would deliver.

Everything was happening at a pace that didn't leave much breathing room. McDonnell Aircraft had a sole source contract to build Gemini, and when its designs were delivered, I was called in to help evaluate instrument panels, control systems, and operational concerns. In many ways, Gemini looked like a bigger version of Mercury. Looks were deceiving. Gemini would be launched by the Air Force's Titan II intercontinental ballistic missile, more powerful than Atlas and now fairly well proven to be reliable. For rendezvous and docking missions, the target vehicle would be an Agena upper stage (already used in several classified spy satellite programs) launched by an Atlas.

North American Aviation got the contract to build the Apollo command and service modules. Grumman Aircraft would build the lunar excursion module. North American's Rocketdyne division, along with Boeing, McDonnell, and IBM, would build the various rocket engines and stages of the Saturn 5 rocket and a smaller version called the Saturn 1 or 1B. Some sooner, some later, I'd get involved with all of them.

With Gemini, my life changed again. It wasn't so much a mark in the sand as an evolution that lifted me slowly away from the nitty-gritty technical details and brought me more into management. It would be at least eighteen months from Gordo Cooper's flight that ended Mercury until we'd fly men in Gemini. I had the new control center to build. Some of my people were poring over the complexities of rendezvous and the intricate orbital-mechanics problems to be solved. Others were developing lunar flight plans. Still others were working on modifications to the worldwide tracking network. All of it required number crunchers for budgets, administrative and personnel specialists, secretaries, and clerks, clerks, clerks. The influx of new people into my Flight Operations Division was like a river rising to flood stage.

Most of them were young and hadn't been around for Mercury. Some new managers came from industry or other government agencies, and the statement "That's not the way we did it in Mercury" carried no weight with them at all. They wanted to do things their way and resented the implication.

We'd grown our own way of doing things from the first days of the Space Task Group. We were colloquial, even provincial, and now the bureaucracy was overtaking us. I'd fight the encroachment for the rest of my career and I'd win some and lose a lot. The weight of bureaucracy is a heavy weight indeed.

The heaviest of it all came from NASA headquarters. Jim Webb looked to the future and saw that managing NASA would more and more mean spending time with Congress, the White House, the Bureau of the Budget, the Department of Defense, and more. He needed help with the day-to-day things and got it, in part, by hiring a TRW vice president to become associate administrator and head of the Office of Manned Space Flight (OMSF). Dr. George Mueller was a physicist who'd been involved in a number of TRW's secret defense and intelligence space contracts, and with the guidance and control of ballistic missiles. More than that, he was a bureaucrat's bureaucrat, and he immediately set out to change things. The layers of management he added only complicated our lives in Houston.

It didn't help that he told Gilruth to get rid of Max Faget and me. Gilruth refused and told both of us about Mueller's demand, and it tainted our relationships with him for years. I did come to appreciate much of Mueller's engineering savvy and his positive contributions to Apollo. But his management style was oppressive, and he usually wanted everything done his way. Nobody is that good, and our deviations from Mueller's straight-and-narrow path frequently opened us to his sharp criticisms.

Pre-Mueller, the Manned Spacecraft Center and Bob Gilruth had major-ity responsibility for the programs, including designing and developing spacecraft, flight operations, and astronauts and their training. We bought rockets and launch operations from the Air Force, and recovery support mostly from the Navy. Gemini started out to be more of the same. But Apollo would be different. Two other NASA centers, the Marshall Space Flight Cen-ter in Huntsville (von Braun) and the Cape Canaveral Space Center (Kurt Debus) were involved. Huntsville was developing the big Apollo rockets. And the Cape was designing and building the new facilities and launch pads it would take to get them and the spacecraft off the ground. Both Huntsville and the Cape were about to consume huge chunks of NASA's growing budget.

One of Mueller's first actions was to bring in General Samuel Phillips as overall Apollo program manager at Washington headquarters. Phillips had been running the Air Force Minuteman ICBM program. Overnight the move put a military flavor on an agency that had been set up by President Eisenhower and Congress to be separate from that taint. The vibrations we got from Washington didn't make us comfortable. But Webb approved and nobody in Congress raised a warning flag.

We'd lived and worked with the military for most of our careers. We could grit our teeth and do it again. But the second impact of Mueller's new organization was much more severe. In Phillips there now resided a whole new layer of management that hadn't previously existed. Mueller in-sisted that he needed the extra oversight. Gilruth, von Braun, and Debus would just have to deal with it. But Mueller didn't rise to bureaucratic heights by being blind to the implications of his actions. He knew that the center directors would be offended. So he created a Management Council of the Office of Manned Space Flight. Its members were Gilruth, von Braun, Debus, and Mueller himself. They would act as a kind of manned space flight board of directors.

I looked up and saw not one, but suddenly two additional layers of man-agement up there in headquarters. The other division directors saw the same thing and we were damned unhappy. Then Mueller compounded his management obfuscation by hiring a newly formed consulting organization called Bellcomm to give him technical support as he oversaw all of manned space flight from his spacious headquarters office.

There was one element missing, and when Mueller saw it, he filled that void by creating the Apollo Executive Committee. Its members came from the top management of companies holding Apollo contracts and gave Mueller a direct channel to each company president. The committee would periodically tour all the major facilities and manufacturing plants to review the program status. That meant pulling people, including me, away from real work to prepare slides, charts, graphs, and presentations that eventually filled file cabinets and boxes in some government storage area.

The growing mountain of Mueller's bureaucracy looked to us working troops as unnecessary, cumbersome, and expensive. It turned out to be all three, but it never went away. In the years after Apollo, management academics wrote volumes about how wonderfully NASA ran such a complex program, with its broad scope and many interfaces. They weren't there to see that simpler was better.

We did the job in spite of Mueller's management schemes. If he'd delegated authority and told us to run the program, it would have reached the same ends, but with far greater economy and rationality. I was in his presence all too often, particularly at management council meetings where I'd report on operations planning, the software in the MCC, and the spacecraft. Mueller invariably gave me a hard time, sometimes with nitpicking observations, sometimes with valid questions. I stood my ground without being insubordinate. When he was right, I did my best to listen and respond. But if we'd followed all of his directions, Apollo would have been a mess. Gilruth, Faget, and I spent many hours discussing Mueller's management foibles as we rode the Gulfstream to and from his meetings in Washington.

We did take many liberties. We had brilliant people doing brilliant work, and when Mueller wasn't looking, or when it was simply the right way to do things, we used our own judgment. The bureaucracy was a weight that we could occasionally shrug off in the name of efficiency or logic. More often than not, when Mueller called us to account, we defended our actions successfully. At the same time, von Braun gave Mueller whatever he wanted, even when it didn't make sense. His proposals for a small space station were exactly in line with Mueller's thinking, even if they weren't technically feasible. Von Braun wasn't happy as a supporting player while Gilruth ran the manned space operations, and he hoped that Mueller would give him a program of equal stature. It never happened.

Of these conflicts was born the Manned Spacecraft Center's reputation for being arrogant and insubordinate. From Gilruth on down, we were guilty. But we got the job done.

The faces were changing all around me. Some of the old hands were disappearing. The new organizational chart in Houston—Bob Gilruth created directorates and I now had the title of assistant director of MSC for flight operations—opened up slots that were filled by people from NASA head-quarters in Washington. Some hotshots from the aerospace industry were recruited and brought in. And all the while, holes were being dug, steel was being put in place, and gleaming white buildings were rising from the salt-flat prairie by Clear Lake, Texas.

I divided my Flight Operations Directorate into three divisions. John Hodge ran Flight Control. John Mayer had Mission Planning and Analysis. And my old friend Bob Thompson plied his expertise in running Recovery Operations. Deke Slayton now had his own empire, too. Gilruth formed the Flight Crew Operations Directorate to watch over the astronauts and their training, and to help develop flight plans and in-flight procedures. The new organization left Walt Williams with only pieces of his former do-main. I didn't know that he and Gilruth were in a power struggle. It wouldn't come to a head for several more months.

But other changes seemed to be happening weekly. James Elms turned in his resignation and George Low moved to Houston to become Gilruth's deputy. Low had proven himself to be a Gilruth ally and was well liked in congress, too. I wasn't so sure about the next headquarters man to arrive. He was Joe Shea, known to one and all as George Mueller's man. He re-placed Charlie Frick—a fractious manager who often alienated his own people—as Apollo program manager at the space center. Shea was an out-sider and an enigma. His close tie to George Mueller made him a stepchild among the Mercury veterans, but he was a good engineer and I could only hope that he'd develop into a competent manager.

The original thirty-six of us in the Space Task Group in November 1958 had now grown to an organization of thousands, with expensive and com-plex programs to carry out. What we could do in Mercury wasn't usually possible in Gemini and Apollo, and some managers who succeeded on the original program sometimes fell short and were replaced. With a few others brought in to handle administrative and procurement slots, the team that would lead the way to the moon was now in place. It was the spring of 1964.

Personality conflicts and management changes were matched by the in-tensity of arguments over design details in the spacecraft. One of the orner-iest came to a head in 1963. The Rogallo wing was supposed to let astronauts land Gemini like an airplane, coming down perhaps at Edwards Air Force Base in California's Mojave Desert. It was an inflatable paraglider rig invented by Francis Rogallo at Langley. It was lightweight and with shroud lines and motor-driven reels, it could be steered to a landing.

While McDonnell was building the basic Gemini craft, North American Aviation got the Rogallo wing contract. Max Faget and I opposed it from the beginning. The paraglider wing forced McDonnell to add ejection seats with built-in parachutes to the Gemini design in case the wing failed. That led to McDonnell and NASA design teams declaring that the ejection seats would be the primary escape system for a pad abort or if a Titan rocket failed shortly after launch. They abandoned the escape tower with its tractor rockets that we'd used successfully in Mercury.

It was a foolish plan. The ejection seats couldn't shoot straight up off the pad or rocket, as they do from an airplane. Gemini's hatch doors opened forward, and the ejection rockets had to be tilted. If the astronauts survived the 20-g kick in the butt it took to escape from Gemini, they were liable to be enveloped in a caustic and toxic gas cloud from Titan's fuel tanks. But the design was adopted anyway.

Then North American had nothing but trouble in developing a Rogallo wing big enough and maneuverable enough to carry Gemini to a safe landing. Test wings crashed or ripped apart. The shroud lines and reels kept fouling. The budget was overrunning. Tempers were flaring. Bob Gilruth called a meeting in the rented Houston building that was our headquarters. This time Max and I didn't just oppose the damned thing. We flat vetoed it. After listening to the growing list of paraglider problems, Gilruth backed us. The paraglider was dead.

But we were stuck with the ejection seats. It was too late to completely redesign Gemini, and tough enough to add a parachute canister that would put recovery back in the ocean.

"Cross your fingers," Max said.

"I am," I said. Those ejection seats were dangerous, and neither of us thought that any astronaut would be foolish enough to use them except in the most dire of circumstances. Events proved us right.

Long-duration space flight was a major Gemini goal and a mandatory requirement in Apollo. Batteries could give us enough power for four or five days in orbit. But the size and weight of a two-week battery bank was beyond our ability to launch. Solar panels weren't an option. Gemini could have carried small ones, but needed more electrical power than they could supply. The solution had to come from virtually untested fuel cell technology. Fuel cells weigh a fraction of their equivalent in batteries and generate electricity by reacting hydrogen and oxygen through a membrane. The byproduct is water, which we could use for both cooling and drinking.

The problem was that nobody had ever made a lightweight, efficient fuel cell that would keep working for fourteen days in space. But after agonizing studies that looked at every option, including a small nuclear reactor, we had no choice. The General Electric Company's design looked best and they got the job. We'd fly two short-duration Gemini missions on batteries to prove out the rest of the spacecraft, then put all our eggs in the fuel cell basket thereafter. The scheduling people had a name for what happened as fuel cell development ran into snag after snag. They called it a "long pole in the tent." It was more than that. Without fuel cells, we'd have no tent at all.

My rendezvous expert was Bill Tindall. His full name was Howard Wilson Tindall Jr., but nobody ever tried to use it. I'd known him slightly in the old Langley days, and with the coming of space flight, he was one of the leaders in developing software that used radar data to do launch and orbit calculations for Mercury. He came to Houston as one of John Mayer's top people in Mission Planning and Analysis, and after Bill organized a rendezvous group, I got to know him a lot better.

Rendezvous was mandatory for Gemini and Apollo. If we couldn't figure out how to launch a spacecraft, then have it catch up with and dock with a ship that was already in orbit, our whole lunar plan would go into the wastebasket. But getting from here to there, I discovered, was not only complicated, but sometimes counterintuitive.

It was pure theory. Nobody had ever done it. Tindall asked around and found a number of smart people who shared his curiosity. A few were astronauts who'd have to do it. Some worked for contractors. Most of them were in my Flight Operations Directorate. They met regularly, pulled out their slide rules and big yellow pads, and tried to figure it out. This week's idea or calculation could be next week's piece of trash, or maybe the mathematical building block they badly needed. Some of their breakthroughs fed back immediately into spacecraft design requirements; maneuvering in space was turning out to be a stranger beast than anyone thought.

One of the new people in the group was Edwin E. (Buzz) Aldrin, an Air Force lieutenant colonel who'd come on board with the third group of astronauts selected. Aldrin was recently out of the Massachusetts Institute of Technology, where he'd written a doctoral thesis on space rendezvous. That made him, in his own eyes, one of the world's leading experts. Before long, the real experts on Tindall's panel were calling him, with a touch of sarcasm, "Dr. Rendezvous."

Gradually, Aldrin started listening and so did the rest of them. He was bright and his contributions melded with the conclusions they all were reaching. Finally in late 1964, they had a series of procedures and calculations that might work. Tindall briefed me regularly and wrote insightful and sometimes funny summaries of the group's work. His "Tindallgrams" became the founding papers of space rendezvous and were required reading for contractors and NASA people alike who were involved in Gemini and Apollo mission plans.

There came a day when I actually understood what he was talking about. It can be explained more simply than it actually happens.

Flying in space is not like airplane flying at all. To catch that airplane up ahead, a pilot pours on the gas to go faster. Speed makes an airplane want to climb, so he maintains altitude with his aerodynamic controls until he catches up and flies in formation.

A spacecraft in orbit is moving about 17,500 miles per hour. Assume that it's going around Earth in a perfect circle one hundred miles up. If you're behind it and want to catch up, you don't add speed. Speed makes you go higher, there's no air for aerodynamic control, and because you're now higher than your target, you're in a bigger circle, so you're going relatively slower. You speed up and your target is farther away than ever.

To catch up, you have to slow down. Your orbit drops into a smaller circle. But because of that, you catch up with the higher-flying craft. At some point after you pass beneath it, you add speed, raise your own orbit, and your target catches up with you.

Those are the basics and it took Tindall's brain trust great effort just to get that far. It's too bad that it wasn't that easy in real life. Orbits are almost never perfectly circular, so there are the altitude vagaries of apogee (the high point of an orbit) and perigee (the low point) to figure in. If you're a little left or right of the target's track—they called it "out of plane"—it takes a lot of fuel and precise calculations to get on track. In Gemini, a quarter mile out of plane was a lot. All the maneuvering fuel Gemini could carry would move it sideways about a mile. That's all. So it wasn't just the in-space maneuvers that were important. The original launches had to be just right, too. When the Russians tried it, their next-day launch was off by about one second. That was enough to put a three-mile gap between their two capsules and guarantee that rendezvous couldn't happen.

Tindall's group worked out the math and the physics of rendezvous. When the time came, it would take a roomful of the best computers IBM could build in 1965 to figure it out for real and let us go do it. Even then, it wasn't easy.

Our optimistic schedule called for flying the first manned Gemini missions before the end of 1964. We came close. There were the normal problems we'd come to expect in developing new spacecraft systems, but the biggest hang-up was with the Titan II booster. It needed some modifications to carry the Gemini spacecraft and to ensure its safety as a "man-rated" booster. But when we loaded one with instruments to monitor its performance in a test flight, we got a rude surprise.

They called it pogo—a violent up-and-down oscillation caused by surges in the propellant flow to the rocket engines. Imagine sitting on a spring-loaded chair and being bounced up and down eleven times a second, with each jolt so severe that you couldn't focus your eyes or make your hands reach for something. When the Air Force and NASA engineers looked back at earlier Titan data, they discovered that pogo had always been there. It just didn't matter when the payload was a robust nuclear warhead because the Titan's overall performance wasn't affected.

But it mattered a lot when the payload was going to be two human beings sitting in a spacecraft full of relatively delicate electronics and plumbing. Pogo at the Titan II levels would wreck everything. The Air force was ready to do what it took to fix the pogo problem. But then the reviews of earlier flights turned up a discrepancy in the Titan's second stage; it was only producing half the thrust it should. Again, that hadn't affected the rocket in its capacity as a weapon. But it meant big problems for Gemini, and the Air Force was less willing to work on that problem. Chuck Mathews and I argued our case in a number of meetings before the secretary of the Air Force ordered it to happen. In the end, the knowledge all of us gained paid off in both Gemini and Apollo, where similar problems cropped up.

Most of the time, the Air Force was a willing partner in working on Gemini problems. It wanted its own manned space program and I agreed. In the flames of the Cold War, there would be distinct tactical advantage to the nation with a continuous manned military presence in orbit. We had no doubt that the Russians understood this and were working to that end. A man-in-space command post would have offered far greater monitoring abilities of enemy activities than robotic spy satellites. The satellites in their various configurations were needed and led to such programs as the global positioning satellites now used by both the military and civilian worlds. But when it comes to reconnaissance, no satellite can equal the instant ability of the human eye and brain.

In the mid-sixties, the Department of Defense started its own manned space program, adapting the Gemini design to a program called Blue Gemini and planning a small space station called the Manned Orbiting Laboratory. Both programs were canceled by President Johnson when the budgetary realities of the Vietnam War overwhelmed virtually everything else connected with the military.

Johnson's support for the moon program begun by Kennedy was never enthusiastic, and he concentrated his energies and the nation's budget on his Great Society programs and on the growing war in Vietnam. Each budget he sent to Congress had less money for NASA than the one before. By the time he left office in 1969, NASA's budget had been cut in half and its vision had been shriveled even more. But we couldn't know any of that in 1964. Our job was getting Gemini off the pad.

While the Air Force was still hoping to get its own manned program, it had an oversupply of young engineering officers. We suggested that some of them be assigned to Houston for training as flight controllers. Through Gemini and Apollo, I had more than one hundred of them working in Flight Operations. They were an invaluable asset as our own workload became more intense, and many of them moved on to flight control assignments for unmanned Air Force programs handled out of the Space Operations Center in Colorado Springs.

We hoped that the second Gemini-Titan launch would carry astronauts. As the Titan problems came to the fore, along with some problems in the Atlas-Agena target vehicle, and with the various Gemini spacecraft design changes, it was obvious that we needed two Titan launches before we'd feel secure in putting astronauts aboard. Engineers at the Martin Company, the Titan's builders, came up with fixes for the pogo and second-stage problems, so we scheduled a test launch in the spring of 1964. The Titan carried a Gemini shell, properly weighted to match the actual spacecraft, and equipped with a variety of instruments, radar beacons, and telemetry gear.

Gemini/Titan-1 was run out of our mission control center at the Cape— it had been renamed the Kennedy Space Center after JFK's death. Our new center in Houston was nearly a year from being finished, but we'd finished the modifications to the worldwide tracking network. GT-1 would test the rocket, give us important data on the Gemini spacecraft, and verify the network changes. It was also our chance to integrate the new Air Force and Martin Company people into an operational Gemini team.

The launch was just what we expected until the Titan's second stage separated, then fired. Then inexplicably we lost all contact with the rocket and the spacecraft. My reaction was echoed throughout mission control: "What the hell . . . !?!"

Then three seconds later, everything was back. We scratched our heads, made some notes, and went on with the mission. Gemini went into the elliptical orbit we'd planned. The Titan's pogo vibrations had been cut by ninety percent, and its second stage worked as advertised. We tracked Gemini around the world, through the network and from mission control, for the next five hours, recording data and seeing nothing that gave us a worry.

After three revs, I called an end to mission control's operation. We'd seen enough to know that Gemini was going to work. The network took over through Goddard in Maryland, continuing to track the spacecraft and practice their procedures until its batteries went dead the next day. On the fourth day, the Gemini shell plunged back through the atmosphere, and the few pieces that survived reentry heating splashed in the South Atlantic between South America and Africa.

By then we'd figured out what happened in those three seconds of lost communications. It was a mini-blackout, similar to the blackouts encountered when Mercury reentered. In this case, separation of the two Titan stages and start-up of the second-stage engine created a momentary sheath of charged ions around the rocket. No radio signals could get through until the rocket outran the ion sheath. It happened on every Gemini launch. If we hadn't seen it early, we might have felt that momentary surge of terror that came whenever an astronaut was involved.

Gus Grissom and John Young were assigned to the first manned Gemini mission. Gus was gradually overcoming his antipathy to the press after the harsh media treatment when his Mercury capsule sank. Young was a Navy test pilot with a subtle sense of humor and incredible engineering and flying skills. Grissom and Young made a great team, and so did their backups, Wally Schirra and Tom Stafford, as we trained for our next step into manned space flight.

The first two buildings at the Manned Spacecraft Center were occupied in early 1964, with the Public Affairs Office in Building 1, a sprawling one-story affair with an attached auditorium, and most of management in the nine-story Building 2 next door. In spite of government regulations, I understand that a fair amount of champagne flowed in Building 1 when our public affairs people and the local press got together for their own brand of ribbon-cutting ceremony.

To keep pace with our activities, I created a fourth division under my command, Flight Support, to manage and operate our control center, and hired Lieutenant Colonel Henry E. "Pete" Clements away from the Air Force to run it. Pete had been with us awhile anyway, as the network officer in the Mercury mission control center. In the bland scheme of naming buildings, mission control was a three-story, virtually windowless structure next to our office building, known as Building 30, and housing the four parts of my directorate. We got into the buildings in the spring of 1964, but installing the computers and mission control hardware would consume most of the year ahead.

The exception to getting into the Building 30 complex was me. Bob Gilruth looked at the site plan, at the intensity of the work ahead on Gemini and Apollo, and then ruled that he wanted all of his top assistants located where he could find them by walking down a hall. I argued to be with my people at mission control, but he was adamant.

I was given a corner office suite on the ninth floor of Building 1, just opposite from Gilruth's office. It was an elevator ride and a six-minute walk to Building 30. Max Faget's offices were next to mine. Deke Slayton was on the corner adjacent to mine, and George Low was in a diagonal corner, adjoining Gilruth's offices. I saw one immediate advantage: My view of Clear Lake and Galveston Bay was more beautiful than I ever thought possible for a government employee. And it turned out that having quick access to all of the Manned Spacecraft Center leaders was a major benefit to getting things done. Gilruth was right again.

I needed more flight directors. Our Gemini plan included flights of four, eight, and fourteen days. A two-shift operation by myself and John Hodge would be worn to exhaustion. Hodge and I talked it over. My first choice for another flight director would have been Tec Roberts, but he'd succumbed to Houston's heat and had transferred to Goddard Space Flight Center in Maryland. We went down our list and picked Eugene Kranz and

Glynn Lunney to go into flight director training and to eventually create their own flight control teams. Kranz was my procedures officer in the Mercury control center, virtually my right-hand man even though he was on my left. Lunney was one of our brightest talents, served as the flight dynamics officer at the Bermuda site, and had a natural talent for leadership. Nearly forty years later, I look back and know we made the right choices.

It took until December 9 to get Gemini/Titan-2 ready to fly. The Titan was in good shape, but Gemini's growing pains kept putting on the brakes. This time we had a relatively complete Gemini spacecraft—it had a fuel cell, but still lacked the computer system for rendezvous, among a few missing systems—and we planned a relatively simple up-and-down suborbital flight to test the spacecraft, its heat shield, and its recovery systems.

Our days in the old Cape control center were numbered. In Houston, our new facility would be ready to support—if not actually control—the first manned Gemini flight in early 1965. We counted down GT-2 and I looked around, filled with memories, wondering if this would be my last time in this historic old room. It wasn't.

Titan's two engines ignited, burned for about three seconds, and shut down. A cloud of red, acidic smoke from the propellants obscured the rocket momentarily on my television monitor; I knew it was red, even though my monitor was still black-and-white. Booster was in my ear with his report. The primary flight-control hydraulics had lost pressure, and when the system had switched to the secondary system, it also shut off the engines.

Damn! I thought. *What next?* We'd fixed every Titan problem we'd found, and now something new had jumped up to bite us. We told the Navy to stand down until after Christmas, then be ready to sail again on the New Year's tides. I went home to Houston to wait for the Titan report from the Martin Company. When I got it, the problem was another of those things that happens when we try to be smarter and more efficient than we really are.

A servo valve in the rocket's hydraulic system fractured. We'd insisted on redundancy in a number of Titan systems, backstops against failures that we thought we could predict. The hydraulic system was not one of them. But redundancy adds weight. Engineers had saved an ounce or so by shaving some metal from the valve. It broke apart when that first pressure spike hit it as the hydraulics activated. In trying to be supersafe, we'd created a failure. I asked for the valve and it sat on my desk for years, a reminder that "safe" and "too safe" are not the same thing.

The redesign was simple. Martin had new valves built, tested, and installed in the Titan for a launch in January 1965. I got to the Cape just in time to come down with the worst case of flu I've ever had. I was in a motel bed for a week, and I'm convinced that only Doc Berry's medical skills and the willingness of the secretarial staff to bring me warm soup and cool drinks kept me alive to get to the final reviews before we tried GT-2 again.

Midway through the countdown, we lost the Gemini fuel cell. Its operation was a secondary goal on the mission. The failure hardly enhanced my confidence in this new technology, and when it went down, I wasn't going to waste any time trying to figure out the problems.

"Take it off-line," I ordered, "and keep the count going."

We launched only four minutes late and everything was working perfectly until the lights went out. We had a total power failure in the Cape's mission control center. I stood there in the dark, shocked momentarily into silence. Then the emergency power kicked in and at least we had some lights. We'd lost all data, but our communications lines had their own battery backups. Power failures were in our training scenarios and we all knew just what to do. I depressed my mike key.

"*Rose Knot*, Flight. You've got control. We've just lost power in mission control."

Rose Knot Victor was our downrange tracking ship. In normal times, it would do little more than track and monitor the Titan and Gemini as it passed overhead. We didn't plan to send any commands up, except in the event that the spacecraft didn't separate automatically from the rocket. My old friend John Llewellyn was on the ship, in charge of the tracking team. His voice went up at least an octave when he replied.

"Rog, Flight . . . we've got it."

He ran a commentary for the next minute. Everything was normal. Then we got our power back at the Cape—the outage was caused when all the network television lights and cameras in the area overloaded the system—and I took control again. There wasn't much to do but watch the data flow in and try not to smile. Gemini went through its paces perfectly, separating from the Titan, dropping off its own equipment modules, setting up for retrofire, then firing its retrorockets. Our second downrange ship, *Coastal Sentry Quebec*, watched it all, held telemetry signals before and after blackout, and confirmed that all was well.

Gemini splashed down just four miles from the recovery carrier, and ninety minutes after launch, frogmen had a flotation collar attached to the spacecraft and it was being winched aboard. We'd decided to control the Grissom-Young flight from the Cape, using it one last time, with the now ready Houston mission control center as a backup.

I lit up my last cigar in the unmanned portion of the Gemini program, took a look around, and marveled at how far we'd come, and how far we still had to go.

13

Space Walk

The old chicken-or-egg question that confronted us was easy to answer and damned hard to do. It took the form of "Which comes first, the simulator or the real spacecraft?" In Mercury, we had the Procedures Trainer. It was hooked up to mission control, and we simulated the connections to the worldwide network. The trainer was a close approximation of a Mercury capsule. An astronaut sat in it, put on his headset, and practiced the relatively simple activities of his upcoming mission. My flight controllers usually worked with him.

The Gemini, and later Apollo, simulators were orders of magnitude more complex. We had to have the simulators, one in Houston and another at the Cape, well before the individual mission spacecraft were ready. We'd learned in Mercury that building spacecraft isn't like building cars. Cars coming off an assembly line may have some differences in optional features, but even those differences are an established part of the system.

But every spacecraft is different from the one before it. Some systems are new or improved. Other systems vanish completely. The design that started on an assembly line as Job One was always modified along the way as we tested systems and subsystems or learned lessons from missions flown. But the biggest learning tool was the training simulator. Astronauts had found some things not to their liking in the Mercury trainer and asked for changes in the capsule they'd fly. In Gemini and Apollo, we knew from the start that these extraordinary machines were not only going to train astronauts and my flight control team, but would also yield a never-ending flow of engineering change orders for the real spacecraft.

Everything was computer-controlled now. The simulators' windows were video tubes, and if an astronaut was supposed to look out and see a Titan booster stage or an Agena target vehicle, that's what he'd see. This was heady stuff in 1964. Computer graphics were barely embryonic then. The video scenes were of actual models, moving on tracks in view of a television camera. All of the simulators' controls, switches, circuit breakers, and sensors were computer-monitored and -controlled. The right things happened or seemed to happen in the visual simulation when an astronaut flipped a switch or made a maneuver with the hand controller. In mission control, the computer fed the information to us just as if we'd received it from radio telemetry.

The simulators were so realistic that the astronauts spent hundreds of hours in practice before we ever opened the links to mission control and did a combined training session. The simulation scriptwriters could make almost anything happen. A spacecraft switch, a thruster, or a heater could be programmed to malfunction. Or maybe the switch would work, but the reading we got in mission control said that it had failed.

When astronauts weren't training in the simulator, engineers might be modifying it to add the latest change in design. Or equally often, something turned out to be more difficult than thought. My flight controllers or the astronauts would recommend a change that ultimately found its way to the spacecraft on the assembly line. When we first used the word *dynamic* to describe the simulators, it meant that they acted pretty much like the real thing. But in practice, *dynamic* also applied to design because we discovered things in simulations that needed to be changed.

We also fell into a routine that became a permanent part of mission training for astronauts and flight controllers. The Gemini IV spacecraft had a number of systems and changes to be made. Our schedule in 1965 was falling into a pattern of launching a mission every eight or nine weeks. In early 1965, the Gemini III crew made its last practice run in the Houston simulator. From then on, their training was in the simulator at the Cape. Within hours, engineers were swarming over the Houston simulator to reconfigure it for Gemini IV. The Cape simulator was held in Gemini III configuration until that mission ended. It might be needed to help work out in-flight problems, if they developed. But then it, too, was rebuilt to match Gemini IV.

That kind of leapfrogging became standard practice. It was the most efficient way to train the astronauts and allowed my flight control teams to stay up with the growing demands of each new mission. Backup astronaut crews had to go through exactly the same training as the prime crews. Everything had to be done twice. The simulators went to a full three-shift operation with twenty hours a day for training and four hours for maintenance, fixing the electronic or mechanical parts that inevitably failed.

Even then, it wasn't enough. Deke Slayton and I argued for more simulators, but NASA headquarters refused our budget requests. Deke was particularly chapped. He was forced to put astronauts on long-day schedules, just to get them enough simulator time to be ready to fly. Many of those hours came in late evening, even in the middle of the night. When I had to start scheduling flight control teams for midnight training sessions, my frustration equaled his. The problem reached ridiculous lows when we had to delay the Gemini V flight by several weeks for no reason other than we didn't have the hardware to get them trained in time. But even that couldn't loosen the Washington purse strings. Lyndon Johnson's presidency was taking the nation in directions none of us foresaw, and the NASA budget was only one of its victims.

By now, Walt Williams was gone. Somehow I'd been oblivious to the power run he was making for Bob Gilruth's job until it was too late. Walt was talking to others, both in Houston and at headquarters, and making an issue over his qualifications to run the Manned Spacecraft Center. When he finally broached the subject with me, he was convinced that he could lead an in-house rebellion to force Gilruth out.

He couldn't have been more wrong. I was stunned and told him so: "Don't do this thing, Walt. It's not going to happen."

But he wouldn't back down. Maybe he'd gone too far by the time he brought me into his confidence. He'd cast his die and the problem landed in George Mueller's lap. Mueller for once didn't hesitate. Walt not only wasn't going to get Gilruth's job, but he had to get out of Houston completely. Mueller offered him the position of deputy for manned space flight operations at headquarters.

"I'll quit first," Williams said that night as Deke and I poured him drinks.

We liked Walt and respected him for all he'd done since joining the old Space Task Group. But he was wrong this time, and neither of us would say otherwise. Our loyalties went with Bob Gilruth. Deke was adamant. "Take the damn job, Walt," he said. "Chris, tell him."

"Take the job at least for a while," I said. "Somebody in aerospace will snap you up from a headquarters slot. But if you look like a quitter, or worse, like somebody who just got fired, you'll be on the outside looking in."

A few drinks later he saw it our way. Within a month, Walt Williams was all smiles in his deputy's job in Washington. And a few months later, shortly after we flew Gemini III, he took a job as a vice president at the Aerospace Corporation in Los Angeles. He did a helluva job for them, too.

None of that impacted our training for Gemini III. Gus Grissom and John Young were first in line to fly, and scheduling the simulators was not on anyone's worry list. Their mission was almost a throwback to Mercury. They'd fly three revs and do a shakedown of the spacecraft. We planned several first-time maneuvers to begin demonstrating Gemini's rendezvous ability.

For a guy with a public reputation as a surly character, Gus could be a free spirit. Most of the astronauts were good friends with Jim Rathman, a famous race driver who owned a Chevy dealership in Melbourne, Florida. One day in the fall of 1964, Gus invited me along on an evening's trip. He was flying Rathman's plane, with Wally Schirra, Gordo Cooper, Tom Stafford, and Rathman in the other seats. We zipped down to Miami, spent the evening with an Eastern Airlines friend, then flew back to Melbourne. I slept both going and coming, exhausted from the work schedule we'd been keeping in advance of Gemini III.

It was long after midnight when we landed in Melbourne. We didn't know that Rathman had told his wife that we were going to a football game—long over—in Tallahassee. She was furious. But she also knew all about the many practical jokes the astronauts and her husband played over the years. So she was sitting in her car watching when we walked to our cars. Rathman's wouldn't start. Gus had his Corvette there and it wouldn't start either. Finally they popped the hoods. She'd pulled the distributor cap wires on both cars and plugged them back into the wrong places. It was one of the few times when Gus suffered a "gotcha" that left him unable to retaliate. I don't know what the Rathmans said when they got home.

A few months later, Grissom created a stir when he named the Gemini spacecraft *Molly Brown*, after the unsinkable lady of the same name. It was direct reference to his Mercury flight. "This time," he said more than once, "nothing is going to sink out from under us." The press had a good time with the story. They'd often insinuated that Gus was at fault for losing his Mercury capsule, and he'd developed hard feelings toward most reporters. But now all was forgiven. Gus was a few years older, a lot wiser in the ways of the news media, and he was willing to pick a name like *Molly Brown* just to prove his point.

It was our NASA bureaucrats in Washington who didn't get the joke. Bob Gilruth refused an order from George Mueller to tell Grissom to change the name. Unable to get his way on Gemini III, Mueller issued an order: Henceforth, no spacecraft would have names. They'd be known simply as Gemini IV, Gemini V, and so forth. I thought that Mueller was completely wrong. But no amount of persuasion could change his mind.

The final few weeks before launch were hectic. No matter how much we tried to close the door, last-minute changes in procedures crept into every mission. Much of that was because we were always doing something new, in a field of endeavor that had little history. Every mission had new goals, new pieces of hardware, and put new demands on flight controllers and astronauts alike.

With Gemini, more and more demands came from the scientific community. Mercury missions were short, the capsule was cramped, and just getting into space and staying there for a few hours stretched our limits. The Gemini spacecraft was bigger. It carried two men, and they were actually a bit more cramped than one man was in Mercury. But Gemini had more stowage spaces, both in the cockpit and in its big exterior equipment bay. The science experiments in Mercury were limited and rudimentary. In Gemini, we consciously invited the science community to play a bigger role in proposing experiments and hardware for the astronauts to use.

It wasn't a mistake to do so, but we were surprised and sometimes frustrated by the Pandora's box we'd opened. Scientists and engineers simply do not think alike. I should have recognized that earlier; physicians can be roughly categorized as scientists, and our battles with flight surgeons almost always came because we looked at the same thing and saw it differently. Now we had a broader spectrum of scientists to work with and placate. Usually a scientist, or science team, provided some kind of hardware for a flight. The hardware had to be operated by an astronaut, and some of it was touchy. Training was involved. So was timing—figuring out where to put this experiment or that one into a flight plan. The simple experiments didn't require anything more than an on-off switch to activate certain kinds of instruments. But more complicated experiments might involve pointing a sensor at a star, taking specific photographs of an area on Earth, prioritizing electrical power use, scheduling the experiment for a particular place over Earth, or a seemingly endless list of other factors.

Getting into space was, and still is, a rare event. Scientists would naturally be excited, and many held emotional and proprietary feelings about their hardware. So they started showing up in the final days before a launch, hoping to spend some last-minute time with the astronauts to go over procedures and to make sure that the astronauts understood just how important their experiment was. In those same final days, everybody's schedule was filled. The astronauts themselves were quarantined against exposure to communicable diseases, so few scientists got through the doors to discuss experiments with the crew. Inevitably there were hurt feelings and even anger. Even after we moved to Houston, I always spent a few pre-launch days at the Cape. I'd end up as a referee, explaining to scientists why it was impossible to give them the astronaut time they thought they needed. I'd also hear the crew side. After months of training, the last-minute changes in systems procedures and the constant demand for "just one more conversation" wore on them. Every debriefing included pleas to stop making changes. We tried. But with each flight getting more complicated, the situation only got worse.

I was one of the few with free access—after getting a physician's approval—to the crew quarters at the Cape. Walt Williams started a tradition in Mercury of having a pre-launch dinner with the astronaut. I continued the custom in Gemini. The astronauts usually opted for a low-residue menu. Bowel movements in space were not comfortably accomplished. With only a short night's sleep between dinner and liftoff, the dinner was the last opportunity for all of us to clear up any unfinished business. These sessions were among the best perquisites of my job. Intensely hard work by all of us was about to pay off. And the knowledge that this was a dangerous business always lurked in the back of our minds. None of us would ever forget the camaraderie at that dinner hour.

Five days before we were to launch Gemini III, the Russians pulled off another space first. I never doubted that they timed this mission to preempt us and to turn the world's attention back to their own space program. They had the advantage of knowing exactly what we were going to do, and when, simply by reading the newspapers. So in the propaganda heats of this space race we were running, the Russians could always count on the element of surprise. I still refused to attend those classified briefings about the Russian program. So what they did on March 18, 1965, was certainly a surprise to me and to the world's press.

They put up their own two-man ship called Voskhod 2. They'd orbited Voskhod 1 in 1963 with three men on board. Decades later we found out that the Voskhods were simply their old one-man Vostok craft with its guts torn out to cram in more people. On the three-man flight, the cosmonauts were packed so tightly that they had to leave their space suits behind on Earth. If Voskhod 1 had sprung a leak, they would have died. Voskhod 2 had other modifications. During its one day in orbit, Alexei Leonov, who did have a space suit, crawled through an airlock and a canvas tube to become the first human to do extravehicular activity (EVA), or what the press commonly called a space walk.

It was an amazing feat. The press and U.S. politicians reacted with shock and dismay that the Russians were once again proving to be superior in space. Leonov's space walk was a risky bid for attention and it worked. Gus Grissom, I heard, had a choice selection of four-letter words to describe his feelings about the event. It put a momentary pall on my own preflight mood before I told my flight control team to ignore the damned Russians. "It's a stunt," I said. Again, it was decades before we learned just how risky it had been for Leonov. His space suit puffed up like the Pillsbury Doughboy and he got stuck in the canvas tube. He had to open a vent valve and deflate his suit several times before he finally crawled back to safety. The Russians, of course, didn't mention any of that when they paraded Leonov through Red Square and pinned a medal on his uniform.

We put Voskhod and Leonov out of our minds and got Grissom and Young into space on March 23. I had my control team at the Cape, and John Hodge backed us up in the new control center in Houston. As I looked around and compared surroundings, things looked small and antiquated at the Cape. We were saying good-bye to a room that was filled with history. But for now, all Gemini authority was here. Hodge and his people were getting the same data we were, though on more capable consoles and with a greater number of displays. But unless something went seriously wrong—perhaps another power failure at the Cape caused by too many television journalists plugging in at the same time—this room was in charge for one last space mission.

The Titan rocket ride was a pleasant surprise for Grissom and Young. It was smooth, with relatively low noise. They went into almost exactly the orbit we wanted, an ellipse about 100 by 140 miles above Earth, with the high point, apogee, over Australia. I heard the numbers and pressed my mike button.

"Tell 'em they're go for three orbits, Capcom."

Almost immediately we got a report back from John Young that the oxygen system was acting up and its instruments were giving strange readings. Young was a sharp engineer. He quickly diagnosed an electrical problem and switched to a backup unit. The instruments went back to normal. Grissom pulled out two science packages, including one that would expose freshly fertilized sea urchin eggs to zero gravity, to see if their initial cell division was affected. When he twisted the sperm-release handle, it broke. So much for sea urchins in space.

Then it was time for the big maneuvering tests, using Gemini's advanced thrusters called the Orbital Attitude and Maneuvering System (OAMS). There wasn't much for any of us to do in mission control except to monitor telemetry readings and tell Grissom that he was "go" for thruster firing. He easily turned *Molly Brown* around so that its OAMS thrusters fired retrograde, into the direction of flight, then over Texas powered on the thrusters for a long seventy-five-second burn. We'd never tried anything so lengthy in Mercury; the thrusters weren't built for long burns and we didn't have the fuel. Now Young did the burn perfectly. As *Molly Brown* passed over the Cape, my flight dynamics officer had some new numbers. I listened with satisfaction, then gave them to the capcom. Grissom had lowered their orbit to an almost circular 93 by 106 miles.

Young told us that their onboard computer gave them almost exactly the same results. I breathed an initial sigh of relief. We'd just crossed a major milestone in getting to the moon. We'd changed a spacecraft's orbit, and we'd shown that the ship's own computer could measure velocity and orbital changes.

A little more than halfway around the world again and Grissom would try the second maneuver, this one sideways to move *Molly Brown* out of plane. That kind of correction was vital to future rendezvous missions, where the orbits of two spacecraft might not be in exactly the same plane.

But first, Grissom and Young had to sample some of the freeze-dried space food sent aloft by the doctors. It didn't taste that good on Earth and had been the butt of many jokes. Now they had to give it a try in space. What Grissom didn't know was that Wally Schirra had slipped into Cocoa Beach the night before and, just before the spacecraft hatches were closed that morning, had handed a package to John Young.

So while Grissom struggled to open a pack of freeze-dried something, Young pulled out his package and unwrapped it. It was a tasty and aromatic corned beef sandwich. Grissom couldn't resist. They each took a bite. It was a great sandwich, but it would cause them considerable trouble later.

Over the Indian Ocean, Grissom set up for the important out-of-plane OAMS burn. *Molly Brown* was facing sideways to the line of flight when he fired the thrusters. Our tracking ship below, *Coastal Sentry Quebec*, watched the spacecraft move one-fiftieth of a degree sideways, almost a full mile. The computer readings on the spacecraft instrument panel said the same thing. I heard the reports and grinned. All that theory about rendezvous was being converted to fact.

The final task was reentry. Max Faget and his engineers had deliberately designed Gemini with an offset center of gravity. By maneuvering carefully during reentry, an astronaut should be able to take advantage of atmospheric lift forces to target his landing with great precision. Grissom tried, but no matter how much he rolled the spacecraft, he couldn't make his instruments converge with the computer-driven display. *Molly Brown* landed fifty-two miles short of the waiting aircraft carrier *Intrepid*. When the data was analyzed, the aerodynamicists found that Gemini's lift was less than predicted. The new data would be factored in for future missions.

On the water, Grissom told us later, he had a brief sense of déjà vu. All he and Young could see through their hatch windows was seawater. He briefly worried that *Molly Brown* might not be unsinkable after all. Then he figured it out. The big Gemini parachutes were still catching wind and were pulling the spacecraft through the water. He released the chutes, Gemini bobbed upright, and there was nothing to do except wait for the recovery forces.

Gemini III was a complete success. But it was tainted by the corned beef episode. Some doctors and a few outraged congressmen complained that Grissom and Young had compromised medical tests with their unauthorized lunch. They hadn't and I certainly saw no problem. No matter how brave or focused an astronaut is, there's a tension in space flight that none

of us on the ground can truly appreciate. A moment of diversion up there is
no bad thing.

Our original plans for Gemini IV were on the scrap heap. My mission
planning people had laid it out as an eight-day, long-duration flight that included the first rendezvous with an Agena target vehicle launched by an
Atlas. But eight days in space would require those fuel cells to generate
electricity. We'd used batteries for Gemini III, and now we got the news
that we'd have to use batteries again because the fuel cells simply weren't
ready.

We went into a series of meetings that produced one undeniable fact:
The best we could do on batteries was a four-day mission. That made the
doctors happy anyway. They still clung to their worries about long exposure
to zero gravity—fearing everything from an astronaut's blacking out to losing muscle control to going blind—and had fought us hard before the original eight-day plan was approved.

Then we got news that the Air Force had run into more troubles in perfecting the Atlas-Agena combination. We not only couldn't have an Agena
in June 1965, but shouldn't count on doing a rendezvous mission until
much later in the year. There were other delays, too. The radar to be carried on Gemini was late in development, and so was the rendezvous software for the onboard computer. Until we had that equipment, there wasn't
much else to do except to extend flight durations.

The major accomplishment of our three-revolution Gemini III flight,
maneuvering in space, was now fact, not theory. But it looked puny to
politicians, and even to some reporters, when compared to what the Russians had done with Alexei Leonov's space walk. Inside NASA, we understood that spacecraft needed maneuvering ability and that the Russian
ships had none. But they got headlines anyway. Extravehicular activity was
on Gemini's agenda. Bob Gilruth asked the question in a meeting of his
senior staff: "With the Gemini IV flight plan in total disarray anyway,
should we include EVA in the rewrite?"

We looked at the implications and came up with an interim answer:
"Yes, if . . ."

We wanted to establish and maintain a schedule for flying a Gemini
every eight to ten weeks. That didn't leave much time to get ready for
Gemini IV. And there were plenty of ifs.

Yes, if the Gemini IV spacecraft hatches could be tested and certified for opening and closing in space. Kenny Kleinknecht ordered McDonnell to do the tests immediately. The hatches passed.

Yes, if the space suits to be worn by Jim McDivitt and Ed White had been qualified for performance in a vacuum. A check of the paperwork showed that they had.

Yes, if Max Faget's engineers could work with McDonnell to design, almost overnight, an umbilical system that would connect to the spacecraft on one end and to Ed White's space suit on the other end. We'd already designed the spacecraft system to handle short periods of high-flow oxygen. But getting it out to the astronaut was the question. "We can do it," Max said. And in a matter of weeks, they did.

The final "Yes, if . . ." involved an operational question. If we just hung Ed White out on the end of an umbilical, it would be a gimmick. But Harold Johnson had been developing a small handheld maneuvering unit to allow astronauts to control their movements on a space walk. It looked like a crossbow and used spurts of carbon dioxide for propulsion. Using it was a legitimate test of a man's abilities to work outside of a pressurized cabin. If the device was ready and White was properly trained, why not try it?

We presented the plan to headquarters and got conditional approval. Then while Ed White practiced using the maneuvering unit, my operations engineers had another idea. The Titan rocket's second stage followed Gemini into orbit. Why not use it as a rendezvous target in the first hours of the mission, before it fell away and reentered? I took the proposal to Deke Slayton and then to Jim McDivitt. "Yes," they said. "Let's do it."

That put an immediate burden on McDivitt. The Gemini simulator wasn't configured for rendezvous, and without onboard radar he'd have to eyeball the whole thing. We found several makeshift training devices around the country, and McDivitt did all that he could to get ready. By mid-May, all of the "Yes, if . . ." questions were answered and we scheduled launch for early June. In addition to all the real-time flight plan changes and new training, we had one more element to consider. This time out, we'd control the mission from Houston. I sent Glynn Lunney and his control team to the Cape where they'd be our backups and be the last ever to use that old-fashioned center during a space mission.

In the last weeks, we scheduled training sessions and simulations almost every day. After my last mission at the Cape, the new facility was even more impressive. We'd been working in a small and even cramped control center during Mercury. The Houston center was spacious, the computers were faster and had much more capacity, the modern intercom system worked, and we were surrounded by support rooms where bright young systems people kept us supplied with every detail we requested. The words *control center* now encompassed all of it. What we used to call the control center—home to our consoles, the world map, and other displays—had a new name, too. It was the Mission Operations Control Room, or MOCR. We pronounced it "moe-ker." I never liked that name.

Around the country, special support rooms and test facilities also were set up by the various spacecraft, rocket, and system contractors. Industry experts stood by to help us, answer questions, or run tests on a moment's notice during a mission. This was a far cry from our early days, and at times I found myself thinking about Mercury as some prehistoric relic. Then I'd remember that we'd flown men in space with that relic, and not long ago, either.

One day in late May, after a long and grueling simulation that involved my flight control team, McDivitt and White, and the worldwide network, a strange feeling hit me as I got into my car. It took a moment before I remembered that I wasn't about to drive to a motel and another in an endless chain of restaurant dinners. I was going home to Friendswood, to Betty Anne and the children, and a home-cooked meal. For the first time, I fully understood that mission control wasn't halfway across the country anymore. It was in my own backyard, and even during a mission I could sleep at home in my own bed. It felt good. When I saw the smiles around the dinner table that night, it felt even better.

We counted down on June 3, and it was so smooth that I felt more confidence than ever about the close timing of our future rendezvous launches. Then we hit a snag on the ground. The erector used to raise and lower work platforms on the launch pad malfunctioned. Three times. A diesel engine carburetor fouled. An electrical wire. Something jammed. We'd had enough built-in holds to let us launch on time. But the Air Force was embarrassed. General Vince Houston, the Atlantic Missile Range commander and the military's designated representative to Gemini, raised the roof. I heard later that his people damn near wore out the equipment in fixing it and testing it to make sure nothing like that happened again. It didn't.

I watched the lights go right up the nominal lines on our new and more elegant screens in the Houston center. McDivitt did a perfect OAMS burn to pull away from the Titan second stage, then turned around, had the stage in sight about 150 feet away, and tried to approach it. It was Gemini's first and admittedly minor attempt at rendezvous. It didn't work right at all.

McDivitt's pilot reflexes took over. He hadn't been able to train in a simulator, with realistic rendezvous hardware or procedures. Everything in his being told him to fly toward that rocket stage. But the more he did, the farther away it got. He was adding speed, thus increasing Gemini's altitude. The higher he got, the slower his ship moved in relation to the rocket stage. His eyeballs and brain told him one thing. The laws of physics dictated something else.

McDivitt wouldn't give up, but I saw that he was using more fuel than I liked. As Gemini IV approached Mexico on its first revolution, I looked at Gus Grissom on the capcom console. "Give it up," I said. "Tell him to save the fuel."

Our first mini-rendezvous was a bust. But we knew why, and we knew that the next time around, the astronaut crew would have the right equipment and the right training. I was only sorry for one thing. If the rendezvous had worked, we would have sent Ed White and his maneuvering gun over to take a close look at the rocket stage. There could have been some spectacular photography. It was a minor point. I told Grissom to pass the word that McDivitt and White should get ready for the EVA. I had my own small worries and I wanted him to get out and then back in as soon as possible.

But that took longer, too. They had to unpack the bulky twenty-five-foot-long umbilical. It was about an inch in diameter, and in addition to a nylon tube for oxygen, it had communications and electrical wires and a braided steel cable to bear the load. White had a small chest pack to wear, too, containing an emergency oxygen supply. By the time the two astronauts were sealed into their space suits and Ed White had the umbilical plugged in, it was like a huge weightless snake on his lap. He had to unstow the maneuvering gun, too, then make sure that all his umbilical and communications connections were secure and in working order. They were nearly around the world again when McDivitt told Grissom that they still had work to do.

"Next pass around," McDivitt radioed.

I nodded to Gus Grissom. "Tell 'em we're happy with that."

So another ninety minutes passed before they were again near Hawaii. This time I gave them a go and McDivitt depressurized the cabin. Not too many minutes later, Ed White tried to push open his hatch. It was stuck.

He pushed again and finally it opened. In mission control, we couldn't say anything to help. Once they switched to the EVA comm system, we could hear McDivitt and White talking. But I couldn't get a message to them, through Grissom on the capcom console, unless McDivitt flipped a switch to let us in. That was a flaw in our thinking that I made certain never happened again.

Gemini didn't have onboard television. I wish that the technology had existed because the scene had to be spectacular. We had only Jim McDivitt's description of Ed White floating out there on his tether, using his maneuvering gun just the way it was intended, and then hanging in space outside the craft. We had still photos and 16mm motion picture footage later, and it was still an awesome sight. But in the back of my mind, I tucked away a promise that one day we'd have television to watch it all in real time. We needed the input in mission control. Even more, we owed it to the American people to give them a front-row seat to their country's finest hours.

Ed White was having the time of his life. Later, I objected when the press said he was euphoric. Still later, I wasn't so sure. He certainly wasn't nervous. And in his place, who wouldn't be euphoric?

Gemini crossed the Pacific between Hawaii and the mainland United States, and I forced myself to concentrate. White's descriptions of sky and Earth and even the Gemini thrusters were mesmerizing. But this was no game. We were strung pretty tight in mission control knowing that Ed White's life depended on the integrity of his space suit, on the oxygen flow through that long umbilical, and on the mental control we all needed to maintain.

Now they were crossing the southern United States and it wouldn't be long before they were out over the Atlantic and running toward sunset. We were behind in the timeline and White showed no signs of getting back in. My gut was tightening with worry about his ability to function in the pitch dark of nighttime space. But that damned communications system wouldn't let us break in to give him instructions.

Finally Jim McDivitt decided to ask. "Gus, this is Jim. Got any messages for us?"

I didn't hesitate in breaking my own rule about who talked to astronauts in space and who didn't. I flipped my override switch and spoke directly to McDivitt.

"Yes," I barked. "Tell him to get back in!"

Grissom looked at me in surprise. Even in our worst-case simulations,

I'd never used that override switch. Then he grinned and shrugged. We listened to McDivitt passing along my order. It took several repetitions before Ed White complied. He'd truly enjoyed America's first space walk, and he'd been in full control of himself for the entire twenty minutes he was outside. He wasn't ready to quit, but he knew an order when he heard it. "This is the saddest day of my life," he sighed. Then he gave it up and went back into the cockpit.

They had a devil of time getting that hatch closed again, particularly with a snaking umbilical trying to wander all over the place. Once it was shut, we decided to leave it that way. We'd planned to let them open it again later to dump out all that EVA equipment. I thought it over and decided that it wasn't worth the risk. McDivitt and White would just have to live with that umbilical and the rest of it for the next three and a half days.

Ed White's space walk had an unintended fallout. He made it look easy. He had a maneuvering unit that worked, and he didn't have to grab on, hold stable, or do any real work out there. We assumed that future space walks would be a piece of cake. We were dead wrong.

Over the next few days, we learned a great deal about long-duration space flight. I'd divided our flight control schedule into three shifts. Glynn Lunney and his team were on call at the Cape if they were needed for backup. A team under Gene Kranz took the second shift in Houston, and the John Hodge team handled the third shift.

Most of the activities took place on my shift. Kranz and his team monitored all the consumables, made sure that systems functioned properly, and oversaw the astronauts trying to sleep. John Hodge and his team would review everything that had happened during the previous shifts and then update the next day's flight plan to make sure we accomplished all of our goals.

At first, "trying" to sleep was the right word. We'd scheduled the astronauts for sleep periods of four hours at a time, but with only one man sleeping. The other would stay awake to monitor systems and be ready for an emergency. It didn't work very well. Anytime the awake astronaut moved, it disturbed the asleep astronaut. They quickly decided, and I concurred, that the ground should watch the systems while they both slept.

At the end of each shift, we met the press. Paul Haney had taken over the Public Affairs Office after Shorty Powers left NASA, and he convinced me that a change-of-shift press briefing was the right thing to do. The flight director would be there, of course, and he'd bring along the controllers responsible for the shift's major activities. Sometimes the sessions were quick and easy. Sometimes they were brutal, with reporters—more than a thousand had showed up in Houston for Gemini IV—pressing for intimate details of the astronauts' medical status, or finagling for a quote that would lead to a headline in tomorrow's paper.

Some reporters seemed unusually concerned about bowel movements in space. The fecal container bags carried aboard Gemini were blue. Almost every day, somebody would ask for a "blue-bag status." It became a standing joke in mission control that we had to be ready to deliver a blue-bag report. The good space reporters, I was happy to discover, never asked.

Paul Haney understood the need to feed the press. When it was obvious that the Manned Spacecraft Center would be inundated with reporters during missions, Bob Gilruth told him to rent an off-site building across the highway as press headquarters. Haney had direct lines run to mission control, and reporters heard most of the air-to-ground conversations between Houston and the astronauts live as they happened. (There was an exception, which created a flurry on later missions.) Haney also hired stenographers to transcribe the air-to-ground conversations, and all press conferences, verbatim. That created a mostly accurate written record of Gemini, and it has become a historical treasure trove. *Mostly accurate* is the correct phrase because the best of the stenographers sometimes got confused with our space jargon and acronyms. Now and then I'd looked at a paragraph and not had the foggiest idea what it meant.

We only had one major glitch on Gemini IV. After about forty revolutions, we asked McDivitt to turn off the onboard computer to save battery power. It wouldn't turn off. We sent commands from the ground and it still wouldn't turn off. Finally we told McDivitt to disconnect the computer's power. That did the job. But when they turned it back on later, it wouldn't work. Somehow its memory had been corrupted.

That was a problem. The computer was an important part of Gemini's controlled reentry. Now McDivitt would have to do a manual reentry, rolling the spacecraft to negate the small lift forces generated by the offset center of gravity. It was a Mercury reentry all over again. In mission control, I put a team to work digging out the exact procedures McDivitt should use. It should be, we decided, not that difficult.

It wasn't. McDivitt and White reentered after four days in orbit and came down under parachutes spotted by one of the recovery helicopters. They were healthy, too. Both had lost some weight and showed a small decrease in blood plasma and bone density. It was nothing that hadn't been expected. The men held up well throughout America's first long stay in space, and so had the hardware. None of the feared medical problems came to pass, and the doctors had no choice but to accept the results.

So we were ready to double flight time again and go at last for that eight-day mission.

14

How We Left the Russians Behind

E d White might have been euphoric during his space walk, but whatever he felt was tame compared to the American public's reaction. The country went wild with excitement over their space program. Ed's space walk completely eclipsed the Russians. We'd announced it in advance, and every minute of it was carried live on radio and television. No TV camera was aboard, but the audio between Ed and Jim McDivitt, Ed's descriptions of what he saw, and the uncontained excitement of the television personalities made the event compelling. Newspapers and magazines followed with some of the most enthusiastic stories I'd seen since the early days of Mercury. For the first time I saw real optimism out there over our chances of winning the race to the moon.

We had our own optimism inside NASA, but it wasn't something we talked about. In one of my Apollo file folders, I had a 1964 tentative schedule that showed manned lunar missions beginning in 1967. It was an internal working document, more of a wish than a "must-do." But it showed that we intended to beat Kennedy's "before the decade is out" deadline by years, not months. In a worst-case scenario, we had plenty of time to recover from serious problems or to accommodate the almost inevitable slips in delivering flight-ready hardware to the Cape.

That schedule wasn't exactly a secret, not in the official sense. If somebody had asked whether or not we had such a thing, I would have said yes. But nobody asked. Nor did we need the added pressure of reporters asking about a schedule that was probably impossible anyway. We had enough pressure when everyone, including Congress and the White House, focused on Kennedy's deadline, or when the Russians pulled off some new spectacular in space.

Strangely enough, the Russian pressure evaporated after Leonov's space walk. We flew Gemini III and IV and were getting ready for Gemini V while the Russians did nothing. We had no way of knowing that the Russian space program was in flux, suffering its own technical delays as they developed a new spacecraft. Nikita Krushchev made a public statement that beating America to the moon was not a Soviet goal, but nobody paid much attention. Nor was any credence given to published technical papers saying that a space station was a better goal than landing men on the moon. As much as anything, the space race was a creation of the press, and the press couldn't let it die. But racing with the Russians didn't dictate our schedule. We were so absorbed in flying Gemini missions and in getting ready for Apollo that we barely noticed the lack of space news from the other side of the Iron Curtain.

We'd made the transition to the Houston control center with only minor problems. In every way, NASA was on the ascendancy, and at least for now, Gemini was the only manned space program in town. The world's press descended on us and we got the benefit, and the problems, of their full attention.

After Gemini IV, I got a call from Paul Haney. The editors of *Time* magazine had made a formal request for a briefing in New York City by Dr. Chuck Berry, the chief flight surgeon, and me. NASA had approved the request. *Time* was one of the world's premier weekly newsmagazines and its pages were read avidly from Main Street to the Oval Office to the halls of Congress. I hadn't been overly impressed with its space coverage. Though it was a sister publication to *Life*, it didn't seem to have the insight or expertise that *Life* writers had acquired in the years since they'd signed that "personal story" contract with the astronauts. This might be my chance, I thought, to tell them a thing or two. Chuck and I packed our bags and flew to New York.

The lunch in *Time*'s offices in Rockefeller Center was small and personal, with only four editors and science writers joining us. It was obvious that *Time* was about to put a lot more emphasis on space. There was an excitement among the *Time* people that I usually associated with meeting movie stars or big-name politicians. Chuck Berry and I weren't just information sources for these people. We were on display.

In all my dealings with newspeople, going back to before Al Shepard's Mercury flight, I'd found that being blunt and honest carried the day. Dick Witkin had taught me that good reporters have a built-in bullshit detector, and I assumed that *Time*'s editors were as skilled as Witkin and his *New York Times* cronies.

"I'm a *Time* subscriber," I said, "and I'm usually frustrated that the magazine doesn't have more savvy about space, or where this technology is taking us."

I chastised the editors for their shallow approach to reporting the real impact of Mercury and now Gemini. *Time*'s stories always left me with the impression that this space race—a race created and fostered by the press itself—was nothing more than the publicity arm of the Cold War. The real payoff from space, I told them, comes from the way we were catalyzing private industry into producing high-technology hardware, software, and services. The space race was changing the way American companies did business, and even we insiders couldn't predict the long-term benefits. But those benefits are there, I said. On the other side of the coin is the inspirational push to the nation's psyche. Space had given America a new kind of pride in itself. In the next few years, I told the *Time* people, we are going to see an American take the first steps in a migration to other worlds.

Dr. Berry was eloquent in describing the medical benefits that were flowing from the space program. The ability to remotely monitor a man's health would become standard in hospitals, he predicted. If we could use telemetry to watch John Glenn's heartbeat in space, a nurses' station could do the same thing for the patient down the hall.

We were pummeled with questions, and over the next two hours, the *Time* people forced us to defend our beliefs. Neither of us backed down. It was a lively and provocative session that we didn't want to see end.

A few weeks later, Paul Haney called me again. *Time* was considering me as a cover candidate. It would mean having one of their reporters follow me around for the next month, and I'd have to sit for a portrait to be painted in mission control. "They've got the wrong guy," I told Haney. "It should be Bob Gilruth on *Time*'s cover, not me."

Haney followed up and called back. They weren't interested in Gilruth. *Time* wanted me and so did NASA's public affairs office. Haney urged me to agree. So it was that *Time*'s Houston bureau chief, Ben Cate, arrived on my doorstep. I quickly figured out that space was only a piece of his domain, along with politics, the oil industry, and Texas in general. He hadn't hired a space specialist for the bureau, and he didn't know much about the intricacies of Gemini and Apollo. He filled that void quickly by putting Jim Schefter, a young local writer who did his homework, on his payroll. Cate spent about six weeks following me around, with Schefter filling in many of the technical blanks, and we became close associates and friends. As Gemini V approached, there wasn't much about my life and my profession that *Time* didn't know.

They arranged with the artist Henry Koerner to paint my portrait as I sat at my mission control console. Koerner was a marvelous character, with an Austrian accent, but he drove me crazy with his intensity and in the way he made me sit motionless while he painted. One night Betty Anne invited Koerner and Cate to dinner at our home in Friendswood. Nobody warned Koerner about a familiar beast, the giant cockroaches (euphemistically called water bugs) that lived in the nearby trees and frequently invaded homes. We were midway through dinner when one of them ran up the dining room wall.

Ben Cate had lived in Texas a few months and wasn't shocked. But Henry Koerner leapt from his chair and backed away shouting, "Unbelievable! Unbelievable!" His accent made the word almost unintelligible. Betty Anne was horrified, but Ben and I burst out laughing. I was relieved a few weeks later to see that the shocking moment hadn't prevented Henry from delivering a masterful portrait that did, in fact, appear on *Time*'s cover on August 27, 1965. It now hangs on a wall at home and is one of my proudest possessions.

We'd put Gemini V into orbit on August 21, after bad weather delayed us several days, and I had my hands full keeping it up there. Our training time, both for the astronauts and for my flight control teams, had been limited because both the simulators and the control center were coming on-line in the early months when training should have started. We'd managed well enough on Gemini IV, and Deke Slayton was finally satisfied, if not happy, with the training for Gemini V.

I felt better about the control teams. We'd run a number of integrated simulations with the astronauts, including some worst-case scripts with astronauts getting so sick in orbit that Gordo Cooper aborted the mission on his own and left my controllers trying to pick up the pieces of when and where the spacecraft landed. The simulations in our new and well air-conditioned center were so realistic that I saw controllers with sweat on their foreheads as they were forced to analyze situations and make critical decisions on a moment's notice.

This was the longest space flight ever attempted by anyone, and the number of things that could go wrong grew incrementally with each new day. We couldn't train for all of them. But my teams were so sharp and well-practiced at handling new situations that I developed a strong confidence in their ability to cope. Cooper and Conrad felt that way, too. There was a confidence that I'd never seen before flowing between engineers at the Cape, the astronauts, the worldwide network, and the flight control teams. The only word I could find to describe it was *professional*. We were good before. Now the Gemini V mission took us to a new level of professionalism that I didn't know existed.

Gordo Cooper and Pete Conrad were on an eight-day flight plan that would surpass any mission flown by the Russians. They had fuel cells aboard to generate electricity and supply water, and a radar was installed. They also carried a radar transponder pod to be released in orbit to serve as a rendezvous target for them.

The troubles started on the way to orbit. After two smooth rides with Titans, this one developed pogo, and it was a rough ride for Cooper and Conrad. Then within minutes after separating from the Titan, Conrad reported problems with the fuel cells. The tank supplying oxygen to the two cells was losing pressure. It wasn't critical yet, but I ordered a constant watch on the telemetry. If the pressure fell below 250 pounds per square inch, the fuel cells could quit working and we'd have about seven orbits—maybe eleven hours—of battery life to get the crew down safely.

But for the time being, I kept things on schedule. Gordo Cooper deployed the lighted rendezvous pod over Australia, then used the Gemini OAMS thrusters to change orbit and move away from it. In the next hour, he started the maneuvers that would bring Gemini V back to the pod. We'd tried a minor rendezvous on Gemini IV and failed. Now finally we were doing one for real, and all the numbers flowing down the telemetry channels told us that Cooper and Conrad were flying it perfectly.

Then EECOM, my environmental systems controller, was in my ear with bad news. Oxygen tank pressure was dropping rapidly. The way to get pressure up is to heat the cryogenic (liquid) oxygen, building up the pressure. The tank heater had obviously failed. The Gemini spacecraft was operating at full power, using all the electricity the fuel cells could put out. As he watched the pressure drop through that magic 250 psi mark, EECOM sounded worried. Our best information said the cells wouldn't work well, or at all, below 200 psi. They weren't restartable, so shutting them down for a while wasn't an option. That became a lesson learned; in Apollo, the fuel cell would be restartable in-flight.

We discussed the troubles quickly in mission control, then told Cooper and Conrad that it didn't look good. I made the only decision possible.

"Capcom, Flight. Tell 'em to abort the rendezvous. Then tell 'em we're going to read up a list of systems for them to shut down. EECOM, give Capcom the list."

Only the most critical onboard systems would stay on-line. Everything else drawing electricity was being turned off. The momentary disappointment of losing another rendezvous was overshadowed by the more urgent business of deciding if we had a mission or if we needed to order reentry.

So far, the fuel cells hadn't failed. But my choices were getting limited. I could bring Cooper and Conrad home at the end of their third or fourth orbit, splashing down in the Atlantic. After that, the prudent landing sites were a day in the future. I could order them down into one of several contingency landing zones around the world. But each of them was remote, difficult for recovery forces to reach quickly. Even worse, we'd probably be out of communications and not know if they'd landed safely or not.

I had ninety minutes to make a decision. The support-room people already had engineers at McDonnell in St. Louis working on a fuel cell, trying to see how it acted under low oxygen pressure. They told us not to expect a good answer for ten or twelve hours. That was no help at all.

Out there where it counted, the oxygen pressure dropped to 190 psi, held for a while, dropped to 100 psi, held again, then dropped and stabilized at 70 psi. EECOM's people told him that it might have enough heat leak to hold there.

"What's the voltage, EECOM?"

"Holding steady, Flight. It seems to be giving them all they need, considering how many things are turned off."

"Purge status?"

"I think we better try it, one at a time, Flight."

As they generated electricity, the fuel cells produced water. At some point, the water had to be purged. If it wasn't, the cell would drown and go dead. I had about sixty minutes to make a go/no-go decision. On the other side of the world, the crew got their instructions. Purge one fuel cell. Wait a minute. Purge the other.

"Flight, EECOM. Looks like the purges worked okay. I'm seeing voltages going up a bit."

"Rog, EECOM."

A few minutes later, McDonnell's engineers in St. Louis got the same results on their fuel cell. But until they had more data, they wouldn't commit to predicting that the fuel cells would keep working. I had twenty minutes to make a decision. I decided not to wait. The fuel cells were delivering enough electricity to keep the spacecraft running. We could gradually turn things back on to see if the cells kept up. The crew was in no danger and this mission might be salvageable. If everything went to hell, we'd just have to risk one of those remote contingency landing sites.

"Capcom, Flight. Tell 'em they're go for now. Give it to 'em straight, though."

Cooper and Conrad were delighted. It had been a long road to get up there and they weren't about to give it up easily. I turned it over to Gene Kranz and his team, confident that we'd make it through the night. He'd keep an eye on things before John Hodge came on duty in the middle of the night. I explained it all to the press at the change-of-shift briefing. This was a complicated business and a lot of the reporters were new. One of the reporters was here for the first time and clearly hadn't done any homework at all. I decided that it was time to take a few minutes and get to some basics.

"Instead of answering your question, young man, why don't I just give you the facts?" The veteran reporters in the room chuckled, but were professional enough not to laugh out loud at the expense of one of their own. Within a minute, most of them were taking notes, too. I think a simple explanation of fuel cell technology and the problems we faced helped them, too.

Then I gave myself the luxury of driving home to relax over dinner with Betty Anne and the children. I knew that the phone would ring within a minute of anything serious happening to Gemini V. The phone didn't ring and I got a good night's sleep.

John Hodge had good news for me in the morning. The fuel cell in St. Louis was working as efficiently at 70 psi as it did at normal pressure. He and his control team had put together a tentative flight plan for the day. I looked it over and agreed. My team moved in and we were back in business.

The batteries in the rendezvous pod were long dead, but Hodge came up with an alternate plan. He called it a "phantom rendezvous." We'd create an imaginary Agena target vehicle in a known orbit above Gemini. Then we'd develop maneuvers to catch it during the next four revs. Gordo Cooper and Pete Conrad thought it was a great idea. So far, they'd been sitting in a poorly lighted, slightly cold spacecraft with little to do. It wasn't exactly boring—they were weightless in space and had a great view of Earth—but they hadn't come this far just to be passive passengers.

We transmitted the phantom Agena's orbital data up to the Gemini V computer. While the astronauts worked out the maneuvering sequence with their equipment, we did the same thing on the ground with our much more powerful computers. Our solutions matched almost perfectly. With Conrad reading numbers and handling the onboard computer, Cooper began a series of OAMS maneuvers. Gradually over the next six hours, Gemini V climbed higher and did all the right things. Without the rendezvous pod, we couldn't use Gemini's radar to get accurate readings on distance or closing rates, but our ground radars tracked them all the way.

At the right time and in the right place, Gemini V arrived within a few feet of that phantom Agena. The theory all worked. In the next day, we activated a radar beacon at Cape Canaveral to give Gemini's onboard tracking a test. It locked on perfectly as they passed over the Cape, giving them both their altitude above the ground and a measure of their velocity. Rendezvous was ours. All that was left was docking with a real Agena. Wally Schirra and Tom Stafford were already training for that on the Gemini VI mission.

The flight plan we'd sweated over in such detail for Gemini V was in the trash can. We'd entered the era of real-time flight planning, with the overnight control teams setting up each new schedule and my daytime team carrying it out with the crew. They had seventeen scientific experiments, and one by one we were getting them done. Then more things went wrong.

The fuel cells were working almost too well, generating electricity and water faster than the crew could use it. The water tank was filling too fast, threatening to flood the cells. This was a problem we hadn't predicted. The tank needed a discharge port to prevent overfilling, and McDonnell's engineers promptly went to work adding one for future missions. But now, all we could do was urge the crew to drink more water. When that didn't work, we had them filling bags with water and storing them around the cockpit. And when that still wasn't enough after five days in space, we were back to square one, curtailing activities and turning off systems to keep the fuel cells from producing so much water.

Each morning, McDonnell's John Yardley arrived at my console with a different way of calculating water production. It was getting confusing. Finally I gave it to him straight.

"John, you're trying to change the chemical formula for water," I said. "I just don't think it's working."

He could only shrug and promise that Gemini VI and beyond would have a relief valve so that extra water could be dumped overboard. Meanwhile, our technicians in the back room kept coming up with innovative ways to control the water.

While we struggled with fuel cells that were too good for our own good, the OAMS thrusters started to fail. One after another, they went out until we finally ordered Cooper and Conrad not only to stop working on experiments, but to simply drift around without thrusters for a few days. I thought seriously about bringing them down early. But there was nothing life-threatening about the spacecraft condition, and a major goal was to get eight days of crew exposure to zero gravity. If they were bored for a few days, that was tough.

On the ground, we already had McDonnell digging into the thruster problems. Something had gone wrong with the OAMS on every mission, and it seemed to get worse as the missions lengthened. Our chief suspects were contaminated fuel or some fuel residue that was fouling the thruster valves. McDonnell had to track it down and find a way to fix it.

Cooper and Conrad were game to go all the way. On the eighth day, they finally got back into action. Gemini carried a separate reentry attitude control system and it activated trouble-free. A hurricane brewing up in the Atlantic forced us to change the splashdown point by several hundred miles and come down one rev early. The reentry was almost the way we wanted it, with the exception of getting the right lift out of the spacecraft. We still didn't have it calculated quite right, and no matter what Cooper did, or how much he tried to follow the computer-generated instructions, all our radars said that Gemini V was coming down well short of the aiming point.

Helicopters and frogmen got to them on the water quickly, and I was about to light up my cigar when some sheepish looking engineers who'd been supporting my flight dynamics people walked into the control center. They owned up to miscalculating a number they called Theta Dot. In simple terms, they forgot that the Earth keeps spinning under the spacecraft after retrofire. I remembered the Broadway show *Stop the World, I Want to Get Off.* These guys had theoretically stopped the world so that Cooper and Conrad could get on. They got on, all right. They were just a couple of hundred miles from where we wanted them.

"I assume that this won't happen again?" I asked.

"No, Flight, it sure won't."

"You know you're going to hear about this at the splashdown parties?"

"Yes, sir."

"Get out of here."

They heard plenty of unmerciful ribbing at the parties that night. And the mistake had to be mentioned at the postflight press conference and in our reports. Theta Dot went into the arcane lingo we used, and for the rest of time, nobody has ever again miscalculated it.

The press, the president, and even some of NASA's Washington people went overboard in analyzing and praising Gemini V. True, we now held the record for long-duration space flight, and the crew came through in pretty good shape. NASA encouraged the notion that we were well along in qualifying man for the eight-day lunar round-trip, and President Johnson sent the crew on a six-country foreign tour to make international political capital.

Though I didn't say anything in public, I kept thinking about the day-after-day decisions we'd made, the real-time flight planning, and the hardware that forced us to curtail so many activities. We still had rendezvous and docking missions ahead, and a fourteen-day flight that now had me plenty worried.

"We're making it look too easy," I told my deputy Sig Sjoberg. "I hope we don't end up paying a price someday for leaving a false impression."

Gemini V forced me into a period of reflection. I was in a privileged position of great responsibility. All of us in the manned space flight program had been handed an assignment. We understood the difficulties, and we marveled then and now that we were allowed to solve the problems with only minimal intrusion. Was there something different about leadership in those days? I think there was.

But did it come from the world situation, the Cold War? Or because almost nobody understood the complexity and risks involved in space flight and thus simply let us do our thing? I'd stand in my backyard at night and wonder if space flight was too much for most people to understand, and by not understanding, they couldn't criticize us? Or was the Camelot of the sixties a final resurgence of frontiersmanship that made America ready to accept one final big challenge?

Or was this the final challenge? I certainly hoped not. After the moon, on to Mars.

The more I thought, the more I appreciated our position in time and space. Whatever the factors that came together to create Mercury, Gemini, and Apollo, they were complex and probably beyond rational analysis. I finally decided it was a great time to be an aerospace engineer and to be allowed to devote my life to being part of what would surely be known as mankind's greatest adventure.

I was lucky enough to remember my family life, too. Between missions, we took trips to the West, to the Colorado Rockies, the Tetons, and Yellowstone. Sometimes we made it as far as my favorite American city, San Francisco. And once a year, I rushed East for a few days of golf at the Cascades Golf Course in Virginia.

At home I was a lay reader in our Episcopal church. Betty Anne was in the Altar Guild and the children were in Sunday school. It was a small parish, sometimes without a priest, and we all pitched in. I tried teaching a Bible class, but I lacked the fundamentalist verve and drove people away when I tried too hard to relate the early church to more modern interpretations. It was hard not to be modern when I spent my working days sending men into space.

Gemini VI and VII shattered our concerns that we were making space flight look too easy. Gemini VI was Wally Schirra's second trip into space, with Tom Stafford in the right-hand seat. I had complete confidence in Wally's piloting skills, and when he told the press that the first nation to do an actual rendezvous and docking would be ahead in the space race, I had to agree.

We'd heard nothing from the Russians since March 1965. Now we were into late October and had an Atlas-Agena on one pad at the Cape and Gemini/Titan VI on another. Schirra's big disappointment was that we decided not to let him fire up the Agena's big engine after docking and then power-climb to a world-record altitude. The Agena was widely used in Air Force programs, including several highly classified spy missions. But we'd modified it for Gemini, and before we trusted a combined Gemini-Agena maneuver, we wanted to see how it worked on its own.

Thus the mission involved launching the Atlas-Agena, followed about ninety minutes later with Gemini. Schirra and Stafford would catch the target, rendezvous, and inspect it, then plug Gemini's nose into the Agena docking collar. After a few revs together, Schirra would back away, and the Agena would be fired by commands from the ground. If everything worked, astronauts on future missions would get the ride that Schirra and Stafford were being denied.

Nothing worked. We were up all night conducting dual countdowns. The Atlas-Agena launched at 9:00:04 A.M. on October 25, 1965. The Atlas did its job, but after the Agena separated, everything went to hell. Agena's big engine didn't fire. Then all radio signals ceased. The range safety officer at the Cape was in my ear reporting that his radar was skin-tracking five separate pieces of Agena. It had either disintegrated or exploded.

A few miles away, Wally Schirra and Tom Stafford heard me scrub their launch. Instead of soaring toward space, they quickly found themselves on the down elevator to the ground.

The Air Force threw manpower into an immediate Agena investigation. On our side, we started looking at options. John Yardley and his boss, Walter Burke, at McDonnell pushed two ideas. First they resurrected an old proposal to build a smaller target vehicle for Gemini. It would have the radars and a docking collar, but no big rocket engine. Their backup was round-filed in an earlier version. Now we gave them a contract to build it. It would take four or five months.

Then they brought forward another idea. "Why not launch two Geminis?" Yardley asked. "Have one rendezvous with the other. You won't get docking, but you'll get rendezvous practice and you'll fill the void in the schedule."

Their mistake was taking the idea to NASA headquarters first. George Mueller and his staff were not enthusiastic. But Yardley and Burke wouldn't give up. They flew to Houston, got a meeting with Bob Gilruth and George Low, and made the pitch again. They saw Gilruth's eyes light up and Low nodding wisely.

"It's an operations problem," Gilruth told them. "We'll get Kraft in on it and see what he says." It was early November 1965.

I said that I liked it, it'll be damned hard to do, and I'll put some people on it to see if we can make it happen. Sig Sjoberg organized an immediate task force and we laid out the problems.

First, we only had one Titan launch pad configured for Gemini operations. It wasn't possible to configure another pad, both for time and money reasons.

Second, we'd have to launch one Gemini, repair the pad, erect a second Titan/Gemini combination, check them out, then launch before the first Gemini ran out of consumables. Gemini VII was configured for a fourteen-day mission. Could we pull it off and launch Gemini VI within ten or twelve days after VII went up? The answer was discouraging: probably not.

Is there another option? It turned out that there was. We'd written a hip-pocket hurricane-contingency plan, but never tried it. In that plan, we had the Titan-Gemini on the pad, checked it out, then took it down to protect it from an approaching hurricane. Afterward, we'd put it back, do a brief check, then launch it. Could we do this with Gemini VI—put it up, check it out, take it down, put up Gemini VII, launch it, then bring Gemini VI back to the pad? We looked at it carefully and the answer was yes.

That left the question of dealing with two manned and active spacecraft at the same time. We'd have to do this in Apollo, when the command module was in lunar orbit and the lunar module was landing or already on the moon's surface. But neither the control center nor the network were yet configured for dual operations. I put another team of flight controllers onto this problem and they found a way. When the remote sites were upgraded to handle Gemini, the Mercury equipment was left in place. We could use it to receive a limited amount of data from one Gemini spacecraft, but the data would have to be sent manually—meaning mostly by Teletype—back to Houston. And if there was trouble, the systems could be switched back and forth quickly, giving us real-time data on our choice of spacecraft. It wasn't a great solution, but it would do.

Now there were only a few more things to sort out. We rewrote the flight plans to have Schirra and Stafford aboard Gemini VI be the active part of a rendezvous with Gemini VII, carrying Frank Borman and Jim Lovell. For the only time in Gemini, we got rid of those official Roman numerals. The new mission was Gemini 7/6. While we worked on the flight plan, McDonnell would have to install a rendezvous radar beacon and antenna on Gemini VII.

All that was left was the inevitable meetings, but sometimes even a growing bureaucracy can move quickly. My people were hot for the mission, and it didn't take much convincing to bring Chuck Mathews and the Gemini Program Office on board. We set up a meeting with Bob Gilruth and George Low, adding Deke Slayton and the two McDonnell executives to the mix. There were risks; we pointed them out and told Gilruth that we were ready to accept them. He nodded. George Low nodded. Then Gilruth put in a call to George Mueller in Washington. It was only three days since Mueller had given the idea the brush-off, and two since Gilruth told me to figure it out. The ball was rolling quickly, and this time Mueller didn't put on the brakes.

In the next day, James Webb was so elated by the idea of two manned spacecraft flying side by side that he called the White House and briefed

President Johnson. The Air Force bought in quickly, too. Our two-Titan plan could be done. Everywhere we looked, there were bright eyes and enthusiasm. President Johnson put the final seal of approval on Gemini 7/6 by announcing it directly from the White House.

From Gilruth's first call to me until LBJ's announcement, only four days had passed. We were going to fly 7/6 in December 1965.

The last-minute plans fell into place, with a few dumb ideas getting the trash-can treatment. One called for one-upping the Russians by having Lovell and Stafford do space walks, then swap seats. Frank Borman quickly vetoed that one, and I backed him completely. It was too damn risky, considering how little we still knew about extravehicular activity and about long-duration space flight. And after the problems we had getting the Gemini IV hatch closed, I didn't want to face the nightmare of having two ships up there with open doors.

Borman carried his concerns a step further. He and Jim Lovell had to live in that Gemini for two full weeks, and he didn't want anything to screw up its systems. So when McDonnell engineers installed the radar, antenna, and some exterior running lights, Borman was looking over their shoulders. Nothing happened without his on-the-spot approval.

In mid-November, Paul Haney of the Public Affairs Office and I flew to Florida for meetings at the Cape. National Airlines knew us well, and this was one of those times when they upgraded us to first class on a DC-8 flight that stopped in New Orleans. As we climbed out of New Orleans, I saw a flight attendant take a young man to a forward toilet. He looked sickly and carried a small paper bag.

"He's acting funny," she said. "Do you mind if I put him in that seat across from you?"

I didn't and agreed to talk to the boy, to see what his problem was. He seemed to be eighteen or twenty and looked nervous. After a few attempted pleasantries, I asked what was in the bag.

"I've got a gun," he said. He pulled it out and pointed it at me. I didn't know what it was, but it looked like a cannon at that moment. Paul Haney said my face turned as white as my shirt when the boy jumped up and demanded to see the pilot. He wanted to go to Cuba.

The flight attendants were well trained and cool. One of them passed the word to the cockpit, and the pilot immediately locked the door. When the boy discovered that he couldn't get in, he pulled out a second gun and went crazy. He fired both guns into the floor of the lounge in front of me. then he was tossed sideways as the pilot put the plane in a high-g turning descent, heading back to New Orleans.

The boy regained his balance, and my optimistic side convinced me that he was firing blanks because I couldn't see any holes in the carpet. Almost immediately, another passenger—I think he was a little drunk—tried to get the boy's attention by showing him a box filled with old coins. It worked. The kid started looking at old quarters and half-dollars, and we all jumped him, me on the bottom with Haney and the other passenger piling on. In a moment, we had the boy subdued and disarmed.

FBI agents boarded the plane back at New Orleans and took a frightened young man into custody. The interviews with all of us took another hour, and while we talked, I watched a mechanic pull up the carpet. At least ten big bullet holes were in the aluminum floor. The kid wasn't firing blanks. Luck was with all of us because the bullets missed some critical hydraulic and electrical lines under the floor.

By now, it was all over the national news. This was one of the first attempted hijackings of an American plane. Back home, Sig Sjoberg managed to get Betty Anne on the phone and explain things before she saw it on television. By then, Paul Haney and I were on another flight to the Cape.

I got a letter from the boy's father later, thanking me for my actions and for not hurting his son. The boy was disturbed and was charged in federal court. I never did hear anything more about him.

More was riding on Gemini 7/6 than I'd let on to the press. This was going to be my last Gemini mission as Flight. Sig convinced me that I needed to spend more time managing the directorate, helping out on Apollo, and getting ready to be Flight one more time on the first manned Apollo mission in 1967. When we'd moved to Houston, Apollo was a foreign country to eighty percent of my people. I'd assigned a few people, particularly John Mayer and Bill Tindall, to do some basic trajectory work for Apollo and to fit our lessons learned in with Apollo's design. The reports I got back said they weren't being listened to by the guys working on the moon program.

Still, I was not anxious to make the change. I was giving up the best job I'd ever had in my life. But I looked forward to 7/6 all the same, partly because of the camaraderie for the crews and partly because I had a gut feel that the rendezvous mission combined with fourteen days of two guys cramped into a Gemini spacecraft was not going to be easy on any of us.

Gemini 7 launched December 4, 1965. Borman and Lovell did their rendezvous practice, catching and flying alongside their Titan second stage. They used a little too much propellant, but we were happy. Then the fuel cells started acting up. This was brand-new technology, and we obviously hadn't fixed all the problems that we'd seen during Gemini V. Pressures dropped, we took some corrective actions, pressures stabilized. But the pattern kept repeating itself, and after a few days Borman and Lovell were getting fed up with playing nursemaid to the balky cells and with the alarm bell and flashing lights that kept going off. It was interfering with their sleep and I fully understood their irritation. We told them to turn off the alarm at night and tape over the lights. It helped some. But Borman was a worrywart and continued to express his concern. He was only partly mollified when we told him that McDonnell engineers had a set of fuel cells running in St. Louis and could duplicate his problems while looking for answers. As it turned out, we needed those engineers badly before the mission ended.

Their other irritation was with George Mueller, or at least an order he'd issued. Fourteen days is a long time to be cramped in a space smaller than the front seat of a Volkswagen, and wearing a space suit, too. We'd designed lightweight suits for the mission and originally intended to let them strip down to nylon flight jumpers after a few days.

Mueller vetoed the plan, insisting that we needed data on how grown men reacted to being suited up for two weeks. We'd gotten him to back off a bit and allow one or the other to go suitless, but not both. Borman agreed to stay suited, but with each passing day, he got more uncomfortable. Space suits are bulky and stiff. It takes effort to bend an elbow or a knee, and when something itches, scratching is a real chore. After a few days in the same underwear, I had to assume that scratching was on Borman's mind a lot.

Despite our agreement to let the press hear all of the air-to-ground conversations, we kept a contingency in our pocket. When the discussions called for a medical conversation between the crew and the flight surgeon, we ordered a "UHF-6" test. That was our code for the Public Affairs Office to go off-line and let us talk privately with the astronauts. Borman's protests during these nightly UHF-6 discussions were getting vehement. He really hated being stuck in that space suit.

I agreed with Borman and put in a call to Mueller in Washington.

"I'm going to let them switch off," I told him. "Borman takes his suit off and Lovell puts his on."

"No," Mueller said. "That ruins the test."

"George, you're a brilliant man," I snapped, "but there's one word in the dictionary that you obviously don't understand."

"What's that?"

"Compassion, George. Compassion."

Mueller sputtered and muttered, then snapped back, "I'm sending Davey Jones down there tomorrow. He'll tell me what the hell's going on."

General David Jones was a Mueller spy. He'd flown with Doolittle's Raiders, bombed Tokyo, and bailed out over Korea. He knew firsthand what discomfort was. He arrived in mission control the next day, more than a little embarrassed by his assignment. We showed him the suits and let him feel how much effort it took to move arms and legs. Then he talked to Deke and some others, including the flight surgeons. Within thirty minutes, he was on the phone to Mueller.

"You gotta let 'em get out of those suits, George. Both of them."

That took the fight out of Mueller.

I passed the word to the capcom. "Tell 'em suits-off is a go." There were two happy astronauts in space that day.

About that time, a few reporters figured out what UHF-6 meant and we caught hell at the next change-of-shift press conference. That damned phrase *the public's right to know* became the issue. Deke and the doctors and I felt otherwise, we argued our case, and we ended up with a compromise. In all future UHF-6 conversations, a public affairs man would take notes and write a summary for the press. This cumbersome solution made no one happy. On later flights, we just said what had to be said out in the open and let the press have it all. It was one more penalty we all paid for living in this fishbowl environment.

The second control room in Houston was staffed with people getting ready for us to start the "6" half of Gemini 7/6. We were about to show the world that none of this was so easy after all. Our plan called for an M=4 rendezvous; Schirra and Stafford had a two-minute launch window to get off the ground, then they'd maneuver during the next four revs, about six hours, to catch up with Borman and Lovell.

Despite its fuel cell and OAMS problems, Gemini 7 was doing okay, and the crew was finally semicomfortable in their nylon jumpsuits. I let the countdown on Gemini 6 get to T-4 hours, then ordered our second control room to take over passive management of Gemini 7 for a few hours. It was before dawn on December 12, and we had a near-perfect countdown going. I could hear Jack King, the public affairs officer at the Cape, counting the final seconds, and instead of a tiny black-and-white monitor on my console, I had a full-color screen, and we had a rearview projection-television screen up front, next to the world map.

". . . three, two, one, ignition. We have ignition and . . ."

That was as far as he got. The customary belch of gray and red smoke from the Titan's propellant combination of unsymmetrical dimethyl hydrazine and nitrogen tetroxide billowed up from the Titan's base. Then it started to blow away and the rocket hadn't budged an inch.

I sucked in a lungful of air and felt my pulse pound. We were a millisecond from disaster, maybe from seeing two astronauts killed on live television. An abort command from the automatic malfunction and detection system had shut down the rocket engines. But the propellant tanks were still pressurized. That Titan II sitting out there could explode before I could blink my eyes. A mission rule that I'd okayed myself required—required!—the astronauts to pull the D ring between their legs and use those damned dangerous ejection seats to get the hell out of there. But if I saw the ejection seats blasting away from the pad, I was afraid that I'd be seeing badly injured astronauts in a hospital later. I held my breath while nothing happened.

I heard Wally Schirra reporting a negative on a liftoff signal. The test director in the Titan blockhouse confirmed it, then gave us our first good news: Pressures in the Titan propellant tanks were decreasing. An explosion was less likely by the second. Maybe twenty seconds had passed and I realized that I was still holding my breath. I forced myself to exhale. Two thoughts came quickly: *Why the hell didn't they eject?* and *Thank God they didn't eject.*

We'd just seen the most dangerous moments so far in the short history of manned space flight. Schirra told us later that despite the mission rules, his own rule was to keep his hands off that D ring and do nothing. He knew just how deadly those ejection seats could be. This time nothing was exactly the right thing to do. If he and Stafford ejected, the mission would have ended right there, and both of them might have suffered life-threatening or at least career-ending injuries. Schirra's experience and intuition saved everything. It would have been easy to chew him out for disobeying a mission rule. But we didn't hire those guys because they always followed the rules. By training and by personal inclination, test pilots react to events, and sometimes that means ignoring the book. So I did something else the next time I saw Wally. I shook his hand and said thanks. It's hard to argue with success.

The pad crews went into emergency mode, and within thirty minutes the rocket and spacecraft were safed and the astronauts were in a van driving away. Now two more questions dominated my mind: *What the hell happened?* and *Can we salvage this mission?*

We got the answers in a matter of hours when Air Force and Martin Company technicians removed the Titan's engine turbines. They found a plastic dust cover blocking a line. It was hard to see and a technician hadn't removed it when the turbine was installed at the Cape. The engine experts started cleaning up the turbine and told us that we could try another launch in two days. I was amazed, but Schirra and Stafford were ready to go. They'd lost a rendezvous chance when their Agena blew up in October. Now they'd survived the most frightening on-the-pad experience we'd ever had. And still, they were going to get that "third one's a charm" opportunity to make history.

It was a charm. I listened to the countdown again on December 15, then watched Gemini 6 lift off only one-half second late. The astronauts' only complaint was a cloud of flame from second-stage ignition that engulfed the spacecraft and left a smudgy residue on their windows. Gemini 7 was 1,227 miles ahead and above and waiting to be caught. Borman and Lovell had now been in space for eleven days, and even without their space suits, they were getting a little cranky and depressed. I didn't blame them, and I was glad to hear a new upbeat tone in their voices as they got ready for visitors.

Down the hall from my console, the support staff to FIDO (my flight dynamics officer) was feeding the latest radar data into our big IBM computers. Those complex programs worked out by Bill Tindall and his trajectory experts were finally getting a real workout. They were ready to calculate maneuvers for Schirra and Stafford for almost any set of conditions that came up. The first maneuvers came up on our consoles.

"They look good, Flight," FIDO told me.

I nodded. "Okay. Capcom, read 'em up to the crew."

Within minutes, Tom Stafford had copied down a full set of instructions—where to point the spacecraft, when to fire which OAMS thrusters and for how long, and what to expect in velocity changes. Wally Schirra was already taking action. After each maneuver, we got fresh radar data from the ground, ran it through the computers, and called up a new set of orders. Tom Stafford wasn't just a stenographer. He double-checked our instructions against his own calculations, using "carpet plot" charts that he carried along. If he'd come up with something different, we would have heard about it immediately. He didn't, and the distance between Geminis 7 and 6 closed rapidly.

We fed them four maneuvers over the next several hours. Schirra's precision in flying Gemini 6 was just what I expected. He'd been masterful in flying Mercury. Now he was outdoing himself, and my respect for him continued to grow. His trickiest maneuver was an out-of-plane adjustment, moving Gemini sideways in orbit. The two ships had been on slightly different tracks around the world. Schirra's steady hand brought Gemini 6 just a little to the south to take care of that.

At 270 miles, the Gemini 6 radar locked onto the Gemini 7 transponder. Now it was a matter of letting the onboard computer do most of the calculations and flying the needles almost like making an instrument approach in an airplane. The maneuvers were different, but the concept was the same. Gemini 7 was still above and ahead. Schirra fired his thrusters again, straight forward, then waited. They were on the dark side of the world when Schirra saw a bright star. He momentarily mistook it for Sirius, then told us that he had Gemini 7 in sight. They were about sixty-two miles apart.

After that, it was easy. Schirra followed his rendezvous radar and computer guides, blipping the thrusters, waiting, blipping again. As he broke out into sunlight and got his first great look at Gemini 7, he was only about 130 feet away. Schirra fired his thrusters again, came to a relative stop, and four astronauts in two spacecraft stared at each other in what had to be a mighty moment of awe.

It was awesome in mission control. This was an event I'd been working toward and waiting for. Rendezvous was ours. Wally Schirra was certainly correct—America could now officially claim to be ahead of the Russians in the race to the moon. They hadn't shown the world a spacecraft that could maneuver, and they certainly didn't have the technology or the rendezvous skills to put two spacecraft nose to nose in orbit. I'd never lit a mission cigar until the astronauts were safe on the recovery ship. But now I pulled two of them from my coat pocket and handed one to Bob Gilruth. We both lit up and grinned like a couple of happy kids.

The Geminis stayed together for several revs. Schirra's magic touch as a pilot was proved over and over as he flew around Gemini 7 as casually as if he were on a walk in the park. He proved to everyone's satisfaction that Gemini could be controlled precisely, even putting his ship nose to nose and only inches away from Gemini 7. He and Stafford could clearly see the bearded faces and tired eyes of Borman and Lovell looking back. They had a few tricks to cheer them up. Schirra, Stafford, and Lovell were graduates of the U.S. Naval Academy. Borman went to West Point. Schirra held up a neatly lettered sign that Borman couldn't miss: BEAT ARMY. Even Borman had to chuckle.

Then they backed off about ten miles and everybody called it a day and got some sleep. Schirra and Stafford came down the next morning. But not before Stafford surprised us all by reporting a UFO. That made me sit up and take instant notice. But I laughed out loud when the next transmission from Gemini 6 was Tom Stafford playing "Jingle Bells" on a harmonica while Wally rang some tiny bells. The "UFO" was Santa Claus. I looked at Deke Slayton in the control room and he was enjoying every moment. When I caught his eye, he nodded. "Yes," he was saying, "don't sweat it, Flight. I approved the contraband. Hey, it's almost Christmas."

A few hours later, Schirra finished a perfect mission. He did a lifting and rolling reentry and brought Gemini 6 down about eight miles from the carrier *Wasp*. We all got to marvel at how much space exploration and satellites had changed things in only a few short years. Now some communications satellites were up there in geosynchronous orbit and a television crew was aboard the *Wasp*. We joined a few hundred million people around the world in watching the first splashdown televised live.

We started all this in 1958, I thought, *not even eight years ago. And look at us now.* It was time for another cigar.

The cigar was finally out, but our work wasn't over. Borman and Lovell were weary, more depressed than ever after waving good-bye to Gemini 6, and they still had two days left up there. Their fuel cells were barely limping along. Much of their onboard electronics was shut down. They'd finished most of their scientific and medical experiments. And it was just damned uncomfortable. We had a UHF-6 conversation that night, and Borman's concerns about the fuel cells made it obvious that he wanted to come down. The next morning I called a meeting. We got on the phone with McDonnell's fuel cell experts in St. Louis and I put them on the spot: "Convince me that the cells aren't going to quit in the next forty-eight hours. Either that, or I bring 'em down."

The conversation went on for about twenty minutes. All of the fuel cell people, mine in Houston and McDonnell's in St. Louis, agreed. Even if performance was down, the cells would keep working. I hung up the phone convinced. Now I had to convince Frank Borman.

That evening it was me, Doc Berry, and Deke Slayton in the private conversation. Borman and Lovell sounded a little more chipper.

"Tell me honestly," I asked, "can you keep going?"

I think the directness of the question had an impact. These guys weren't quitters and they damned sure didn't want the world to think they were. Their answers were emphatic: "Yes, we're go." But they wanted to know more about those fuel cells. If they failed suddenly, Frank would be doing a quick reentry to a contingency landing site, and that was not on his list of favorite things at the moment. A long wait for recovery, in their weakened and fatigued state, wouldn't be good.

"Frank, we've gone over it and over it down here," I said. "Those cells are going to hang in there."

"That's the best bedtime story I've had in a long time," Borman answered. "Okay, we're staying."

A day later, we transmitted wake-up music to Gemini 7. It was Dean Martin singing "Going Back to Houston." In a matter of hours, they were packed up, in retro attitude, and on the way down. Borman and Schirra had a bet about who'd land closer to the carrier. Even after fourteen days of what Jim Lovell later called "living in a men's room," Borman's touch was perfect. He did Gemini's second lifting and rolling reentry. We watched again on live television. He landed a mile closer to the carrier than Schirra and won his bet.

We watched them walk a bit shakily and hunched over from the recovery helicopter to the ship's sick bay. I was surprised that they could walk at all after what they'd been through. I lit my last cigar as Gemini Flight and waited. Within an hour, we had the first reports from the medical team. Borman and Lovell were in remarkable physical condition. At first look, they were in better shape than Cooper and Conrad had been after eight days in space. I wasn't sure that the comparison was valid, but I carried it with me to the postflight press conference and reported it to the world.

By year's end, the verdicts from the public and the press were unanimous. America was clearly ahead of the Russians, they said, and was gong to have no trouble getting to the moon. That set me to brooding again. One more time, we'd made it look too easy, and now our critics were our biggest fans.

"It's not going to be that easy," I told myself, "so don't start expecting it."

I made some notes about the troubles we were seeing in Apollo development and set my mind toward a new mission for myself in the next few years—getting more involved and getting ready for that "easy" trip to the moon.

15

Out of Control

It didn't take long to understand why Sig Sjoberg had badgered me to come up for air, turn over my Gemini Flight duties, and start paying more attention to Apollo. The bad vibes from Apollo were everywhere, and Bob Gilruth sided with Sig and started needling me to get more involved with the moon program. It was obviously suffering from the most common bureaucratic disease known to government or industry, the not-invented-here syndrome. My people had been fighting it for a year, and I'd been too busy to notice that they were losing.

Now I saw symptoms of the disease everywhere. The Apollo Program Office wasn't listening to our pleas either over the kinds of information we wanted to see in mission control, or for some basic changes in the two moon spacecraft that would improve chances for success. Even simple changes such as adding brackets to switches and fuses to keep them from being accidentally moved in zero g, or following the software verification procedures we'd developed in Mercury, were ignored.

That stubborn attitude filtered down quickly to North American Aviation, building the command and service module, and Grumman Aircraft Corporation, building the lunar excursion module. We'd learned a lot during the first half of Gemini and would learn a lot more on behalf of Apollo. That was Gemini's whole purpose. But our lessons learned didn't matter.

Some of the North American problems had gone away by the time I got actively involved. To many of North American's top management, Apollo was both a money machine and an excuse to party. I'd gone to a management meeting at North American's facility in Downey, California, in 1964, and the company threw a party on the night we arrived. They put us up in a fancy Hawaii-themed motel that felt like anything but home. It was the biggest damned party I'd ever seen, outside of a splashdown party. The food was piled high on tables. The booze was flowing. The women were flowing. I don't know if some of those women were full-time North American employees or only part-timers hired for the night. I do know that I'd never seen anything like this from the McDonnell people in St. Louis, where the entertainment highlight of a management meeting might be a boxed lunch in a conference room with Mr. Mac himself.

Later that night one of the women knocked on my motel room door. She was a pretty brunette and not too happy when I asked her bluntly, "What the hell is going on here?" She didn't have a good reply, but I gave her credit for knowing when to take no for an answer. By the next morning, I was fuming. Then came the real shock. Apollo management turned out to be dysfunctional in the extreme.

At the time, Charlie Frick was our Apollo manager in Houston. An ex–Air Force colonel named John Paup had the mirror-image job at North American. I'd heard stories about the friction and arguments between Frick and Paup. I wasn't prepared to watch two overage bullies venting in public. This wasn't just friction. It was venomous. I sat behind Bob Gilruth in amazement as Frick and Paup turned a review meeting into a shouting match. One of them didn't like this. The other didn't like that. They were screaming and cursing at each other, each blaming the other for whatever problem was on the table.

I leaned forward and whispered to Gilruth. "This is idiocy, Bob. How can these two guys survive? How can Apollo survive?"

Gilruth's eyes were slits and his cheeks quivered with anger. "They won't survive," he said through gritted teeth. "Count on it."

They didn't. Joe Shea moved from Washington to Houston, taking over the Apollo office after Frick was fired. At North American, Dale Myers replaced Paup.

But Shea brought some new problems and I started running into them almost immediately in the form of new guys. At both NASA and the contractors, key executive and engineering slots were now filled with people who had no manned space flight experience. They didn't have the background and the gut-level experience to understand that sending men into space was not like flying airplanes or building satellites for the Air Force. The operational requirements were infinitely stricter and more detailed. When their machine went to work, it would carry men off to do dangerous things, men who would live or die depending on the skills of the men who stayed behind.

The other problem was the schedule. The new guys fervently believed that landing on the moon "in this decade" was the overriding factor, and that everyone should accept whatever risks the schedule demanded, even if there was a better way to design the machine. Their dedication was honest, but their fervor was misguided. They were not just accepting risks. They were creating risks.

We'd opened the door to some of the problems ourselves when we'd hired the TRW Corporation to work for John Mayer's division on analyzing and calculating the trajectories needed to get to the moon and back. These fed into our mission planning function and into developing software for our mission control computers. TRW was so good at forward-thinking engineering that the Apollo Program Office soon asked for help, too. I didn't recognize that I was letting the camel's nose under the tent when I said yes. Before long, Apollo engineering was an add-on to my mission planning contract. Almost overnight, TRW had four hundred people at work and its local engineering organization was close to a mirror image of the Manned Spacecraft Center.

It wasn't a happy situation. Joe Shea had brought in many of the new guys. One of them was Bill Lee to oversee design and development of the lunar excursion module. I remembered Lee only too well from our encounter years before at Bell Labs when his dismissive approach to designing mission control led us to drop Bell and hire Philco. Lee's understanding of manned spacecraft design was even less.

Then there was the jealousy. When I surfaced from being Flight for Gemini, I discovered that living in the public eye, holding press conferences, and even my being on the cover of *Time* hadn't endeared me or many of my people to the Apollo staff. They reacted with disdain, and if open war didn't break out, it was sometimes close.

TRW was caught in the middle. Companies have their own bureaucracies and politics, and TRW reacted accordingly. On almost any issue, TRW sought out advice and recommendations from my people, from Faget's people, and then from Shea's people. We were pitted against each other, and the TRW position became the one that best served TRW and Shea's Apollo Program Office, not necessarily the position that was in the overall program's best interest. From the earliest days of the Space Task Group, Bob Gilruth had fostered a cohesive management system that worked toward consensus. Now we were bickering and distrustful.

I took my observations to George Low and he had a straightforward response: "Go see Shea." What Low didn't tell me was that a mini-cabal of Shea and George Mueller was causing a rift with Bob Gilruth. The whole thing was fraught with side issues, and on a terrible day still nearly a year away, the whole thing would end with tragedy followed by angry and intense management confrontations.

The animosity between my people and Shea's was so intense that I decided to get to know Shea before confronting him. He was a good athlete, despite his penchant for pink or red socks and terrible puns. Several of us used the astronaut gym after work, and I got him to teach me squash. He was good and I didn't have to let him beat me. We grew a little less tense with each other, went to dinner a few times with our wives, and when I found an opening, I told him what I thought about our troubles at work. He listened.

Now my camel's nose was under his tent. It had taken much of 1966, but Shea and his Apollo engineers, even Bill Lee, began to pay attention. Some of the problems were amazingly basic.

Unlike McDonnell in Mercury and Gemini, the Apollo contractors refused to supply systems information and schematic drawings to my flight controllers. Then they refused to send people to the meetings—admittedly a lot of them—to thrash out those all-important mission rules. It was in their contract, but they wouldn't spend the money or manpower. With no previous manned flight experience, Shea's group didn't understand the importance and hadn't backed us up. Now they did.

Deke Slayton was having the same problem. North American and Grumman weren't giving out the data and drawings needed to train the astronauts. Our growing rapprochement with the Apollo Program Office finally got that done, too.

These were breakthroughs we needed in process and procedure. We also found that our other skills were needed. The entire flight control team, through its Mercury and Gemini experience, was perhaps the best systems engineering organization in the world. When we got into those schematics, we made some frightening discoveries.

One of the most dangerous was that Grumman omitted redundant valves in the LEM's fuel manifold. A single valve was in the line, and if it leaked, corrosive propellant would drip onto the top of the Saturn rocket's third stage. That was the dome of its huge liquid oxygen tank, and the metal wasn't thick. The result could be catastrophic during a launch.

Despite my warming relations with Shea, Bill Lee wouldn't back down. Redesigning the fuel manifold was expensive and time-consuming. We kept after him. I was on the verge of demanding that Shea step in personally when other valves in other parts of the LEM started leaking. Lee finally understood. He ordered the redesign and listened more carefully when my people pointed out other flaws. There were a lot of them.

I missed being Flight, but this was more important. I was appointed to positions on two of the most important review boards at North American. One made sure that improvements and changes ordered by our Manned Spacecraft Center engineers were actually done. The other board had final authority on accepting the command and service modules for shipment to the Cape. Earlier boards had approved the first shipments in 1965 and early 1966, for unmanned test flights. When the spacecraft arrived in Florida, they were not only unfinished, but NASA engineers complained about shoddy workmanship. Exposed wires lay on open trays in the command module or ran in bundles across the bulkheads. Electronics boxes and systems were changed so often that wires were chaffed, scarred, or even bared.

We fought for changes and got them. But it was slow going. Covers on wire trays? That meant added weight, more time consumed, more money spent. We finally divided Apollo work at North American into Block I and Block II spacecraft. The early manned missions would still be Block I, and the first of those wasn't very good when we shipped it across country from California to Florida.

A few weeks before Gemini VIII, we lost two astronauts. Elliott See and Charlie Bassett were the prime crew for Gemini IX. They flew into St. Louis in bad weather in their T-38 two-seater jet. See executed a missed-approach maneuver in bad visibility, grazed the building at McDonnell Aircraft where their spacecraft was near completion, and crashed. Elliott had done a number of shifts in mission control as Capcom, and I liked his dedication and his engineering skills. Charlie sat in at times and I was just getting to know him. Now both of them were gone.

Gemini went on without me standing at my old familiar flight console in mission control. John Hodge and Gene Kranz had their control teams in place for the 1966 missions, working twelve-hour shifts, and I assigned Glynn Lunney and Cliff Charlesworth to form new teams of controllers. We'd work them into the schedule as Gemini reached its climax and thus have four teams ready for the Apollo transition.

They got a workout. We had five missions on the schedule for 1966 and every one turned into an adventure. I found myself sitting at my console with Bob Gilruth and Chuck Mathews in the mission control back row, sometimes gritting my teeth to keep from interfering. It wasn't that Hodge or Kranz did anything wrong. I'd trained them and they were good. It was simply that mission control was my home, and when things got exciting on one mission after another, I wanted to be there at the flight director console giving the orders.

That feeling of wanting to rush in and take charge first hit on Gemini VIII. Neil Armstrong and Dave Scott were going up to rendezvous and—at last—dock with an Agena target stage. We'd lost the Agena on Gemini VI, putting the Air Force under pressure to get it fixed and get it right. At the same time, we'd given McDonnell a contract to build a smaller vehicle they called the Augmented Target Docking Adapter, or ATDA. Either the Agena or the ATDA could be used for docking, but only the Agena could be fired up to push Gemini into a higher orbit.

We decided to accept the Agena only a week before Gemini VIII's launch on March 16. Armstrong and Scott were trained to a fine edge for this first docking in space. But we all had qualms about the Agena. Over and over, one thought was drilled into the astronauts: If something goes wrong, get the hell out of there. Fast.

This time Agena made it to orbit. It was in almost perfect position for Gemini VIII's launch ninety minutes later. Rendezvous was no longer a mystery. Wally Schirra's precision in bringing the Gemini 7/6 combination nose to nose had us feeling confident. Neil Armstrong and Dave Scott did just what we expected. A little more than six hours after launch, with help from their own rendezvous radar and onboard computer, plus radar and computer help from the ground, they coasted up to the Agena and took a look. It was a beautiful sight that we all shared later in the photos they brought home. I leaned back and chatted with Gilruth while John Hodge gave them the word we'd all been waiting to hear: "Go for docking."

Armstrong's touch was as fine as any astronaut's. He blipped Gemini's thrusters, and at a speed of 0.75 feet per second—barely more than a half mile per hour—nudged the spacecraft nose into the Agena docking adapter. The two vehicles locked together and we'd done it. Rendezvous and docking both were American firsts in space. More important, we'd just proved that a maneuver vital to Apollo wasn't all that difficult. With a number of revs coming up where the astronauts were to be mostly out of contact with our ground stations, John Hodge had Jim Lovell on the capcom console pass up a message. It was another reminder about the Agena's touchy behavior. If they had trouble, Lovell said, back away. They'd no more than passed out of radio contact when all hell broke loose.

We didn't know it, of course. Gemini VIII was on a long Indian Ocean and trans-Pacific/Atlantic pass with no ground stations to keep track of things. Not until they came within range of our Tananarive station off the east coast of Africa was the alarm sounded. "We have a serious problem here," Scott reported, and tried to give us some quick details. It wasn't much and his voice showed stress. They'd had a control failure while docked. Thrusters on both Gemini and Agena were firing, and Neil Armstrong hadn't been able to get things stabilized.

I heard Armstrong's voice, amazingly calm, tell us that Gemini was rolling to the left and he couldn't get the thrusters turned off.

Both Armstrong and Scott had that reminder from Jim Lovell that included the warning to get away if something happened. The natural assumption was that any problem had to be out there in front, with the Agena. So Armstrong undocked. That only made things worse. Scott's reports were clipped, but concise. As soon as they undocked, the Gemini spacecraft began a wobbling spin—it was twisting in both the yaw and roll attitudes—and that spin was getting faster. Then they went over the horizon and we lost contact again.

Gilruth and I huddled in conversation in the back row of consoles while Mathews took action as Gemini program manager and went off to talk with some of his technical experts. "It's not the Agena," Gilruth said, and I agreed. We watched the clock as Gemini VIII approached our tracking ship, the *Coastal Sentry Quebec*, stationed in the Indian Ocean, and speculated on what was going on up there.

Then we heard Dave Scott again, his voice relayed through the CSQ. Their wobbling spin was twisting them so fast that his words weren't completely clear. Our CSQ controllers reported that the spin was making it impossible to receive any telemetry information. What little we knew came from Scott, and only later did we learn that he and Armstrong were being tossed around, had banged heads a time or two, and were beginning to suffer from grayed-out vision. It was clearly a life-threatening emergency in space, the worst we'd ever encountered.

Gilruth's hands were on his knees and his knuckles were white as we tried to catch Scott's words. They'd turned off the OAMS, the main thruster system, and had activated one of the two thruster rings in the reentry control system (RCS). It sounded as if Armstrong still couldn't get the hand controller to work as they passed again out of radio range.

"Mission rule, Bob," I said. "If they activate an RCS ring, there's no backup. They have to come down."

"I know," Gilruth answered. "The question is where. And that assumes that they get the damned thing under control."

The flight plan included an ambitious space walk for Dave Scott. Now that was gone. Ed White's short EVA on Gemini IV had been our only walk, and though he made it look easy, we knew that we had a lot to learn. But that would have to wait for Gemini IX.

We speculated on the hand controller. Was it sending false commands to the thrusters? Had it failed completely? Chuck Mathews was in a back room with his engineers. I had a few brief chats with John Hodge and listened more than I talked. It was important that John know he was in charge and that I wasn't looking over his shoulder and taking notes. As the day wore on, Gilruth and I talked with Deke and shared his worry about the crew's safety during those periods when they were out of contact. Chuck Mathews could only tell us that support engineers were looking at possibilities; they had damned little real data to study.

Gilruth must have sensed my frustration and gave me a half smile and a nod. For the first time I began to understand how he'd felt all those years. He'd picked the people for all the key jobs, including me. He'd watched us train and learn. He'd rarely intruded, but he was there when we needed him. Then when we started to fly, he had to sit back and let us do our jobs. This was a guy who'd crammed himself into a single-seat airplane with a test pilot and gone up to see things for himself. It wasn't easy to watch when you wanted to jump up and get involved. He ran the show, but he'd kept his seat through all of our space flights. And now I had to keep mine.

It was almost forty minutes to the next contact, the tracking ship *Rose Knot Victor* in mid-Pacific. I forced myself to sit back. Bob and I talked about the reports, kept our minds busy with analytical guesswork, and watched the Gemini spacecraft symbol on the big tracking screen slowly climbing toward the acquisition circle that had the letters *RKV* in the center. It seemed to take forever. Then we heard the calm, clear tones of Armstrong's voice. They had it under control. They were stable.

John Hodge didn't waste any time. He ordered them down into a contingency recovery area in the far western Pacific. The mission rule was inviolate. Once Armstrong activated that RCS thruster ring—and he had to activate it to regain control of his spacecraft—the only option was to come home, and fast.

Gilruth and I stayed to the end. Bob Thompson briefed us quickly on the reentry procedures. He wasn't worried about their safety on the way down, only about how to get to them quickly once they were on the water. It was a beautiful reentry, right on target. There was no live television because they were far from any of our normal recovery ships, so we heard much of it through the static of Gemini's high-frequency radio. Hodge's team had them pinpointed all the way. Within forty-five minutes of splashdown, a search-and-rescue aircraft flying out of Guam was overhead, and frogmen fully equipped with Gemini gear jumped into the water. It hadn't been pleasant, riding the ocean swells and waiting for rescue. Both of the astronauts were seasick. But they were safe.

Engineers tore into the OAMS system when the spacecraft got home and found a short circuit that made one thruster fire continuously. We learned an important lesson—never put electrical power to any system unless it's supposed to be on. The OAMS was rewired so that a short circuit would always give us a dead thruster, not one that kept firing until a circuit breaker was opened by an astronaut.

As we walked through the chilly March evening across the space center campus and back to our offices, Gilruth was already looking ahead. He'd help Mathews form an investigation team to pinpoint exactly what had gone wrong, then make sure that it could never happen again. In his office, he found a bottle of brandy in a credenza drawer and we shared a reflective moment.

My mission control team had done everything right, I decided. John Hodge hadn't faltered, not even for a second, when hard decisions had to be made. I'd left things in good hands.

"But you know, Bob," I said, "I still wanted to be doing it myself."

Bob Gilruth looked me in the eye and raised his glass. "I do know, Chris. I do know."

16

Blinded in Space

With only four missions left on the Gemini schedule, and less than a year before we planned to put the Grissom crew into orbit on the first Apollo flight, it was time to do some hard thinking. My mission planning people were most comfortable with rendezvous during the fourth revolution, M=4, but we were getting some reasonable arguments from the Apollo side to do it faster. Ultimately they'd like Gemini to demonstrate a quick M=1 rendezvous because that was the target when a lunar excursion module launched off the moon, heading back to the orbiting command and service module.

Both sides had good arguments. John Mayer's people were supporting the software development for the Apollo rendezvous techniques, but were stretched thin when it came to modifying the existing Gemini software. Still, it had to be done. We brokered a compromise for Gemini IX. Tom Stafford and Gene Cernan would do an M=3 rendezvous, and we'd look seriously at trying even shorter times on the final Gemini missions.

That satisfied the Gemini and Apollo program offices, but we still had some Gemini IX fights with the Air Force. Gene Cernan was set for a long space walk, and part of the plan was for him to use a maneuvering device developed for the Air Force "Blue Gemini" program. This Astronaut Maneuvering Unit (AMU) was a jet-powered backpack with its own environmental control system. Little of the internal controversy leaked to the press, but still it was intense. The astronauts, their trainers, and the flight control team thought the AMU was too dangerous. Its hydrogen peroxide thrusters could damage Cernan's space suit. Its system didn't have enough redundancy or safety margins. And while the Air Force wanted Cernan to fly free, nobody at NASA was willing to risk his life out there without a tether.

The solutions didn't make anyone happy. The suit people added some insulation to Cernan's space suit, making it bulkier and more uncomfortable. Just getting from the cockpit to the backside of Gemini where the big AMU was stored meant adding Velcro to Cernan's gloves, and installing both handholds and footholds so he could crawl back to the AMU, then get it on. Nothing on space walks was easy, we were finding. Everything was trial and error. We also insisted on that tether, in the face of Air Force arguments that it could tangle up and put Cernan into even more danger. To counter that, our engineers devised a reel system that would keep tension on the tether. In the end, none of it mattered at all.

We were ready to go on May 17, 1966. Once again, I was in the back row with Bob Gilruth, an uneasy watcher instead of an active participant. We watched the Atlas-Agena combination lift off after a smooth countdown. We watched the plot boards as it climbed along its programmed trajectory. And we watched it turn sharply from its course, head off in the wrong direction, and destroy itself. A spurious signal drove one of the Atlas engines into a "hard-over" position, and we lost it about sixty miles from the beach.

Tom Stafford was famous for his colorful language and unquotable quotes. He'd been strapped into Gemini capsules three times now, waiting for launch, and twice there'd been problems.

"What do you suppose Stafford's saying about now?" Gilruth asked with a wince.

"I don't know," I said, "but you can bet that it isn't 'Oh, shucks.' "

We didn't waste any time figuring out what to do. That Augmented Target Docking Adapter built by McDonnell was ready for us. So was a backup Atlas rocket, which the Air Force had promised and was ready to launch. Convair engineers quickly determined that the previous Atlas went off course because a wire was pinched during assembly, then broke. They checked the backup Atlas wiring and it was good.

Tom Stafford and Gene Cernan were on the pad two weeks later, on

June 1, with their ATDA safely in orbit, when the Stafford Jinx hit again. Three minutes before launch, communications between the ground computers and the onboard Gemini computer failed. The launch was scrubbed for two days while a radio receiver was replaced. Once again, Stafford's comments were not recorded for history.

While that was happening, controllers reported problems with the ATDA. Telemetry seemed to say that its conical shroud hadn't deployed. If that was true, Gemini IX couldn't dock; the docking mechanism was under the shroud. We decided to launch anyway. An M=3 rendezvous was too important to miss. This time it was perfect. Stafford and Cernan pulled up to within about one hundred feet of the ATDA in a perfect three-revolution rendezvous. The two halves of the ATDA shroud were still there, but gaping wide open.

"It looks like an angry alligator," Stafford radioed, and thus Gemini IX got its nickname. The problem turned out to be another human error. The shroud's lanyards had been hooked up backward. Instead of turning loose when explosive bolts fired, they held on. Could Cernan cut them during a space walk? That turned into an intense argument with backup astronaut Buzz Aldrin claiming loudly and not too politely that it was a safe and reasonable plan.

The rest of us disagreed. Bob Gilruth turned back each of Aldrin's arguments by pointing out the dangers to Cernan if the shroud flew off suddenly, and by listing the difficulties Cernan would have in stabilizing himself to do the cutting and in just getting over to the ATDA in the first place. We'd already proved on Gemini VIII that docking wasn't difficult, and we'd have three more Gemini missions to practice it. Aldrin wouldn't budge, but the decision was obvious. It turned out to be the right thing to do. Aldrin never forgot that day. Years later, he was convinced that Gilruth and I faulted him for his aggressive opinions and that affected his future flight assignments. It wasn't true, but I don't think Buzz was ever convinced.

Cernan's space walk turned into a near disaster. Imagine the feeling of being encased in a hard and heavy, little-yielding balloon. It was fine for Ed White, floating around with a handheld maneuvering unit. But now Cernan had to do real work, starting with clawing himself hand over hand with Velcroed gloves to get to the back end of Gemini. His body floated this way and that, each time requiring strong arm and shoulder muscles to pull himself along and maintain some kind of normal position. By the time he made it to the AMU, his heart rate was soaring and he was sweating more than his thermal underwear could absorb. He backed himself into the AMU, lodged his feet into footholds, and tried to catch his breath. Then his helmet visor started to fog over.

He rested awhile, on orders from the capcom and flight surgeon, but it didn't help much. When he went back to work, trying to don the AMU, his face plate fogged over completely and he was effectively blind. The best he could do was try to get himself into the AMU seat by touch and feel. He finally did it and was strapped in. But this experiment was going no further. It would be foolhardy to send him out, tether or no tether, when he was exhausted and mostly blinded by fog. Any work he did would only make things worse. With the concurrence of Stafford, the flight controller team, and some unhappy Air Force people, the exercise was terminated.

By the time Cernan crawled blindly back to the cockpit and got inside, his visor was so foggy that Stafford couldn't see his face. Then they had problems getting the hatch closed. Stafford had to lean over and help before it was down and latched.

I asked myself how we could be so dumb. The only truthful answer is that we blew it. We'd been lulled by Ed White's euphoria during his space walk, and we didn't foresee the heavy workloads needed to move around free-floating in space. Gene Cernan paid the price for our lack of understanding. We not only put him into a precarious situation, but we ultimately deprived him of a space walk that might have become legendary. Cool thinking by both Stafford and Cernan kept things from getting worse than they did.

Gilruth sat with his chin in his hand, glumly assessing what had happened. I chatted with the controllers, then brought him their thinking. "We've got a lot to learn about EVA," I said.

"It's only eight months until we start Apollo," he answered. "We better do some hard thinking about what happens on the rest of Gemini."

None of us were prepared for the external reaction. The press said that Gemini IX was one failure after another, starting with the errant Atlas and ending with the aborted EVA. NASA demanded an investigation to satisfy our critics. We took a strong position, pinpointing the hardware problems, showing how they were fixed, and then taking the time to explain once again what these expeditions into space were all about. And we found a chemical spray to keep condensation from fogging the visor.

The risks are there, we said, and unknowns will always exist. Our job is to recognize those facts and deal with them. The press called them failures on Gemini IX, but we called them learning experiences. Failures today show us how to achieve the routine missions we wanted to see in the future. If flying in space was easy, we would have done it a long time ago. Bob Gilruth put our philosophy into but a few words: "If you can't accept the risk, don't go to the pad."

The responsibility of accepting risk meant that we had to understand our failures and not repeat them. In examining Gemini IX, we noticed something good. All those failures with OAMS thrusters on earlier missions had disappeared. McDonnell engineers had learned along the way and made the right fixes. Gemini's thruster systems worked perfectly, but we were so intent on the rest of what was happening with Stafford and Cernan that we hadn't realized how good the OAMS had become. The fix had been one of those simple things: A sonic cleaning in a liquid bath removed thermal scales caused by welding. The scales had been clogging the thruster filters in space.

With the critics backing off, if not silenced, my mission planning people went to work on the plans for the rest of Gemini, and I turned back toward working on Apollo problems. But I couldn't stay away from mission control when the countdowns started. I was getting used to my role as observer, and when I stopped to think about it, I knew I had one of the best seats in the house. We'd learned our lessons well, we'd keep learning with each mission, and the last three Geminis were a joy to watch.

John Young had command of Gemini X and wouldn't allow corned beef within a mile of his spacecraft. Mike Collins was his partner and their mission plan was ambitious. We went back to an M=4 rendezvous, which went off without a hitch. Young docked easily with the Agena target vehicle, and they set to work. Over three days, they did a pair of EVAs, fired up the Agena's big engine to set a new altitude record of 468 miles, and did a rendezvous with the Gemini VIII Agena left behind by Armstrong and Scott.

The only real problems came on Collins's second space walk. He was tethered to Gemini with an umbilical hose and pulled himself over to the docked Agena. There he removed a micrometeorite measurement package and brought it back to Young. So far, so good. He was slightly tired, but he'd been thoroughly briefed by Gene Cernan about how to move around and conserve energy. The next job wasn't easy. Collins used a handheld maneuvering gun, similar to Ed White's, to carry a replacement science package back to the Agena.

This time the umbilical whipped and snarled in space, wrapping itself around Collins and making life generally difficult. After an exhausting battle with the hose, he and Young aborted the space walk. Getting back to

Gemini was a full-time chore, with Young leaning out to unwrap the umbilical from Collins's leg. Once Collins was inside, they closed the hatch
easily; its mechanisms had been modified and the new design worked as
planned. We'd learned a lot about space walks, it appeared, but not
enough. That left two Gemini missions to get things right.

Pete Conrad and Dick Gordon's job was to satisfy the Apollo rendezvous
requirement. Pete and I had talked often about it, with Pete pointedly saying that he'd do it on Gemini, then wanted the chance to do it again on an
Apollo moon flight. Ninety minutes after launching on September 12,
1966, Pete radioed a message to me from orbit. "Mr. Kraft," he said, all but
laughing in that high-pitched voice of his, "would you believe M=1?"

I believed it because I was in my usual watcher's seat in the back row of
the control center. Conrad and Gordon were docked with their Agena as
they passed over the Cape for the first time. The rendezvous they'd just
done was a duplicate of getting a lunar excursion module and a command
and service module docked in the shortest possible time at the moon. I
went home happy that night. The next day things went to hell again.

This time, Dick Gordon was to get out of the spacecraft, then attach a
long tether between the Gemini and the docked Agena. This "tethered
spacecraft" experiment was important to our understanding of how best to
keep spacecraft flying closely together in orbit. But it was obvious that our
understanding of space walking was still not good. Gordon couldn't keep
himself steady, and attaching the tether was a backbreaking job. The more
he tried to force himself into a stable position, the more his body moved
one way or another. His strength was fading rapidly and his frustration was
growing apace. At one point, I thought he was going to have a panic attack.
He struggled against the one thing that shouldn't have required a struggle—weightlessness. But weightlessness imposed its own rule and wouldn't
cooperate. Gordon was close to incoherent when he finally got the tether
attached, and Conrad ordered him to cancel the rest of his workload and
get back indoors.

"Thank God," I said, and both Gilruth and Mathews beside me agreed.

"We just don't have a handle on EVA yet, do we?" Gilruth asked. It was
a rhetorical question and it would nag us for years to come. Dick Gordon
found himself in a situation that none of us anticipated. The handholds he
needed were not right, and neither was his training. The best we could say
so far was that we were learning a lot of what not to do.

The rest of the mission was a charm. Pete Conrad used the Agena to set another altitude record, taking the docked combination to 851 miles above Earth. The view was terrific. I wished I could be sitting in Dick Gordon's seat and bantering with Pete about the state of the world below.

Back down at a lower altitude, they tried the tether exercise. This meant undocking and backing away from the Agena until the tether was just taut. Then Pete maneuvered to set the whole combination into a giant spin. It took some adjustments, but it worked. Gemini and Agena rotated under control, proving a new way for two spacecraft to stay together in orbit. It was an interesting experiment for the time, but we never needed the technique in any future manned space program. It was used, however, in several science satellite missions, though with mixed results.

The end of Gemini was upon us. With all of the training we could muster, Gemini XII set out to prove once again that man could overcome the nuances of working in space. The mission was planned and replanned almost up to the moment of launch, with Jim Lovell and Buzz Aldrin sitting on the pad as the best-trained astronauts we'd ever flown.

That EVA proof was critical to our Apollo plans. The CSM for the first Apollo flight was at the Cape (and still in rough shape), looking toward a launch in February 1967. Not long afterward, we figured to be flying both the lunar excursion module and the command module in Earth orbit. If something went wrong with docking and two astronauts were stuck in the LEM, the only way home was to do a space walk to the CSM and climb in through the exterior hatch. So far, all we'd really proven on Gemini was that even getting out to the docked Agena was a real chore.

By now, rendezvous was old hat. Less than a year ago, it was a mystery and a major hurdle. Now we'd done it in four, three, and one revolution, plus the extra rendezvous with the Gemini VIII Agena. Docking with the Agena was routine, too. Lovell and Aldrin docked and undocked a number of times, coming at their target at slightly different angles, yet always completing "capture" with perfection. Our confidence in the Apollo system was getting higher all the time.

That left the big question of EVA. We'd learned big lessons from the last few space walks, and now astronauts were training in a big underwater tank, the Neutral Buoyancy Facility, which came as close to simulating what it was like in zero gravity as we could manage. Their space suits were weighted just right and they could spend hours underwater working out procedures and practicing their moves. The concept of underwater training is still used today and is a vital part of training for every astronaut who has to do a space walk from the Shuttle.

My mission planners had worked with the astronauts on the smallest details of Aldrin's space walks. After the earlier Gemini experience, they made Aldrin's first EVA easy. He started with a "stand-up," opening the hatch and just standing there to get used to the feel of his pressurized suit. Then he lifted his camera—moving around even this little bit gave him more time to adjust to things—and took some pictures. Then he leaned out and installed a number of additional handholds, and a handrail that he'd use the next day. Finally he practiced installing a movie camera that would record everything the next time he went out. His stand-up lasted just over two hours, and when it was finished, Aldrin felt refreshed and confident. But the next day was the real test.

This time we got it right. All the evaluations, debriefings, and long hours of training paid off. Aldrin went outside with ease, moved along the handholds and rail, then attached tethers both to himself and to the Gemini-Agena. (Another tether exercise was planned for the next day.) The movie camera he'd mounted caught it all. It didn't look exactly easy, but it was obvious in the films and from the real-time reports that Aldrin was in full control. At one point, he fixed constraints to his space suit, then used several space tools to loosen and tighten bolts, mount clips, and do other construction/assembly work. Without the constraints, Aldrin would simply have rotated himself when he tried to use his space wrench. Instead he was stable and kept up a calm and steady flow of reports on what he was doing.

In mission control, our mood was getting better with every passing moment. After the problems of earlier space walks, this was our last chance on Gemini, and everything was working. He was out there for three hours and he did it all. When Buzz Aldrin climbed back into Gemini and shut the hatch, we sat back and enjoyed a feeling that had been missing on previous space walks—real satisfaction and pride.

The rest of Gemini XII was anticlimactic. When they splashed down and I lit up my cigar, it was hard to believe that only twenty months had passed since Gus Grissom and John Young had taken *Molly Brown* up for three short turns around the world. In that time, we'd flown ten Gemini missions, proven out all the required techniques for rendezvous and docking, flown two spacecraft and four astronauts in space at a time, set endurance records of eight and fourteen days, done spacewalks badly and then perfectly, and proven that the American space program could handle failures such as rockets blowing up and target vehicles malfunctioning, then come back in a matter of weeks and do it again right.

In all that time, the Russians did not fly even a single cosmonaut in space. Their mission a week before Gemini III was their last, and in early 1966, the mysterious head of their space program, Sergei Korolev, had died during botched surgery. Until that announcement, we hadn't even known his name. Now we heard that the Russian program was in disarray and that competition to succeed Korolev was hurting their chance of going to the moon. We also heard that they'd abandoned their moon program in favor of developing a space station. But nobody on our side knew whether that was right or wrong. The Russians didn't announce advance plans or hold press conferences to discuss the future. We only found out what they were doing when they did it. So we had to assume that this space race we were running still focused on putting men on the moon.

Gemini seldom gets proper credit in the eyes of space historians, except for those few who were there to see for themselves. We were so focused on the operational aspects of flying one mission while getting ready for the next that we only saw the larger context when we had time to look back at what we'd done. My view is simple: Gemini bridged the technology gaps that made Apollo possible. Without Gemini, the Kennedy goal of landing a man on the moon and returning him safely to Earth by the end of that marvelous decade would not have been accomplished.

SECTION IV

THE APOLLO MISSIONS

17

"We're on Fire!"

I was back home in mission control, working with my guys as Flight on the first manned Apollo mission. We were a month from launch and doing a full checkout of the spacecraft and its interface with the ground communications on January 27, 1967. It wasn't going well. Gus Grissom, Ed White, and Roger Chaffee were buttoned up in the command module at the Cape, sealed tight in a cabin pressurized to 16.7 psi of pure oxygen.

Eventually I was satisfied with the telemetry we were receiving in Houston, but the voice checks were still poor. Gus was getting frustrated. "How do you expect to communicate with us in orbit if you can't even talk to us on the pad?" he complained, and he was right. Somewhere in those bundles of wires spread across the floor, and more wires running along the cabin walls, something was putting static and garble into the voice communications.

When the Cape engineers called a hold in the test and went looking one more time for the problem, I walked back to my office a few hundred yards away and buried myself in paperwork. The crew stayed buttoned up in their spacecraft. By 4:00 P.M., with no word from mission control and bureaucratic boredom setting in, I went back to my console. Better to wait in mission control, I thought, where at least I could tune in multiple comm loops and listen to the troubleshooting. I had the flight director's loop, the Cape test conductor's loop, and the astronaut voice loop all in my ear when the pad technicians were finally ready to go "plugs out" and remove external power supplies from the command module. After that, the spacecraft would be on its own internal power, and the test could roll on to its conclusion.

Suddenly I heard confused and loud talk in my ear. People were yelling and I thought I heard muffled screams. I stood bolt upright, a cold shiver running up my spine, and I heard someone yell, "We're on fire!" It was Gus.

My God! I thought, and before I could think more I heard a voice yelling, "Get us out of here!" I felt sure that it was Roger. The sounds of confused scuffling and movement echoed in my headset, along with reports of a fire in the "white room"—the clean room surrounding the spacecraft. I started praying, harder than I'd ever prayed in my life.

At the same time, I kept my finger off my mike button. I desperately wanted to ask the test conductor for an update, but I knew that he had his hands full and I'd only be an intrusion. Around me in mission control, there was an awful silence. Every member of my team sat white-faced and rigid, every ear tuned to the terrible sounds coming from the Cape. We were nine hundred miles away and helpless. This was not an emergency that mission control could handle.

The test conductor, George Page, came on the line. His voice was shaky and filled with emotion. There was fire in the spacecraft and the white room was full of flame and smoke. It was clearing and they were trying to get to the spacecraft hatch. I heard his words and I heard despair. He said no more.

I watched two minutes go by on the clock, then asked the worst question I could imagine: "Test Conductor, Flight. What about the crew?"

His answer jolted us to our souls. "Not much hope, Flight. We'll have the hatch open in a minute."

About that long went by before a voice came on the loop. I couldn't tell if it was Page or somebody else: "The crew is dead."

My stomach lurched and I felt sick all over, so weak and drained that I almost collapsed into my chair. Fire and asphyxiation had killed three good men. I reached for the phone to call Bob Gilruth and George Low. They already knew. Both had been listening on squawk boxes in their offices. They'd heard everything.

I'd dealt with flight-test death before, but this was different. Memories flashed through my mind about the responsibility of sending a pilot up in an experimental airplane to perform maneuvers that test the limits. I remembered Herbert Hoover dying in a B-45 crash in 1955 at Langley, and the guilt and grief we all felt. It was only mitigated by the fact that Herb was a professional who knew the dangers of the game, and the limits of his abilities. That time the dangers caught up with him when the B-45's wing structure failed and his escape system wasn't good enough to get him out.

Now we'd put three astronauts into harm's way and made their escape impossible. They were dead and we knew that it was our fault. The fire on the pad and its consequences would be profound.

Deke Slayton and Joe Shea thought about sitting in the forward well of the command module during the test to see the problems for themselves. Gus said they'd just cause more problems and told them to go home to Houston. Shea did, but Deke was sitting in the launch control center with Stu Roosa at the capcom console when the fire broke out. He went to the pad, took a look, and then did his job. He'd set up procedures and immediately sent astronauts and their wives to the Grissom, White, and Chaffee homes. It was rough duty, but everyone knew that the families needed all the support and understanding they could get on this fateful night.

The next morning, Saturday, I drove Bob Gilruth to each of the homes. We could only offer our sympathies and the full support of the Manned Spacecraft Center and NASA. I found the right words hard to come by. It would have been easier to simply hug the widows and kids and let the tears run free. Bob was in the same condition, shaken and filled with sorrow. But our feelings were nothing compared to those of the wives and children who'd just lost husbands and fathers. Somehow we managed to keep our voices steady and to let them all know that they weren't alone. We drove back to the center in complete silence. There was nothing to say until we got to the ninth floor and Bob put it simply: "We've got work to do."

The world reaction was equal to ours. President Johnson, members of Congress, and the press expressed sorrow, followed by grave concern over Apollo's future. The Russians sent their sympathy to the astronaut families, but the Russian press criticized the United States for an overzealous rush to send men to the moon. The thought crossed my mind: *Are they right?* It was a question that we'd face and be forced to answer in the months ahead.

I got a call from Betty Grissom. She wanted Bob Gilruth, Walt Williams, and me to be pallbearers for Gus. He and Roger Chaffee would be buried at Arlington National Cemetery. Ed White would be buried at his beloved West Point. We all flew to Washington by NASA airplane and spent the night before the ceremonies at the Georgetown Inn. We sat up late talking about Gus, Ed, and Roger with the families, and sometime after dinner, Vice President Hubert Humphrey came into the room. Hubert was a friend of the space program and we'd all met him on numerous occasions. Now he sat with us and drank and reminisced about the astronauts. He was a comfort to Betty Grissom and Martha Chaffee, and his concern and caring were so strong that it was 2:00 A.M. before we finally broke up and went to bed. Hubert somehow made things a little better for everybody that night.

I'd been to Arlington Cemetery as a boy to see the Tomb of the Unknown Soldier, and I'd watched the funeral and burial of President Kennedy on television. But that morning, walking behind the caisson carrying Gus Grissom in a casket, was something I never wanted to do again. The cold, gusty, and penetrating day made our task all the more uncomfortable. The eulogies, military regalia, rifle salutes, and the bugler playing Taps are familiar to everyone. But when that caisson carried my friend, and I had this aching feeling of responsibility for putting him there, it made me not only cry, but resolve that it would never happen again.

This determination to make sure these men did not die without cause, I believe, gave us all the strength to continue our job of landing men on the moon. It also brought us all closer together and made our responsibilities crystal clear. For some, it was more than they could bear. The fire on the pad took its toll beyond the deaths of three brave men.

Several things happened. Jim Webb and Bob Seamans had the good sense in NASA headquarters to appoint an accident review board filled with highly skilled engineers and associates who appreciated the problem they faced. And Congress had the good sense to stand down and keep from interfering while the investigation went ahead. Webb and others would keep the appropriate oversight committees fully informed, but there would be no congressional hysteria or politicking allowed.

Floyd Thompson, director of the Langley Research Center, headed the board. He was an inspired choice. We all respected him, and the aerospace industry recognized him as one of the best engineers and administrators in the country. His board members included Frank Borman, representing the astronauts, and Max Faget, whose engineering expertise was unquestioned. Other board members came from industry, other NASA centers, the U.S. Air

Force, and the president's Science Advisory Council. This blue-ribbon group was determined to get the answers, then make the right recommendations.

Nothing inspires introspection more than failure. Nor is anything more clear than twenty-twenty hindsight. I thought about the day a month or so ago when we were running a training simulation and Gus called me into one of the support rooms off our mission control center. He handed me a dime encased in Lucite.

"It's from *Liberty Bell 7*," he said. "I've been saving it for you."

With the small gift, Gus was saying that we were friends and partners. He trusted me to run his mission. And along the way, we'd put Gus and his crew into a pure oxygen environment that surely was a major cause of their deaths. I have that dime where I see it every time I sit at my desk. It's a reminder that we hadn't thought of everything in time to save the guy who carried it into space on Project Mercury.

Why did we use pure oxygen, under high pressure, in that test on the pad? Any spark would ignite an explosive fire. The answer is that we were trying to be perfectly realistic. In space, the cabin would be filled with pure oxygen, but at 5.2 to 5.6 psi, not at 16+ psi. Outside the cabin the pressure was zero—the vacuum of space. The command module was designed to handle a positive pressure differential of about 8 psi from inside to outside, but only a 1 or 2 psi negative differential from outside to inside. If the outside pressure was 2 psi higher than inside the cabin, the pressure hull could implode.

Sitting on the pad, the whole Apollo system was at sea level, about 14.7 psi of normal atmospheric pressure. If we evacuated oxygen from the cabin, to duplicate its pressure in space, it would indeed implode. So for those kinds of pad tests, the cabin was pumped up to exceed sea-level pressure. What we didn't factor in was the danger from fire. We knew it intellectually, but so far, we'd never had a problem. We'd done such tests during Mercury and Gemini without incident. But the Mercury and Gemini hardware was cleaner and more secure than this Apollo spacecraft, sitting there with its exposed wire bundles and trays.

The review board ordered a series of laboratory fire tests. The results were astounding. We'd run combustion tests at 6 psi of pure oxygen as part of the routine for qualifying materials to be used in the spacecraft. Some of them burned, even rapidly, but could be contained at 6 psi. At 16 psi, that same material exploded and gave off a poisonous gas. The ethylene glycol/water solution used in the cooling system burned, too, with a thick, black smoke. The Velcro placed on almost every open surface to help the crew keep things from floating away in zero gravity burned fiercely. Even a solid bar of aluminum burst into flame when ignited in a 16 psi oxygen atmosphere.

To everyone's horror, that high-pressure pad test was a fire waiting to happen. All it needed was an ignition source. There were plenty of candidates. There were frayed and even bare wires after the command module reached the Cape, all the result of shoddy workmanship at North American. Wire bundles on the floor were often stepped on by technicians or the astronauts. Wires in open trays were jostled and bumped during training sessions, potentially leading to frays in insulation. The exact ignition source was never identified. The fire almost certainly started below and to the left of Gus Grissom's couch.

But which wire? We would never know. Except as an intellectual exercise, it didn't matter. Every wire and every bundle would be wrapped and protected before Apollo could fly. Fixing the wire problem was mandatory. Sloppy workmanship was a thing of the past.

That wasn't all. The command module hatch came in for sharp criticism. Ironically, it was a design that flowed from the old explosive Mercury hatch that sank *Freedom 7* and nearly drowned Gus Grissom. The hatch was designed to never, never come off accidently. Once it was locked in place, an astronaut had to grab a latch handle, insert it into an inside slot, then turn the handle repeatedly to unlock it. It was tough work, requiring a lot of strength, and it was Ed White's job to do this. Once the hatch was free, White would have to pull it inward and drop it to the floor. He was desperately trying to do all that when he died.

The conclusion was obvious. The hatch had to be redesigned, both for easier release and so it could be removed quickly from the outside in any emergency.

As the investigation continued, Joe Shea got increasingly melancholy. John Yardley, still working for McDonnell but volunteering his services to NASA, tried unsuccessfully to console him, reassuring him that this wasn't his fault. Shea wasn't buying it, and before long I heard from Max Faget that both Shea and Yardley were getting stupefied drunk every night.

Who can say how much or how little blame Joe Shea deserved? He'd been getting closer and closer to the rest of us in the months before the fire, and he was beginning to do many of the things we wanted. But his blind spot was with Bob Gilruth. Shea would talk to Max and he'd talk to me. But he wasn't keeping Gilruth informed. His formal reports went directly to George Mueller at headquarters, and Mueller wasn't passing them back to Gilruth, either. We were all getting too compartmentalized. Max and I didn't know how much in the dark Shea was keeping Gilruth. And without the kind of information that Shea was withholding, Gilruth had no basis for intervention.

Floyd Thompson was a shrewd observer. I was at the Cape for a meeting when I got a call from his secretary. Could I meet Floyd that night at the Holiday Inn? He was alone in his room when I got there. Straight out, he asked my opinion: "How do Shea and Gilruth get along?"

"Shea's not very helpful to Gilruth," I answered honestly. "He's not keeping him well informed and sends everything to Mueller at headquarters."

Thompson nodded. "That's what I thought." Whatever technical problems we had with Apollo, Thompson had figured out that we had a management problem, too.

Six weeks later, it came to a head. Thompson had a report ready and was taking it around to brief the various NASA centers. We met in one of our big conference rooms at the Manner Spacecraft Center, with General Sam Phillips running the meeting. He introduced Shea and gave him the floor. What happened next shocked us all.

Joe Shea got up and started calmly with a report on the state of the investigation. But within a minute, he was rambling, and in another thirty seconds he was incoherent. I looked at him and saw my father, in the grip of dementia praecox. It was horrifying and fascinating at the same time. *I've seen this before,* I thought, and tried to look away. But I couldn't. After our early problems, Joe had become a friend. Now he was falling apart in front of me. Whatever was happening in Joe's head, it all came out in a jumble of mixed words and meaningless sentences. Sam Phillips listened only a few more moments before he stood up and put his hand on Joe's shoulder.

"Thanks, Joe," he said. "Why don't you sit down while we get on with the agenda."

That was the end of Joe Shea. Within a week, he was transferred back to Washington, and in a matter of months he'd resigned and taken a job at Raytheon. He was out of the NASA loop and never really recovered.

The final review-board report cited a host of technical problems and offered a series of well-conceived recommendations, and we accepted all of them. Congress let its unhappiness with NASA be known, but wisely left it to us to do the right thing. Still, there was a terrible moment for Jim Webb during testimony.

He was asked about reports written by Sam Phillips that criticized North American and its subcontractors for shoddy workmanship and other problems. Webb was taken aback. He knew of no such reports and said they didn't exist.

But they did. They'd gone to Webb's deputy, Robert Seamans, who'd failed to forward copies to Webb. Back at the office, Webb discovered that the reports were extremely critical of North American and even included threats to cut North American's fee and possibly even terminate its contract with NASA. Now Webb was both embarrassed and angry. He immediately informed Congress and took the flak. But his reputation was damaged almost beyond repair. At the same time, Webb lost all confidence in Seamans. Before long, Seamans had resigned.

George Low took over the Apollo Program Office in Houston, replacing Joe Shea. He was perfect for the job, a close confidant of Gilruth's and a highly skilled administrator whom we all respected. His first job was to set up a high-level Configuration Control Board, filled with our senior managers. What Low said to me is what he said to all of us: "You personally are on this board. Not your deputy. You."

Then he told us at the first meeting, "This is our opportunity to fix Apollo. We have the time to redesign those things that have to be redesigned. But we don't want to overdo it. I want another meeting in two days and I want everyone's suggestions, in writing, with justifications."

We put 125 things on that list for the command module. Armored sleeve joints on fluid lines. Fire resistant Velcro. Switch guards. Nonflammable cooling liquid. Flame retardant paper for onboard flight plans and documents. Redesigned hatch for quick opening. And all the rest. Within six months, before the end of 1967, we'd done all 125. Then we did the same thing for the lunar module. It was no longer the lunar excursion module; George Mueller thought that *excursion* sounded too frivolous and ordered the word dropped from our lexicon.

George Low called me in to talk about mission software, particularly for the moon flights. "We can't get the software moving out of MIT," he said. "So I'm giving you responsibility for making it happen."

I sent Bill Tindall to the MIT Instrumentation Laboratory—now called the Draper Laboratory—which was writing our software, to find out what was wrong. The legendary electronics expert Stark Draper himself was running the place and welcomed Bill with open arms. Software was different then. It was encoded onto electronic strips called ropes, and it was being overloaded with instruction sets. Tindal came home with some thoughts, then took me to lunch in our cafeteria. We sat down at a table with astronauts Dave Scott, Rusty Schweickart, and Dick Gordon, and the conversation naturally turned to business. It was an eye-opener. They needed this. They wanted that. The software should do this and must do that, and we can't do without this other function either.

We walked back from lunch convinced that better is the evil of the good. We were trying to get Draper to write software ropes that were all things to all people. It couldn't be done. So I set up my own Configuration Control Board for software.

"They've got one month," I told Tindall. "After that, there will be no software changes unless I personally sign off on it." Tindall's grin as he went off to call Stark Draper told me all I needed to know. Within that month, Draper had an operating program that we declared our benchmark. From then on, I was the bottleneck to approve or disapprove changes.

Suddenly our Apollo software opened like a flower. Teams everywhere had been waiting for it—trainers and simulators at Houston and the Cape needed it to operate accurately, Grumman and North American needed it for their own training and for spacecraft missions.

In a single month, everything changed on the software front. And I didn't do anything beyond making a management decision that worked.

While the command and lunar modules were being brought up to our new standards, George Low called us together to lay out a road map to the moon. He asked a simple question: "What objectives can we attack in a group of flights?"

Within a few days, we had plans labeled A through G. We didn't know if each set of objectives would be one flight or several, but at least we codified the process. An A mission would be unmanned in Earth orbit, testing a command and service module. A G mission would land men on the moon. In between there'd be missions in Earth orbit with the lunar module and missions to the moon without landing.

As all this was happening, management changes came to North American, too. Bill Bergen took over as president and brought in Bastian "Buzz" Hello, a program manager on the Gemini-Titan program, to run North American's operation at the Cape. Then he named John Healy to run the Block II command and service module program. That was now our manned spacecraft, and all of us thought that Bergen was putting the right people in place. Low also sent Frank Borman to North American as his personal alter ego, and for the time being, Borman reported directly and daily to Low. Then later, Low sent one of Wernher von Braun's top deputies, Eberhard Rees, out to California to be NASA's in-house man overseeing the spacecraft assembly and its upgrades. We finally had control of things and were back on track to the moon.

The Saturn V moon rocket was an impressive monster. It stood on the launch pad taller than a thirty-story building, and any man working at its base could be lost from sight as no more than a golf tee in a pine forest. Wernher von Braun built a masterpiece. Its statistics should never be forgotten.

The first stage was thirty-three feet in diameter. It had five rocket engines, each producing 1.5 million pounds of thrust. Those engines would burn 4.5 million pounds of kerosene and liquid oxygen in just 160 seconds.

The second stage was thirty feet in diameter. It was powered by five slightly smaller rocket engines that burned 1 million pounds of hydrogen and oxygen in the next six and a half minutes after staging.

The third stage sat atop a tapered adapter and was almost twenty-two feet across. This was the S-IVB with one J-2 engine, and it had to be fired twice, once to get the spacecraft—CSM and LM—into Earth orbit, then again to power them away from Earth and toward the moon. Its tanks carried 192,495 pounds of liquid oxygen and 39,735 pounds of liquid hydrogen. Put all three stages together and those numbers still boggle the mind.

Above the S-IVB, another tapered adapter hid the lunar module inside. Finally the service module and command module sat on top of it all, with a laticework escape tower mounted on the command module's nose.

All we had to do was make it fly.

I was at my console in Houston mission control when the first Saturn V lifted off November 9, 1967. No one anticipated the true power of that rocket. More than three miles from the launch pad, ceiling tiles fell on television reporters at their anchor desks. The corrugated metal sheets covering the press viewing stands came loose and flapped dangerously. People said that they felt the Saturn's power through the ground and up into their chests before the sound reached their ears. Then all they could do was lean into the pressure of the crackling sound waves and watch that rocket climb.

I never saw a Saturn fly. I never saw any rocket lift off, except on television, in Mercury, Gemini, or Apollo. My place was in mission control, where the action was. Maybe I missed something. But I couldn't be in both places at once. I'd made the right choice and never regretted it.

That flight was to test both the rocket and the spacecraft. It seemed to be perfect. After orbiting Earth twice, the S-IVB fired a second time and sent Apollo nearly ten thousand miles out into space. At its peak, the command and service module turned; the big Service Propulsion System (SPS) engine carried in the service module fired and rammed the whole thing back down through the atmosphere. The thermal conditions would equal the worst case possible during a return from the moon. It could hardly have been better. The command module's guidance system controlled it through some complex entry maneuvers, its heat shield gave it more than enough protection, its parachutes opened, and it splashed down less than ten miles from our aiming point. Barely nine hours after liftoff, the command module was aboard a recovery ship and heading for home.

In the glow of success, we all missed something. There was a resonance problem in the Saturn V, that old pogo effect come back in giant size, that nobody noticed until the rocket flew again.

Along the way, we had to fly a legless lunar module on the smaller Saturn 1B rocket. It was the LM's first time in space, and with no moon to land on, all that mattered was that we test the machine's two stages—the descent stage, which carried the rocket that would power astronauts down to a lunar landing, and the ascent stage, which was their cockpit and their ride back into orbit to rendezvous with the waiting mother ship.

The LM was fragile. It was designed for lunar conditions, where gravity is only one-sixth as strong as it is on Earth. So we used the lightest possible materials in building it. It was lightweight inside, too, with the most delicate of wires that could carry the needed current, pipes so thin that leaking coolants had been a constant problem, and electronics packages stripped to their bare essence. Two astronauts would stand inside to fly it. There was no provision for them to sit or lie down. When they needed to sleep on the moon, they'd curl up on the floor. Otherwise, they stood and looked out triangular windows.

It wasn't a pretty machine, except perhaps to astronauts and aerospace engineers. The angular lower part was the descent stage, with four long legs and a rocket engine pointed down from its middle. Each leg had a dish-shaped pad, big enough for a man to stand in, that would distribute weight on the lunar surface. A ladder ran from the upper stage's hatch down one leg, ending about 18 inches above the pad. Various external compartments in the descent stage carried equipment that the astronauts would use on the moon. After the first landing, we planned to add more and more cargo, most of it scientific. But on that first landing, the stowage compartments would carry only the minimum gear needed.

The upper part of the LM was the ascent stage. It would be the only part that left the moon, the descent stage serving as a launch pad. The crew lived and worked in the ascent stage. Under the floor was the ascent engine. It had to work or they'd be stranded on the moon. Directly over the astronauts' heads was the docking port and circular hatch that opened into the command module. The command module pilot's job included docking with the LM, then once the two craft were sealed tightly together, removing the docking mechanism to clear the tunnel between CSM and LM.

The concept sounded simple. But the engineering needed to make it happen was extraordinary. One failure, in something as minor as one of the latches that held the LM and CSM together, could threaten a mission's success and maybe even the astronauts' lives. Everything had to work.

The first LM flight was naturally unmanned. We sent it up on January 28, 1968, and transmitted all the necessary commands to put it through a harsh exercise in orbit. The descent engine was fired and refired. The ascent engine was fired in an emergency mode to separate from the descent stage and did the maneuver just as we expected. This lunar module was proving itself an able space ship. We fired that ascent engine twice more, running it out of fuel and bringing it down into the Pacific Ocean near Guam. At the end of the day, LM got nothing but high marks. There was a lot to be done before men could fly it. But we knew that the design and the lunar module systems were solid.

That left a second test of the Saturn V. We got it off on April 4, 1968, and it went bad almost from the start. This time the pogo effect was obvious in the real-time data. That big first stage was twitching like a leaf. It staged all right, but the pogo picked up again in the second stage. This time the abort system forced two of the shaking engines to quit. The other three fired under guidance control until the propellant tanks ran dry and the third stage took over. It didn't have the juice to get all the way to the circular orbit we wanted and to put the CSM into an elliptical orbit. We thought we could work around the problem with the service module's SPS engine. But first we needed a second burn on that third-stage J-2. It didn't ignite.

I stood there in mission control, absorbing the shock of seeing nothing happen. Then we put our contingency plan into effect. Flight controllers ordered the CSM to separate from the S-IVB stage, point toward the sky, and fire its SPS engine. It was good, but not perfect. We got the spacecraft

high enough to turn it around and do another lunarlike reentry. The veloc-
ities didn't quite get where we wanted. But we did the maneuvers and
brought it home about fifty miles from the recovery ship. Combined with
the results of the first high-speed reentry a few months earlier, we were
ready to declare the CSM man-rated for a moon flight. We weren't so sure
about that Saturn V.

It took some damn good engineering detective work by von Braun's engineers to figure it out. All three Saturn V stages had troubles. The first
stage solution turned out to be easy. When they fired an F-1 engine on a
test stand, engineers discovered that liquid oxygen was hammering and
surging in the lines. A pressure system that pumped helium at the oxygen
valves quieted things down.

The upper stages used identical J-2 engines. We were all afraid that
some generic problem was hiding inside and could be a bear to fix. But
again, von Braun's team was brilliant. They discovered that frost built up
on the propellant line while the Saturn V sat in Florida humidity on the
pad. When the lines vibrated during launch, the added weight broke one
or more. Stronger supports to keep the lines from vibrating fixed the
problem.

But there was more. One of our own structures engineers found a problem that von Braun's people missed. I got a call to join Bob Gilruth in his
office for a briefing from Don Wade. The I beams holding the center engine on the second stage had deflected as much as eighteen inches during
the last mission and came perilously close to structural failure. If a beam
broke, the entire second stage would fail catastrophically. It would explode.

"We've got the best structures people in NASA here in Houston,"
Gilruth said. "I'm afraid that the Huntsville people won't see just how serious this is."

I'd done my share of structures work at Langley, and it was obvious that
Wade was onto something serious. Gilruth sent us to Huntsville to show
them what we had. Von Braun's people looked at us like we were some alien
life-form. We showed them the data and they didn't understand. At least
one hundred people were watching and listening, and one of them got so
irritated with me that he lapsed into German, berating me, I think, for
being a dummkopf. I let him rant and suddenly the room burst into laughter. Maybe half of the engineers understood what he was saying, and the
rest couldn't hold back at seeing and hearing the one-way torrent of Ger-
manic abuse.

It broke the tension. We all went back to speaking English, and now the room was listening carefully. Our recommendation was simple: Put a helium damping system on the second stage, virtually identical to the first stage fix. That I-beam would be protected and the danger of it breaking apart midway in flight would be gone. We flew back to Houston that night with a commitment from the von Braun team to make all the fixes and then test them on the ground in another controlled rocket firing.

Nobody thought that we'd get away without another unmanned flight of a Saturn V. We were wrong.

18

Suddenly the Moon Comes Up

By midsummer of 1968, we were close to putting men back into space. The A and B missions were behind us, testing the unmanned command and service module, then an unmanned lunar module, in Earth orbit. Even the Saturn V fixes looked so good in ground tests that we were now seriously thinking that the third big moon rocket could carry men.

Wally Schirra, Walt Cunningham, and Donn Eisele were assigned to the newly renumbered Apollo 7 on the first C mission and it looked as if we'd have them up there by late October. But the Schirra who flew such meticulous missions in Mercury and Gemini and was so easy for all of us to work with was strangely missing. This new Schirra was grouchy, angry, and hardheaded.

It was particularly tough to watch his relationship with Frank Borman disintegrate. Borman was George Low's pick to ride herd on the command and service modules, and nothing he did was good enough for Wally. They ended up at each other's throat on almost every issue. We shipped the spacecraft from California to the Cape ahead of schedule and over Wally's objections; he bitched and moaned about everything to the point where some of us thought that he'd crossed the line and was actually afraid of the damned thing.

His fears were misplaced. The CSM was in the best shape of any space-craft ever when it arrived in Florida, and it was going through systems checkouts and a run in the vacuum chamber with flying colors. We couldn't say the same thing for the lunar module, and all thoughts of fly-ing a D mission with all the spacecraft in Earth orbit before the end of 1968 were fading. Grumman was still having trouble with fluid leaks, with elec-trical systems, and with getting the first LM ready in general. At the same time, I was riding herd on Stark Draper's MIT operations to get the first LM software ready for the mission, and the December deadline didn't look possible.

That was when George Low showed his genius. He spent a few days re-viewing lunar module progress at the Cape in late July and came home with bad news showing on his face. He usually hid his feelings. Not this time. I'd never seen him so down.

Then he snapped out of it. A day or so later, he stuck his head in my door. "Chris, I've got an idea about the moon. Do you have time to see Bob with me?"

We walked down the hall to Gilruth's office. Bob was expecting us, and George started by going through a laundry list of LM problems. He didn't see any way that we could have an all-up stack—Saturn V topped with a command and service module and a lunar module—before February or March 1969. There'd been some problems with the big Saturn V in the two unmanned test flights, too, but those seemed to be under control.

Finally he got to the point. "What do you think about flying to the moon without a lunar module? If Apollo 7 does okay in October, can we juggle the schedule to fly a crew to the moon and back in December?"

His idea was a shocker. Gilruth leaned forward, elbows on his desk, and asked for more details. "What kind of lunar mission? Just out and back, a circumlunar trip? "

"Yes, that's what I'm thinking," George Low said. "It would ace the Rus-sians and take a lot of pressure off Apollo. And we have to go there sooner or later anyway. I don't think it makes much sense to fly twice in Earth orbit if we can go around the moon."

Gilruth reached for his phone. "We've got to get Deke in here." The four of us—Gilruth, Low, Kraft, and Slayton—had become an unofficial committee that got together often in Bob's office to discuss problems, plans, and off-the-wall ideas. Not much happened in Gemini or Apollo that didn't either originate with us or have our input. Deke showed up within minutes and George put it to us plainly.

"Chris, can you get the control center, the controllers, and the software ready for December? Deke, what about a crew? Can you give us a trained crew in four months?"

Deke was both quick and careful. "Yeah, I think so. But Chris and I need to look at it real careful. Give us a day?"

"Take two," Gilruth said.

In the hallway, Deke was almost bouncing with anticipation. "I think I'll have to do some crew-swapping. I gotta make some calls."

I hit my desk and made calls of my own. My head was abuzz with all the things we'd have to do. But it was one hell of a challenge. If we could pull it off, it would be absolutely pivotal to landing men on the moon. An hour later, I had five of my best, including Bill Tindall, John Mayer, Ted Skopinski, John Hodge, and Rod Rose in my office. Their eyes lit up like Christmas trees, particularly Tindall's and Mayer's. This was a challenge they wanted. They were gung ho guys, ready to do whatever it took. I sent them off to look at everything from getting mission control ready to having operational trajectory software that would take three Americans around the back of the moon and bring them home again.

I got in early the next morning, but there was already a message from Tindall and Mayer. They showed up fifteen minutes later. John Mayer started: "It's difficult, but we think that challenging the system is the right thing to do. And we have another suggestion."

"You know about the Ranger probes to the moon and lunar orbiter missions," Tindall said. "Well, the math model of the moon's mass isn't very good. We still can't get a good orbit prediction around the moon."

"If we're going all the way to the moon," Mayer jumped in, "let's go into lunar orbit and put the guys into the same orbit as we're going to fly on the landing missions. That'll give us a good measure of the orbit, an empirical measure, not a math model."

"You're talking a different thing," I answered. But I was getting even more excited. "The risks are bigger, going into lunar orbit and getting back out."

"We know," Mayer said. "But we can do it."

"Put it together," I said. "We need details." Then I called George Low. His silence went on forever, but when he finally spoke, I heard only good sounds.

"Okay," he said. "Let's look at it. Tell Deke and then see if we can take it to Bob tomorrow morning."

My guys had a full package waiting for me the next day. I listened to them as I skimmed through it. They were convinced that MIT and IBM could have lunar software ready by December and that our flight control teams would be ready to go. A few minutes later, Low, Slayton, and I were back in Gilruth's office. I told Gilruth what I wanted to do.

"What a great mission that would be," he said with real enthusiasm.

"We're willing to give it a shot," I said.

"We can have astronauts ready," Deke said. "Hell, yes, we can have them ready."

"We've got to get Wernher on board." Gilruth was reaching for his phone as he said it. He turned on the speaker so we could all listen.

"This is Bob Gilruth for Wernher."

"He's in a meeting," said a secretary. "Can he call you back?"

"No, this is important. Can you pull him out?"

Moments later, von Braun was on the line. He and Bob had come to terms over the years and their relationship was now cordial, if not friendly. Gilruth told him that he was with Low, Kraft, and Slayton. That got Wernher's attention immediately.

"We've got a proposition on a new flight and we need to talk to you," Gilruth said.

"Go ahead."

"No, in person. What are you doing this afternoon?"

Now von Braun was really curious. If Gilruth wouldn't talk on the phone, it must be important. "I can be free this afternoon."

"Then we'll fly over and talk to you in your office. Wernher, you're going to like this one."

Slayton immediately called Base Ops to order sandwiches for our Gulfstream. He wanted it ready for takeoff in thirty minutes. Low looked thoughtful and, with Gilruth's permission, called Sam Phillips out of a meeting at the Cape. Phillips was curious, too, and agreed to have his own plane on the way to Huntsville immediately. We flew to Huntsville, eating our sandwiches and mapping out who would say what when we got into the meeting.

Gilruth opened with a discussion about changing our plans on going to the moon. That got everybody's attention immediately. Low followed with a brief summary of the lunar module problems and the likelihood that no LM could fly until early in 1969. So far, von Braun and Phillips were silent. They knew about the LM and the delays in getting it ready for manned flight. Suddenly there was a sense in the room that something big was about to happen.

"We were thinking that we could do a circumlunar flight if Apollo 7 is perfect," Low said. "But Chris said, no, if we go at all, we might as well go into lunar orbit."

Von Braun had been leaning forward at the end of the conference table. Now he sat straight upright and looked at me. So did Sam Phillips. I had the floor. We could send a crew with a CSM, no lunar module, into lunar orbit in December, I said. It was a daring mission, maybe the most important in NASA's brief history. There were risks. But it could be done. It should be done.

Deke had a crew plan ready. Jim McDivitt, Dave Scott, and Rusty Schweickart were supposed to fly Apollo 8 with a lunar module in Earth orbit. The next crew in line was Frank Borman, Bill Anders, and Mike Collins. Deke didn't want to interrupt McDivitt's LM training, but the Borman crew had barely started their LM work. He would move them forward to fly to the moon. McDivitt would still fly the LM in Earth orbit, but in 1969 instead of 1968.

The excitement in that room was flush. I walked over to von Braun and said, "Wernher, we need you to commit to your next Saturn V flight. It has to have men on it and it's going to the moon."

He didn't hesitate a moment. "It's a great idea. We just have to see if we can get everything together. You don't give us much time."

Sam Phillips, head of the Apollo Program Office at headquarters, stood up and asked for silence. "This is the best idea I've heard yet. I'm proud of you guys for coming forward with this. But hear this: We have to keep this a secret. There's a lot of risk here and we need more information before we make a decision."

Jim Webb and George Mueller were in Vienna for an international meeting. Phillips told us to take two days to find any showstoppers. If we were still go, he'd bring Webb and Mueller into the loop.

On the way home, Deke put in a call to Frank Borman in California: "Drop whatever you're doing and come home. We gotta talk to you tomorrow."

Borman was mystified until I sat down with him and some of my best planning people showed him what we wanted to do. He was stunned by the idea of going into lunar orbit and needed to talk to his crew. Later Borman told me that it was one of the greatest meetings of his life. We gave it to him straight. He had the option on the backside of the moon to make a free return to Earth if everything wasn't right. "The choice will be yours," I said. "Your call."

Borman was convinced. He'd take the mission. The rest of us looked for showstoppers and didn't find any. For two days, the phone lines hummed between Houston, Huntsville, the Cape, and headquarters—many of our conversations originating in our wire-cage rooms that couldn't be bugged. This was something that we didn't want leaking to the press, or to anyone else.

George Mueller, of course, was furious when Phillips finally called him. He was convinced that we'd waited until he was out of the country, then replanned the Apollo program behind his back. When he spread that story to Jim Webb, the administrator was even worse. He backed Mueller in ordering that no decision be made until they were back in Washington on August 22.

At the same time, we were already meeting with Deputy Administrator Tom Paine at headquarters. Time was short and we needed all the support we could get. Mueller's recalcitrance shocked me, and we all understood that he was the biggest roadblock to be overcome. To Paine's credit, he understood the same thing. We thrashed out the whole Apollo 8 concept, along with its alternatives, its consequences, and its risks.

Finally Paine stood up and looked us all in the eye, one by one. "I want to hear from each of you. Are you in favor of a lunar orbit mission or not?" Paine pointed to Wernher von Braun first.

"It doesn't matter whether the mission is Earth-orbital or lunar-orbital." There was little difference in what Wernher's Saturn V rocket would have to do. "I say lunar orbit."

"I see all the positives and no negatives," Deke Slayton said. "This is our only chance to get to the moon before 1969. Let's do it."

"It's not the only path to the moon, but it certainly is the one that enhances our chances," Bob Gilruth said. He held Paine's gaze and his voice was firm. "We're in a risky business and I think this is the path of less risk. I vote in the affirmative."

"Flight operations has the toughest job here, to get ready, and we can do it," I answered when Paine looked at me. We'd been training for this moment from the beginning. "It won't be easy, but we want a lunar orbit mission on Apollo 8."

George Low put his hands flat on the table. "This is the only thing to do," he said forcefully. "The only thing. If Apollo 7 is successful, we have no other choice. This is not a question of whether we can afford to fly to lunar orbit. It is a question of whether we can afford not to."

By now, Tom Paine was leaning forward and I was waiting for him to smile. He did the next best thing. "I congratulate you all for not being prisoners of old plans and for having the courage to be bold. I think you're right. This is what Apollo must do, and I will personally work with Webb and Mueller to show them why."

I wanted to start applauding. We had a counsel at the top of NASA who was in our court. I folded my notes, put them into my briefcase, and made a vow that we wouldn't fail. There was too much at stake.

Sam Phillips did the only thing he could; he put it all on paper and sent a courier to Vienna with the package. Whatever he wrote, it made sense. Webb and Mueller calmed down. But Mueller still insisted on details when they got back to Washington. Webb listened only briefly, then turned his authority over to Tom Paine, while Mueller wanted every damned detail in writing. It took two weeks to convince him that Apollo 8 could go to the moon.

By mid-September, we had a go. It was tentative, all contingent on the success of Apollo 7. But it was a real plan. Jim Webb took it to Lyndon Johnson, just something we're thinking about, he said. Lyndon liked it and didn't say no. So Webb ordered his Public Affairs Office to put out a news release, simply saying that we were looking at options beyond Apollo 7 that included a long-duration flight in high orbit, or maybe a circumlunar or lunar orbit flight. We were all amazed when the press didn't pick up on our hints. The sharpest space reporters in the world missed one, and we avoided the questions, interviews, and press conferences that we'd thought were inevitable.

The biggest thing in our way was Apollo 7. Wally Schirra was still raising hell, bitching and hollering about everything. Then just two weeks before liftoff, he announced that he was quitting when the flight was over. That pissed me off completely. His timing was terrible. As things went on his flight, it was worse than terrible. He made himself look foolish.

After the fire, I'd rethought my own plan to be Flight on this mission. There was too much else going on and the moon preparations consumed me. I named Glynn Lunney as prime flight director, with Gene Kranz and John Hodge handling the second and third shifts. I'd have to be content to sit there again and watch. It was like having a ringside seat at the Wally Schirra Bitch Circus.

They got off the pad in fine form on October 11, 1968. Their Saturn 1B rocket was perfect and so were their orbital numbers. Even their first maneuvers, separating from the Saturn's upper stage, turning around to simulate docking with the lunar module that wasn't there, and then firing the service module's big engine, the SPS, went perfectly. That SPS burn was important. We needed that engine to fire right every time. It would be the rocket that slowed Apollo into lunar orbit and the rocket that kicked Apollo back out of orbit on the return trip home. It was the only rocket they had for those jobs. So when Wally fired it up and moved into an orbit that would let him rendezvous back with the rocket stage the next day, we relaxed a bit in mission control.

It was too bad that Wally and his crew didn't do the same. On day two, Wally came down with a cold. His brittle mood cracked under the pressure of a drippy nose and sneezes, and before long he'd infected Walt Cunningham and Donn Eisele with the same symptoms, including the bitchiness. When the time came for the crew to unpack their small black-and-white television camera, Wally refused. Even Deke couldn't get him to back down. Schirra was exercising his commander's right to have the last word, he said, and that was that.

The camera had been a bone of contention anyway. For a long time, Deke opposed including television on Apollo. It was a burden in training and a burden in flight that his crews didn't need, he said. I finally sent him a short memo reminding him that we owed this to the American public. "I can't conceive of this country 'sending' 3 men to the moon and not being allowed to see the lunar surface and the sight of a U.S. LEM on the moon," I wrote. "I believe you should reconsider your point of view."

Deke was too smart to stay stubborn. He reversed course immediately and became a television supporter. I think he wanted to see it all with his own eyes, anyway. I know that I did.

Schirra finally relented, and as the flight control teams wrote and rewrote the flight plan each day, he brought out the camera and put on an entertaining show from space. But his overall mood didn't get much better. At one point while listening to updates in the flight plan, he broke into the transmission and demanded, "What idiot in the MCC came up with that request?"

I looked at Gilruth and said that Schirra was getting paranoid. I repeated the remark a few years later in an interview and was quoted in the press. Wally was not the same old easygoing pro that I'd known on earlier missions. But my comment was out of place. Wally took exception and he was right to do so. Whatever was eating him up to and during Apollo 7 remains a mystery. He's long since gone back to the original fun and skilled Wally Schirra I knew from the beginning.

On one count, we had no complaints. Schirra and his crew did it all, or at least all of it that counted. They fired up that SPS engine over and over, doing a neat re-rendezvous with their rocket stage and proving to everyone's satisfaction that the engine was one of the most reliable we'd ever sent into space. They operated the command and service module with true professionalism, and if Wally had been worried about his spacecraft's safety on the ground, he was more than happy with how it performed in space.

But he had to have one last confrontation. The crew would not, repeat not, wear their space helmets for reentry. Damned if Schirra cared that it was a mission rule. Cunningham and Eisele backed him up. Schirra's reasoning was spurious; he was afraid that the pressurized helmets and their ongoing head colds would cause eardrum ruptures during reentry. The flight surgeon disagreed and Slayton himself got on the horn to argue with Wally. It was no use. We all talked it over and decided to declare that a helmetless reentry was acceptable. The other option was to declare the crew insubordinate and air all the dirty linen in public.

But it was insubordinate. I told Deke straight out that this crew shouldn't fly again. With Schirra, it was no problem; he was leaving. But Cunningham and Eisele had me worried. I didn't want to see either of them in a command position, and in my talks with Deke, he felt the same. Later I heard Cunningham's side of things. He was in a tough position up there with a headstrong and angry commander. I told him straight that he'd gotten on my wrong side, but maybe he should have another chance. It was up to him to take his chances and make things right. But he left NASA before that could happen.

So Wally and crew came home. The SPS was also their retrorocket, and it still had a perfect record. They splashed down in front of the television cameras, only one mile from their aiming point. The first manned Apollo flight was over. From a hardware and software standpoint, it was everything we wanted. At the regular postflight debriefing, Wally walked up to me and stuck out his hand.

"Congratulations, Wally," I said, and we shook on it. He'd been a pain, but he'd done a good technical job and nothing could change that.

"I wasn't sure you'd still be talking to me," he said. I nodded, but didn't know what else to say. We've never touched the subject since. Besides, something more important was on my mind.

Now it was time to give Apollo 8 the final word. Go. Or no-go.

Jim Webb quit as NASA administrator before Apollo 7 flew, tendering his resignation to the lame-duck president, Lyndon Johnson. Some people thought he was being magnanimous in clearing the way for a new president to appoint his successor. But I always wondered if the Apollo 8 decision wasn't too much for him. He felt betrayed by the fire and by the critical reports that had been kept from his eye. I lean toward the belief that he didn't want any responsibility for something going wrong on a moon flight.

Thomas Paine took over as acting administrator and leaned in our direction on the decision. After the huge success of Apollo 7's hardware, if not its crew, we were ready to make a formal and final pitch for approval. George Mueller's support was evident, but sometimes seemed lukewarm. But he knew the tide was flowing toward the moon, and in early November he did what we knew he'd do. He called a meeting.

The Apollo executives were there, all of the presidents or CEOs from the prime contractors. So were all the key NASA leaders. Sam Phillips opened and put it to them bluntly: We'd outline exactly what we wanted to do, then we wanted their advice and consent on the mission we were now calling C-Prime (C'). It took several hours, with presentations from von Braun's people on the Saturn V readiness, from George Low on the spacecraft, me on the operational plans and mission control status, Deke Slayton on the crew (he'd replaced Mike Collins with Jim Lovell when Mike needed shoulder surgery), and Rocco Petrone on the Cape's ability to do the launch. I took copious notes and thought that George Low summed it up best with this statement:

"The risks are no greater than those that are generally inherent in a progressive flight test program, and we believe that the probability of success of the ultimate lunar landing mission will be greatly enhanced."

Around the room, men were listening and taking their own notes. This was a moment of truth for Apollo and everyone knew it. Sam Phillips summed it up, then turned the meeting back to Mueller, and we waited to hear the verdict. George Low wrote down the reactions of the nation's most important aerospace executives.

"The S-IVB is ready to do any of the missions listed; however, McDonnell Douglas feels that we ought to fly a circumlunar flight instead of a lunar orbit mission in order to minimize the risks." That was Walter Burke of McDonnell Douglas, and I thought he was surprisingly timid.

Hilly Paige of General Electric was blunt: "GE would like to go on record that we should go ahead with an Apollo 8 lunar orbit flight."

"The guidance-and-navigation hardware is completely ready," said Paul Blasingame of AC Electronics. "Generalizing to the mission as a whole, when we risk the lives of people, we ought to get something for this risk. A lunar orbit flight looks like the right-size step to make."

I expected the words from Stark Draper of MIT: "We should go ahead with the mission."

IBM's Bob Evans was with us: "The program is in good shape, and the instrument unit is ready to go."

So was George Bunker from Martin Marietta. "The presentations made a persuasive case to fly a lunar orbit mission," he said thoughtfully. "The risk in lunar orbit is certainly greater than in earth orbit, but in assessing the risks for a lunar landing mission on a cumulative basis, it appears that the lunar orbit mission now will lessen the overall risk. I am for a lunar orbit mission."

T. A. Wilson, known simply as T, put a slight hedge to Boeing's position: "There is every indication that the lunar orbit mission is the right thing to do."

"As manufacturers of the spacecraft," said Lee Atwood, North American Rockwell, "our motivation to take chances is no higher than Frank Borman's, but we are ready to go."

"I have no reservations in supporting the complete mission," said Bob Hunter, Philco Ford.

Tom Morrow from Chrysler was wistful, then forceful: "We have no hardware on this mission, but we wish we had. We strongly feel that we ought to go for it. We must take steps like this one. We cannot move forward without progressing on each step. I vote yes."

"It is difficult to quantify the risks. I am impressed by what I heard," Bill Gwinn of United Aircraft said. "I support the recommendation to proceed with the mission."

Joe Gavin of Grumman was with us: "Since we have no hardware on this flight, our interest is only with respect to the overall program. The mission makes a lot of sense. I have no reservations."

Bill Bergen of North American Rockwell's Space Division, George Stoner of Boeing, and Gerry Smiley of General Electric all agreed. The only naysayer had been Walter Burke from McDonnell. American business leaders, I thought, showed the courage and fortitude that day that is the real foundation of our nation's industrial strength. It was done of necessity in private, but those men cast votes that would have very public impact on their companies, and more important, on their country.

The lunar orbit mission overwhelmingly carried the day. The key NASA people met twenty-fours later, on November 11, 1968, with Tom Paine. The decision was in the balance.

This time the recommendations came from the government side. The Department of Defense favored the mission. George Mueller's consultants from Bellcomm were opposed. The president's Science Advisory Committee leaned in favor, but made no recommendation.

Mueller argued from a statistical position, saying that the least risk came from flying the least number of missions. Sending Apollo 8 to the moon, he said, was not a positive step.

Bob Gilruth scoffed at his reasoning: "It's like saying that the faster you drive your car, the safer you are, because your exposure is less."

Everyone in the room understood the animosity between Mueller and Gilruth. There was a tense silence, finally broken by Tom Paine: "George, your argument just isn't valid. We're going to be in the flying business for a long time to come, and we will fly on all of the Saturn V's, whether we use them in the lunar program or not."

Future events disproved that statement, but in 1968 it seemed to be a fact. Paine listened a bit more, then called for a smaller meeting between his own senior staff, George Mueller, Sam Phillips, and the NASA center directors. While the rest of us waited, a third meeting followed with only Paine, Mueller, and deputy administrator Homer Newell attending. Then Paine came out and gave us his decision.

Apollo 8 would be a lunar orbit mission. It would be announced in a Washington press conference the next day.

I was fascinated by the entire procedure, not surprised by any of the comments made, nor surprised by the people who opposed the lunar orbit flight. Tom Paine obviously wanted to give everyone the chance to comment, but it was also evident to me that he was a little miffed that some present were not as emphatic as we from the centers.

I saw that he didn't want to commit himself, but was still in favor of C-Prime. Later I heard from George Low that Paine had made sure the major supporters in Congress and President Johnson were given a chance to override the decision, and that they had given their consent without question. None of us knew when he consulted with them, but we went away happy and anxious to get on with the job. The press, of course, took up hours of our time in the next few days. Apollo 8 to the moon had captured the world's imagination, and it was one of the biggest stories in a year of bad news from Vietnam, student demonstrations, riots at the Democratic National Convention, and a bitter presidential campaign just won by Richard Nixon.

We hadn't started our Apollo 8 plans by thinking about its impact on the outside world, but now that impact was blaring from television sets and newspaper headlines. Whatever press attention we'd gotten before was about to be dwarfed by a news media maelstrom.

We had six weeks to meet our schedule. We'd be around the moon on Christmas Eve, and the crew would splash down in the Pacific a few days later. That raised a big problem. The Pacific Fleet had already issued orders for a stand-down during Christmas week. Its carriers would be in port at Pearl Harbor, and the Navy planned to give as many sailors as possible leave in Hawaii or stateside.

The two-star running recovery operations at the Cape, Vince Houston, dumped the problem in my lap. "John McCain is CINCPAC and his son is in a Vietnam prison," Houston said. "I'm not willing to be the one who tells him that we need part of his fleet at sea over Christmas. Chris, you should go out there and make the pitch."

"I'll do it," I said, "but you're coming along. And you set up the meeting."

Houston called back in a matter of hours. "CINCPAC will see you a week from Thursday at ten a.m.," he said. "I'll pick you up next Wednesday in my Sabreliner and then we'll go commercial from Los Angeles."

The following Wednesday night we had dinner in the bachelor officers' quarters at Hickam Air Force Base in Honolulu. The next morning I put on my best suit, a new shirt, and gathered up the slides and charts I'd asked John Mayer to prepare. At McCain's office, General Houston and I were met by two rear admirals, who offered us coffee. I had tea. The admirals knew what we wanted and were completely noncommittal. Finally they took us to an amphitheater conference room, nicely paneled and with about a hundred seats. We sat down front as the damnedest bunch of brass I've ever seen—all captains and admirals and even a couple of four-star Army and Air Force generals—filtered in and filled every seat.

At 10:30 A.M. sharp, someone yelled, "Attention!" and in walks four-star Admiral John McCain, commander in chief of the Pacific Fleet, who's a couple of inches shorter than me and full to the brim with military bearing. Somebody said, "Be seated," as McCain took his chair right down front and looked me in the eye.

"Okay, young man," he barked in a loud voice, "what have you got to say?"

Neither before nor since have I stood in front of that kind of high-powered audience. McCain's eyes only left my face to look at my charts and graphs. I don't remember many of my exact words. I simply ran through the mission and told him what we wanted to do. I know that I stressed the importance of the flight and its risks, and that the greatness of the United States of America was about to be tested in space.

Then I got to the real point—we had to land the crew in the Pacific near Midway. I'd memorized exactly how I wanted to put it:

"Admiral, I realize that the Navy has made its Christmas plans and I'm asking you to change them. I'm here to request that the Navy support us and have ships out there before we launch and through Christmas. We need you."

There was complete silence in the room for maybe five seconds. Mc-Cain was smoking this big long cigar, and all of a sudden he stood up and threw it down on the table.

"Best damn briefing I've ever had," he said, again loud enough for everyone to hear. "Give this young man anything he wants."

And he walked out.

Apollo 8 was go.

Back home the next night, I stood outside mission control and looked at a waning three-quarter moon. *By the time I see that moon phase again,* I thought with a stunning clarity, *we will have been there and back.*

First we had to go to the White House. Lyndon Johnson threw a dinner honoring James Webb and Apollo 8, and Dr. Chuck Berry did his best to stop it. He didn't want Borman, Lovell, and Anders exposed to all those germs and viruses just before flying to the moon. Berry lost that fight. "The docs weren't happy with me," Lyndon joked during his remarks that night, "but I prevailed on them."

We were all there. Betty Anne was at the vice president's table. I sat with Al Shepard to my left, Mrs. Clark Clifford to my right. Next to Al was Charles Lindbergh. That was an exciting surprise. I saw Lindbergh land the *Spirit of St. Louis* at Byrd Field in Richmond, Virginia, when I was four, and he'd been one of my idols ever since. We immediately got into a discussion about Apollo and the coming flight to the moon. Lucky Lindy kept drawing the conversation back to Al's Mercury flight, wanting to hear every detail from America's first man in space. I think Lindbergh was as impressed with Al Shepard as we were with him.

It was hard to be polite and to spend time talking with the others at our table. I knew that being this close to Charles Lindbergh was a rare event. But so did the others, and Lindy managed to talk to all of them. Stark Draper was with us, and Betty Grissom and Mrs. James Van Allen. So were

Mrs. Kurt Debus, and Jim McDivitt's wife, Pat. It was a night to remember and nobody caught a cold. My table all signed our menu cards, giving us a treasured memento of the evening.

The next day we were in another of George Mueller's "final" reviews at NASA headquarters. "For God's sake," Bob Gilruth wrote in a sharp note, "you're killing us with meetings when we should be in Houston working." But Mueller got his way. He wanted a final piece of paper from everybody, including the Apollo executives, signed almost in our blood and agreeing to put Apollo 8 into lunar orbit. We didn't prick our fingers, but we all signed and were glad to get these meetings finally behind us.

Apollo 8 went into lunar orbit on Christmas Eve, 1968. I was in my usual watcher's seat in the back row of mission control, and Cliff Charlesworth ran the mission as my prime flight director. With the demands of Apollo, we'd added flight control teams under Charlesworth, Gerry Griffin, Pete Frank, Milt Windler, and Phil Shaffer. Along with the now seasoned teams run by Gene Kranz and Glynn Lunney, they'd all have their turns, and their moments to face critical decisions, in the years ahead.

With only a few changes in the outbound flight plan, the Apollo 8 trajectory was perfect. It was the crew that wasn't so good. Unlike Gemini VII, where Frank Borman and Jim Lovell were confined to their seats for fourteen days, the crew was free to float around in the Apollo command module. Borman got sick first, almost certainly a form of motion sickness that included vomiting and diarrhea. He hid it during his live conversations with mission control, but recorded everything on a voice tape that we downloaded in the next few hours. By then, Lovell and Anders weren't feeling so hot, either.

Motion sickness was always a possibility in space. It still is. The doctors prescribed one of the onboard medications and it helped. Well before they reached the moon, Borman and his crew were looking out the windows at a gloriously blue and white Earth receding behind them and at the growing gray-white mass ahead. It was a profound experience for them, and their descriptions had an effect back on Earth that would grow to mystical proportions. Their television images weren't so good; Earth showed up as a glowing ball without much detail. But the astronauts' enthusiasm for what they were seeing was infectious. In the enclosed container of mission control, we had no idea of what was going through the world's collective mind. But we'd soon find out.

We had one short dissension in mission control. The first midcourse correction to the moon was planned to use Apollo's big SPS engine. It had worked perfectly during Apollo 7, but now some controllers wanted to abandon that plan and use only Apollo 8's smaller thrusters. The big engine, they feared, might disrupt the free-return trajectory that the spacecraft was following and make it difficult to return to Earth in an emergency. As the discussion grew more heated, even some of my own mission planning people began to get nervous.

I had to step in to settle the argument: "We need that SPS engine to get into lunar orbit and I want to see it work before they go behind the moon. Stick to the plan."

They did. The only anomaly was a slightly higher buildup in thrust. We quickly traced that to helium accidently left in the fuel lines before launch. It was no big deal. We knew that the SPS worked and that it would do its job of getting Apollo 8 into lunar orbit and back out again.

The crew wives and children came to mission control, and a mythical story arose about dissension between Susan Borman and me. According to several television shows in the years since, she and I had heated discussions about sending Frank out to face so much danger. Those conversations never took place. I was honest about the risks and she was concerned, but never critical. Hell, I was concerned. We'd just sent three men out where no humans had ever gone. Jim Lovell later said that Susan didn't speak to me for years thereafter. He must have missed our many chats, both surrounding the Apollo 8 press conferences and at the social events we all attended.

Apollo 8 disappeared behind the moon. For the next thirty-three minutes, we could only sit and stare at the big display screens in the front of mission control and imagine that the SPS engine was firing on time to drop our guys into lunar orbit. There was hushed conversation, and many of the controllers dropped their headsets on their consoles, made long-overdue pit stops, and grabbed cups of coffee or cans of soft drinks. I settled for my usual tea while Bob Gilruth and I talked quietly about how far we'd come in ten short years.

"Ten years and a month," Bob reminded me. "That was when I put out the memo establishing the Space Task Group. There were thirty-six of us on the list, including you, Chris."

"And now we've got three men out of sight behind the moon." I looked at the displays and sipped my tea. "If we weren't sitting here, I don't know that I'd believe it."

"I wonder what Lindbergh is thinking." Bob had known the man for thirty years, back to Gilruth's earliest days as a young engineer at Langley. "He sure seemed excited at that White House dinner."

"His mind is out there with the crew," I suggested. "Just like yours and mine."

There was a stir in the controller rows below us. Both of us held our headsets to our ears to catch what was happening. It was scratchy and garbled, but there was Jim Lovell's voice coming from nearly a quarter million miles away. It sounded as if they were . . . yes, they were in lunar orbit, and the room erupted with cheers and hollers. We were all standing and waving our arms. I could barely hear Paul Haney on the public affairs console a few feet away screaming into his microphone, "We got it! We got it! Apollo 8 now in lunar orbit!"

It was glorious pandemonium, and through the mist in my own eyes, I saw Bob Gilruth wiping at his and hoping that nobody saw him crying. I put my hand on his arm and squeezed. He'd led us all to this point, a brilliant and decisive leader whose reputation was largely unknown and whose legacy would too long be ignored. More people knew Chris Kraft than knew Bob Gilruth, and that was how he wanted it. He lifted my hand from his arm and shook it strongly. There were no words from either of us. The lumps in our throats held them back.

The rest of Christmas Eve and Christmas Day became the stuff of legends. Haney's stenographic pool recorded and transcribed every word. Thousands of print and radio/television reporters spread the stories to every part of the world. Hundreds of millions of people saw live images transmitted by television from the moon and heard the astronauts describe the scenes. One of the most vivid in my memory was from Bill Anders, looking out the window, then saying the words that sent shivers up my neck:

"Earthshine is about as expected, Houston."

I shook my head and wondered if I'd heard right. Earthshine! Yes, it was true. And a few hours later, here was Bill Anders again describing Earthrise over the lunar horizon and snapping off a series of photos that would become some of the most famous pictures ever taken. The sight of a full-color Earth hovering over the craters of the moon—it had been unimaginable until now.

It was the afternoon of Christmas Eve in Houston. Frank Borman was a religious man, a lay reader in the Episcopal Church, and he read a prayer in lunar orbit. Then he and the others took naps, Lovell and Anders only giving up their posts at the windows when Frank ordered them to get some sleep. In mission control, we were all in the grips of an exhilarating exhaustion. No one wanted to go home. There was more to come after dark.

The words were Frank Borman's: "This is Apollo 8, coming to you live from the moon." On any other Christmas Eve, Americans and much of the world's citizens would be celebrating quietly at home or in church. Tonight they were glued to television sets and I was sitting again in mission control. We listened in awe as Borman, then Lovell and Anders, took turns pointing their television camera at each other and at the moon, describing what they saw, and still feeling so calm, so cool, yet so reverent at being out there around the moon. In the final minutes before they went behind the moon again, they read from Genesis: "In the Beginning, God created the Heaven and the Earth . . ."

No eye was dry in mission control. Nor, I'm told, in the newsroom packed shoulder to shoulder with reporters covering the greatest story of their lives. And maybe not even anywhere that televisions and radios were playing on Earth. Frank Borman had the last words: ". . . from the crew of Apollo 8, we close with: Good night. Good luck. A merry Christmas. And God bless all of you, all of you on the good Earth."

I went home drained, to a late Christmas Eve with Betty Anne and the children. There had never been a Christmas like it for the Krafts, nor would there be another to equal it. We weren't alone. Christmas Eve and Apollo 8 were uniquely inspiring to Earth's citizens, wherever and whoever they were.

The emotions didn't quite make us forget why we had men out there. John Mayer and Bill Tindall had called me down to the computer center where they had a series of displays ready. One showed the true radar data we were getting as Apollo 8 tracked around the moon. Another showed how the tracks looked in apolune and perilune—two more new words in our vocabulary—when they were calculated from the math models. Still another showed distance errors plotted in a straight line.

I saw it immediately. Those straight lines gave us precisely accurate measurements of the real orbit around the moon. It was exactly the information we'd wanted, and now we had it, for Apollo 8 and for every future mission that flew to the moon. The math model errors had been corrected with live data. On that basis alone, we were vindicated in putting Apollo 8 into lunar orbit.

"Great news," I said, shaking hands all around, "and great work. Now let's get them back out of orbit and on the way home."

That came about seven hours later. Apollo 8 was behind the moon again, firing the SPS engine while we sat in mission control with tight stomachs and tensed muscles. Jim Lovell's sense of humor made it all wash away when Apollo 8 appeared exactly on time from behind the moon and on course to Earth.

"Please be informed," Lovell radioed with a chuckle, "there is a Santa Claus."

Apollo 8's parachutes opened almost right over the U.S. Navy carrier *Yorktown*. It had done a skip reentry that bled off excess speed and energy, then dropped through the atmosphere and splashed down a few miles away in full view of television cameras.

We'd done it. The crew had done it. The Navy was there and picked them up. *Time* magazine named Borman, Lovell, and Anders its Men of the Year. They deserved it. In early January 1969, we all sat down and drew up a list that was the perfect measure of what we mean when we evaluate risk versus gain. Apollo 8 has to be history's prime example of both.

- Apollo 8 was the first manned flight of the Saturn V.
- It was the first manned vehicle to leave the earth's gravitational field.
- It was the first use of a computer that, combined with the navigation and guidance system, provided total onboard autonomy in inertial space.
- It was the first manned vehicle in lunar orbit.
- It allowed man his first and closest naked-eye view of another planet.
- It was man's first exposure to solar radiation beyond the earth's magnetic field.
- It was the first vehicle to rocket out of lunar orbit.
- It was the first manned vehicle to reenter the earth's atmosphere from another planet.

It was something else, too, that we didn't put on the list. But we all knew it.

It was great.

19

The Last Few Steps

We were accelerating toward landing men on the moon. The Apollo command and service modules worked. They were so good that I remain convinced that the tragic fire in fact gave us the tools to reach the moon. If we'd somehow avoided the fire, we would have flown an imperfect spacecraft and found flaws. We would have slowed down to fix them, flown again, and found more flaws. As delay piled on delay, I have the most serious doubts that we could have met Jack Kennedy's end-of-the-decade goal.

Instead we dove deep into the spacecraft troubles and fixed them. It took a seventeen-month stand-down before Apollo 7 could fly. But then it was only ten weeks until we flew Apollo 8, in what was surely the gutsiest and most important decision of the entire program. And less than twelve weeks later, we had Apollo 9 in Earth orbit with a lunar module up there being tested by Jim McDivitt, Dave Scott, and Rusty Schweickart. The moon was no more a stranger. All we needed was the proof that our LM was as good as we thought it now was.

Good men died to force these results. Good men lived to see it through. I woke up at night and wished for the godly perfection to revisit those days and make it right without losing Gus, Ed, and Roger. That power isn't given to humans. We could only go on and hope that somewhere they watched and understood and forgave.

It barely crossed our minds that we'd be able to go through our flight test program without repeating a mission. But the overwhelming successes of our first two Apollos had most of us thinking that maybe, just maybe, we could pull it off. It was up to the McDivitt crew to make the critical manned tests with their lunar module in Earth orbit. But if something went wrong, a conservative few in NASA were ready to pounce and insist that the first lunar landing be unmanned. George Low and Sam Phillips kept them at arm's length while we watched the lunar module from mission control.

Apollo 9 was scheduled for my forty-fifth birthday, February 28, 1969. Deke was born two days after me—we bracketed that magic date of February 29, 1924—and he never let me forget who was the old man between us. Then the whole of the McDivitt crew came down with head colds and the launch missed both of our birthdays, finally getting off the ground on March 3.

There'd been a change in thinking at NASA headquarters since the end of Gemini. The astronauts again had the privilege of naming their spacecraft. For Apollo 9, the call signs were *Gumdrop* for the CSM and *Spider* for the lunar module. I chuckled when I heard the choice. We needed a bit of levity balanced against our serious natures.

Now they were up there, getting ready to dock with their LM and pull it free from its storage shed on top of the S-IVB rocket stage. Dave Scott, flying the command and service module, immediately ran into a problem. Some of his thrusters weren't working. It made docking with the LM harder, but still he pulled it off. On the ground, I listened to controllers troubleshooting the problem. It turned out to be an easy fix. Somebody had bumped a switch and turned off one set of thrusters. A quick instruction from the capcom turned them back on. I made a note: "Add cover guards to switches." My own experts and engineers at North American didn't need my reminder. They made their own notes and switch covers were added to all future spacecraft.

A few hours later, spring-loaded clamps released *Spider*, and for the first time, the full Apollo spacecraft combination was flying free in space. It was a remarkable moment for us, but it was just the start. Because it was designed to land on the moon, where gravity was one-sixth as strong as on Earth, and because we needed to save as much weight as possible, the lunar module was fragile. Now it was docked to the CSM, and we'd soon find out if it could handle the jarring force of that big SPS engine firing from the service module. It did. Scott only fired the engine for five seconds, but it was enough to let our engineers on the ground get the readings they needed. The LM held up in fine shape. Over the next two days, they fired the SPS again and again, even simulating the long burn needed to go into lunar orbit.

Then we hit another problem. This time it wasn't hardware, but human. I was listening to the communications loop when Jim McDivitt suddenly asked for a private conversation. That meant the press wouldn't be listening live and he could talk freely. Rusty Schweickart was sick. He'd been space sick, in fact, almost from the beginning and was weak from vomiting.

My second reaction, after concern for Schweickart, was anger. Why the hell had McDivitt waited so long to let us know? We were coming up on crucial parts of the mission. He and Schweickart were supposed to crawl into the lunar module and check it out. They'd separate from Scott, leaving him alone in *Gumdrop* while they flew *Spider* freely for hours. And later, Schweickart was supposed to don a big backpack—a portable life-support system—and do a space walk that included getting out of the lunar module, then crossing over to and entering the command module through its main hatch.

Now McDivitt wanted to curtail the space activities. Given the timing, it was the right decision. But if he'd reported on Schweickart a few days earlier, the flight surgeons would probably have prescribed medications that could have eliminated his symptoms. It didn't happen, I'm sure, because McDivitt was trying to protect Schweickart, and because relations between the astronauts and the medical corps had never been good.

The whole thing had an unfortunate fallout. The astronaut image had always been one of physical strength and fortitude. Now Rusty's space sickness made him seem weak, and his reputation never recovered. Deke didn't help. He exercised his sole authority to keep Rusty from flying again. That kind of authority bothered me. If I saw any flaw in Bob Gilruth, it was the way he turned over full control of the astronauts and crew selection to Deke. I thought that it was too much power to be vested in one person. Deke's flaw was that he accepted it all and acted accordingly. He seldom had to justify his actions to Gilruth or anyone else. Among the few times I knew him to reverse course were on the Apollo television decision and when he was directly challenged by the news media to make astronauts more available to the press. Even then, Deke frequently shielded his boys from interviews and photo sessions, and it sometimes took strong arguments and even threats from Paul Haney to convince him otherwise.

So here we were, hoping to be only a few months from landing on the moon, and simple pride over reporting a sick astronaut could be blocking our way. The questions were about Rusty's true condition and how to recover. Was Schweickart healthy enough to work at all? Yes, he said. McDivitt's concern was putting Rusty into a helmeted space suit, then having to contend with more vomiting. Could McDivitt and Schweickart go ahead with powering up the LM and starting its tests? Yes, they could do that. We talked it over in mission control, then gave them a go. The space walk questions would be settled later.

While Scott manned the CSM, McDivitt and Schweickart moved into the LM and got to work. The next hours could hardly have been better. There were the now expected moments when systems and switches had to be reconfigured, and then McDivitt fired the LM thrusters to control the whole stack of modules. They weren't just adequate. They were up to the job in every respect. Then came the big test, firing the LM's descent engine to move the docked modules through space. Again the firing was perfect, with McDivitt using the throttle-down capability near the end. Reducing power on that engine was absolutely mandatory in the final seconds before landing on the moon, and I held my breath when McDivitt called out his move. The plots on our big boards in mission control showed the engine responding exactly as it should. Another milestone was behind us. The descent engine design was sound and so was its operation.

Rusty felt better the next day. Jim McDivitt talked to us frankly, and we decided to take the EVA one step at a time. Schweickart was supposed to crawl out onto *Spider*'s front porch and retrieve some science samples placed there. Then after Dave Scott opened the command module hatch, he would have crossed over and gone into the other ship. That was the only part they didn't do. Schweickart was tiring, and McDivitt called it off after Rusty had been outside for nearly an hour. It was a good decision and we concurred.

Only the hard part of the mission was left. On day five in space, McDivitt and Schweickart closed up their LM, backed away from the CSM, and were on their own. Gene Kranz was Flight for that shift, and after hearing Dave Scott's description of the lunar module looking so good—his amazing photographs proved that point to all of us a few days later—Kranz said the magic words: "Capcom, tell 'em they're go for separation."

That was what McDivitt was waiting to hear. He fired the "dips," the descent propulsion system, and in moments the LM was moving quickly off into the blackness. Radars on both ships locked on and tracked the departure. So did our radars on Earth, with high-speed data flowing into mission control. Everything was in sync. Out to fifty miles, Scott could see the LM as a bright star against the sky, then as it moved on to a full one hundred miles away, he lost it to the eyeball, but not to his instruments. Again I had that frustrating feeling of wishing I could be there. As good as it was, my seat in mission control was second best to those seats in outer space.

The most dangerous moment of the flight was upon us. McDivitt and Schweickart would hit the switches to separate the LM's ascent stage from the descent stage, then fire the smaller ascent engine to begin the rendezvous maneuvers that would bring them back to Dave Scott. That separation always worried us. It simply had to work.

And it did. In three different places, eyes, brains, and computers set to work on the upcoming rendezvous maneuvers. In the CSM, Dave Scott worked his radar and computer to calculate the path he'd have to follow to rescue the LM if something went wrong over there. In the LM, Jim McDivitt and Rusty Schweickart did the same thing in reverse. It was their job to fly back to the CSM, just as if they were coming up from the moon. And in mission control, some of the world's smartest engineers were running the world's biggest computer center to do their own calculations double-checking everything.

Before long, Scott had the bright star that was *Spider* in sight again. Rendezvous in space was so routine now that we didn't expect anything less. Finally, with *Spider* sitting in perfect position outside, Scott nudged *Gumdrop* forward and the docking mechanism snapped into place. The last piece of our complex moon plan had just been proven. I went home that night knowing that we could actually do this thing. We could fly those complicated spacecraft to the moon. We could land. We could launch again and rendezvous. The machines all worked. So did the humans, in mission control and among the astronaut corps.

We'd put together one helluva team. It was time to go back to the moon.

There were things to fix, of course. *Spider*'s performance was more than good enough, but still there were nagging problems with some of its instruments and with some of our procedures. Flight controllers had improvised over and over again, passing along instructions to McDivitt and

Schweickart to do this or that differently, or to work around something that wasn't quite right.

I wasn't unhappy with how it worked out. We flew for exactly those reasons— to find the bugs, come up with creative solutions in-flight, and then to fix them for next time. My flight control teams were filled with young people at the top of their game. The average age in mission control in 1969 was twenty-seven or twenty-eight. They looked at me as the Old Man, and if I didn't feel old at forty-five, I understood their perspective. I appreciated their youth and their quick grasp of situations more than I was able to say. This was the new generation of space controllers, and their performance on Apollo 9 told me all I needed to know about flight operations. We were ready for anything. As it turned out, we got plenty of it on the next two missions.

The decision was made. Apollo 10 would go to the moon "all-up," and Tom Stafford and Gene Cernan would fly their LM to within fifty thousand feet of the lunar surface, doing a near duplicate of the maneuvers needed to go in and land. The press speculated that Stafford might try to land, but that was never in the cards. His lunar module was the last of the early series built, and it was simply too heavy. He knew it, we knew it, and nobody argued for an Apollo 10 landing.

While the Cape, Houston, and the astronauts were getting ready, the old television question raised its ugly head again. Somehow George Low got the idea that putting a television camera on the first lunar landing wasn't really that important. It added weight and it would take some of the astronauts' valuable moon time to set it up. Scientists with instruments and experiments bound for the moon were ambivalent. Television didn't add much to their research, and if they weren't against it, they told Low that they couldn't really justify it either.

"Chris, I want this TV thing resolved," Low told me. "Get everybody together and settle it."

"We can settle it by just keeping the camera," I said. Low knew my position, but he wanted more. The usual process had to be followed. I called a meeting. Naturally I tried to stack the deck. Max Faget was as strong for television as I was, so he'd be there with some technical people to talk about the system. Hell, we both wanted to see those first steps live and in living black and white; a small color camera still wasn't available.

Public Affairs had to be there, too. I knew how badly they wanted television from the moon. And it would be their task to give the bad news to the networks if the camera was left behind.

Finally I asked my own Flight Control Division to work up a formal— and fair—presentation on the pros and cons, then to give their view of having TV on the first landing. I should have given them better direction.

We filled the largest conference room in Building 2. It looked like the first landing crew would be Neil Armstrong, Mike Collins, and Buzz Aldrin. They were there when my lead communications engineer, Ed Fendell, made the final presentation of the day. He was good, outlining the lunar surface timeline, showing examples of how the camera could be used, and discussing its technicalities. Then he put up his summary slide. He made a recommendation, all right. He recommended that television was a nicety that should be left behind.

Damn! I thought. *My own guy!*

The room erupted into discussion. Public Affairs was aghast and so were Faget's people. I kept my mouth shut until the noise dropped to tolerable levels, then stood up. If the look on my face didn't bring silence to the room, the tone of my voice did.

"I can't believe what I'm hearing."

Max interrupted with almost identical words.

"*We* can't believe what *we're* hearing," I corrected. "We've been looking forward to this flight—not just us, but the American taxpayers and in fact the whole world—since Kennedy put the challenge to us.

"Now you're willing to exclude the people of Earth from witnessing man's first steps on the moon? I don't believe it, and if you think about it, I don't think you'll believe it either."

It might have sounded trite, but I didn't care. I sat down, damned angry, and Max stood up to make almost exactly the same points. If anything, he was more forceful than I'd been. The looks around the big table and around the room changed from stolid to sheepish. One by one, there were nods from everyone in the room. Most of them worked for either Max or me, and when they saw our determination, any thought of arguing disappeared. Neil Armstrong and the crew nodded, too, and Neil's "Yes" was the final word.

We had television for the first landing. We all hoped it would be Apollo 11.

May 1969. Tom Stafford, John Young, and Gene Cernan were in lunar orbit with the CSM they named *Charlie Brown* and their lunar module,

Snoopy. For the second time, I appreciated the humor that had come into spacecraft names. We had plenty to be serious about, and when a few carping critics in the press took exception, I told my people to consider the source.

After a night's sleep in lunar orbit, Tom and Gene put on their space suits, crawled into *Snoopy*'s stand-up cabin, and got ready to do a last dress rehearsal that could clear the way for a moon landing in July. A few hours later, *Snoopy* and *Charlie Brown* were miles apart. Alone in the CSM, John Young mounted a television camera in a window and let us watch the moon's cratered surface flowing by below.

Aboard *Snoopy*, Tom Stafford put his hand on the engine controller and fired up the descent engine. Within minutes, *Snoopy* was in a new orbit with a high point of sixty-nine miles and a low point of only about seven miles. He and Gene Cernan were about to come closer to the moon than any men in history. That was when static and garble overwhelmed their radios.

"Aw, this fucking comm," I heard Stafford say. The static cleared up just in time to let his voice reach Earth loud and clear. I looked up into the glass-walled viewing room behind my console. Some of NASA's top officials were there and a few looked embarrassed. But there were smiles, too. Nobody really thought that all astronauts were noncursing Boy Scouts. In the next couple of hours, we'd all get several more earfuls from the moon.

Gene Cernan's happy calls told us that *Snoopy* was rapidly dropping toward those cliffs and craters below. "We is go, and down among 'em," he radioed with an ebullience that had me sitting upright and cupping my earphone to hear every word.

It had to be exciting, and radio protocol simply disappeared. Coming up on a huge crater, I heard Stafford holler, "I've got Censorinus right here, bigger 'n shit!"

Cernan was so excited that all he could say was "Son of a bitch! Son of a bitch!" As they passed just forty-seven thousand feet over another crater, he called out, "God damn! That one looked like it was coming inside!"

Bob Gilruth laughed out loud and said, "Listen to them. A couple of old pros and they're as excited as kids." If they'd been his own kids, he could not have been more proud at that moment. In fact, they were his kids, and they were out there doing man's work at the moon. If a cuss word or three slipped out, well, who the hell cared anyway?

Then it was time to dump the descent stage, fire up the ascent engine, and get the heck back home to *Charlie Brown*. Up to that moment, we'd had a perfect mission, and a landing on Apollo 11 was only a few decisions from being ordered.

The next thing we heard was Cernan, and this time he wasn't some awestruck kid. He was a test pilot in trouble. ". . . son of bitch!" he shouted, and a second later we saw it in mission control's telemetry readings from the moon. The lunar module *Snoopy* was jerking about and all but out of control.

"Let's make this burn on the AGS, babe," Cernan was shouting. "Make this burn on the AGS."

AGS was the abort guidance system, an emergency backup that nobody ever wanted to use. Stafford chose to fight the beast with the primary guidance as it rolled hard to the left, and in moments they were free and away from the descent stage, back under control. We traced the problem to a switch setting that was wrong on their checklist. Glynn Lunney was Flight on that shift, and before he faced the press a few hours later, he had it all figured out. Yes, he said, it had been an exciting moment. But it was never dangerous, though Cernan and Stafford surely got their adrenaline jolt for the day on that one.

Stafford's reflexes and his complete knowledge of his lunar module turned it into an almost routine maneuver as he piloted *Snoopy* back to orbit and their rendezvous with *Charlie Brown*. Still, that was a checklist that got double-checked twice during the rest of Apollo. There was always enough excitement in going to the moon without adding an extra thrill by misediting an instruction.

Apollo 10 made the home trip from the moon seem routine. When they splashed down only a few thousand feet from the Navy carrier—after crossing a quarter million miles of space—I lit my cigar, put my feet up on my console, and let a new feeling pass over me.

"The practice is over," I said to nobody in particular. "Next time we land."

20

. . . And a Giant Leap

No mission equaled Apollo 11. I put four flight control teams on the job, run by Cliff Charlesworth, Glynn Lunney, Gene Kranz, and Gerry Griffin. Then I spent every spare minute of my time in the back row at mission control, watching them train, simulating every aspect of the flight over and over and over, and letting it all soak in until I knew it well enough to watch Apollo 11 unfold in my sleep.

There'd been some arguments getting ready. The science and medical community had raised a fuss years ago about the possibility of contaminating Earth with some alien organism brought back from the moon. It was so far-fetched that only Hollywood could turn it into a script. But nothing succeeds in this world like a few scientists crying wolf and flying the flag of fear to Congress and the press.

So now we had a quarantine facility. Armstrong, Collins, and Aldrin would be handed sterile jumpsuits and hoods while they floated on the Pacific. They'd have to struggle into them, then on the aircraft carrier go through a tunnel into a sterile trailer that would be loaded onto a C-140 and flown home to Houston. Once back home, they'd be confined for weeks. At least we could debrief them through windows, and the doctors, quarantined with them, could do their examinations. But whatever samples they brought home from the moon would be under seal and kept away from humanity until the scientists were certain that the astronauts weren't about to unleash a plague upon the world.

It was stupid, disgusting, and politically mandatory. We went along with the game because we had to. The same thing will happen, I'm sure, when men and women first return from Mars. Hysteria cows common sense every time.

Another worry plagued us in the month before Apollo 11. The lunar module was a hard machine to fly. We'd built the procedures simulators and connected them to mission control. But the strange sensations of flying a rocket that moved like a helicopter couldn't be duplicated. The result was a dangerous flying machine called the Lunar Landing Training Vehicle (LLTV). It looked like an open-air flying bedstead, with the astronaut controlling its flight from a seat in one corner. To simulate the one-sixth gravity of the moon, a down-pointing jet engine removed five-sixths of the LLTV's weight during landing maneuvers, and its four corners had small thrusters to maintain its balance. It was as close to a real lunar module in flying qualities and handling as could be done on Earth.

Neil Armstrong ejected from a malfunctioning LLTV in 1968. The frightening films show that he escaped death by just two-fifths of a second. Winds were gusting that day, something that can't happen on the airless moon, but Armstrong was fully in control for the first five minutes. He took it up several hundred feet and was ready to practice a nearly vertical descent and landing. Then the machine suddenly dropped. He steadied it and climbed back up another two hundred feet. Then the LLTV began to bounce around in the sky. It pitched down, then up, then sideways. It's stabilization had failed and it was clearly out of control. A ground controller radioed Neil to bail out. He activated the ejection seat with only a fractional second of margin. Neil's parachute opened just before he hit the ground. He wasn't hurt, but the LLTV was demolished in a fireball.

The LLTV's control system was modified, but then our chief test pilot, Joe Algranti, had to eject when the new version went wrong. Bob Gilruth and I were ready to eliminate it completely, but the astronauts were adamant. They wanted the training it offered. After more changes, a new LLTV was ready in mid-June. Neil flew it for three days while we held our breath. Then Buzz Aldrin went through the same training.

Nothing went wrong, but I was sure that we were pressing our luck. I asked Neil to stop by and we talked about the LLTV.

"It's absolutely essential," he said. "By far the best training for landing on the moon."

"It's dangerous, damn it!" I snapped. I really wanted him to give a negative report.

"Yes, it is. I know you're worried, but I have to support it. It's just darned good training."

So I gave in. But I didn't give up. Either Bob Gilruth or I grilled every returning lunar astronaut, hoping to find some way to get the LLTV grounded forever. We lost every time. The astronauts wanted it. To a man, they said it was the best training they received and was essential to landing on the moon. So with our fingers crossed, we let them keep it.

We flew to the Cape on July 14—Gilruth, Low, Faget, Slayton, and me—for the final flight-readiness review on T-2. Armstrong and his crew were already there, hidden away from the world's germs in the regular preflight segregation period. But we'd all been checked by the doctors and were cleared for one last visit.

Neil was Neil. Calm, quiet, a gentle smile, and absolute confidence. The whole crew knew that they were in this historic position as much by chance as by careful consideration. When Deke put Neil, Buzz, and Mike together as a crew, nobody knew that Apollo 11 would land on the moon. If things had gone differently on earlier missions, that first landing might have been Apollo 12 or 13 or 14. Deke knew his pilots, and he assigned crews that would be compatible and would complement each other's skills. It was a real feat because he was dealing with a group that included world-class prima donnas.

We all knew that any of the mid-Apollo crews could have landed on the moon. So we didn't look at this crew as being specially chosen to make history. Nor did anyone dispute that Armstrong, Aldrin, and Collins were a great choice. Each of them was an outstanding pilot. Each had flown in space under rigorous circumstances. They were qualified for the job, and the luck of the draw handed them the Apollo 11 assignment.

In the last month, we'd had Neil in mission control to go over the rules for lunar descent, landing, surface operations, and takeoff. Mission rules could leave the ultimate decision to the astronaut, but that wasn't something we encouraged. Now I wanted to make certain that all of us understood exactly where we were. We got down to the finest details— descent-engine performance, computer bugs that we knew about, landmarks on the lunar surface, even talking through the most unlikely events we could imagine during the landing.

The computer and the landing radar got particular attention. We'd be sending last-minute updates to the computer on the lunar module's trajectory, its engine performance, and location over the moon. Until *Eagle* was about ten thousand feet high, its altitude was based on Earth radars, and its guidance system could be off by hundreds or thousands of feet. Then the LM's own landing radar was supposed to kick in and provide accurate readings.

That led to some heated discussions. Neil worried that an overzealous flight controller would abort a good descent, based on faulty information. "I'm going to be in a better position to know what's happening than the people back in Houston," he said over and over.

"And I'm not going to tolerate any unnecessary risks," I retorted. "That's why we have mission rules."

We argued the specifics of the landing radar, and I insisted that if it failed, an abort was mandatory. I just didn't trust the ability of an astronaut, not even one as tried and tested as Neil Armstrong, to accurately estimate his altitude over a cratered lunar surface. It was unfamiliar terrain, and nobody knew the exact size of the landmarks that would normally be used for reference. Finally we agreed. That mission rule stayed as written. But I could tell from Neil's frown that he wasn't convinced. Everything in his experience had taught him to trust his own judgment. I wondered then if he'd overrule all of us in lunar orbit and try to land without a radar system. Those conversations came back to me when I saw Neil a few days before the launch.

"What can we do?" I asked Neil. "Is there anything we've missed?"

"No, Chris, we're ready. It's all been done except the countdown."

He was right. If there was anything undone, none of us could say what it was. Apollo 11—its Saturn V, ground crews, mission control, and astronauts—was the most ready space mission ever. I looked at Neil, then at Mike and Buzz, with a sudden tension and envy that filled me with an awe I'd never experienced. These men were about to go to the moon and land there. Our work, our dreams, our failures, and our successes all rode with them. We had come at last to this point, and for a moment I felt my legs shake.

George Low broke the moment: "Have you thought about what you're going to say, Neil, when you step off that ladder?"

Neil was quiet for a moment. It wasn't the first time he'd been asked that question. So far, he'd stayed enigmatic and not given an answer. He was true to form: "Sure, George, I've been thinking about it." Then he changed the subject. "Tell everybody thanks from all of us. We know how hard everybody's been working."

We didn't press him on his historic quote. Whatever he said would be good enough, I thought. He'd come up with something that none of us were likely to ever forget. He'd say the right thing.

But words weren't enough. A high-level committee had decided what special items should go to the moon. Most important was a plaque on the lunar module leg, to be unveiled by Neil. "Here men from the planet Earth first set foot upon the moon. We came in peace for all mankind," it read. And it was signed by Armstrong, Collins, Aldrin, and President Richard Nixon.

An American flag and a microminiaturized photo-print of goodwill letters from heads of state would be left on the moon, too. Miniature flags of the states, the United States and its territories, and the United Nations would be carried in the lunar module and brought home to Earth. In the command module, there would be a stamp die, a stamped envelope to be canceled en route by the crew, and two full-size American flags that had flown over the House and the Senate.

The crowds were enormous. NASA invited more than twenty thousand VIPs, setting up bleachers, portable bathrooms, water tanks, and refreshment facilities at a site a little more than three miles from Pad 39 where the Saturn V stood tall and awesome. The Reverend Ralph Abernathy showed up to lead demonstrators against "this foolish waste of money that could be used to feed the poor." He had no effect on the mood of the crowd, either inside the Kennedy Space Center or among the million-plus people who had gathered on the beaches and roads of eastern Florida. Sometimes the common understanding of human greatness exceeds the bitterness of dissent. Apollo 11 was one of those times.

Between Houston and the Cape, thirty-five hundred journalists wore NASA accreditation badges. The press conferences were packed to overflowing, and no one counted how many languages were being spoken by reporters. Even the Russians had reporters there, and we were glad to have them.

We had our last meeting at the Cape on the day before launch. The night before, every room within hundreds of miles was taken, so I stayed in the crew quarters. It was one last chance to talk to Neil, Buzz, and Mike, and once again, mission rules were a hot topic. The flight plan called for Neil and Buzz to grab some sleep on the moon after landing. Neither of them thought that was possible.

"I think we're going to be too excited," Buzz said. "Maybe we should just get on with things and sleep later."

Aldrin was probably right, I thought, but it was too late to change the official plan.

"If that's what's best," I said, "we'll all decide together. But not until it happens. For now, plan on getting some sleep and sticking with the plan."

At home the night before launch, there was nothing on television but Apollo. In the minds of tens of thousands of people around Clear Lake and the Manned Spacecraft Center—NASA people, contractors, journalists, and the guy who ran the 7-Eleven—there was nothing else to think about, talk about, and store away in memories so that on future days we could all say, "I was there."

I slept fitfully that night. By 4:00 A.M. on July 16, I was on the road to mission control. We'd planned a Flight breakfast, my flight directors and me in the control center on this historic day. We'd watch liftoff from the control room floor, know that things had started well or not, then my men would go their own ways to get ready for the work shifts and long days ahead. It wasn't the finest breakfast ever served—cafeteria bacon and eggs and sausage and toast, with coffee for most of them and tea for me. But we sat there together and talked quietly about days past and especially this day present. I looked at the four of them—Cliff Charlesworth, so confident and competent; Gene Kranz, now chief of the Flight Control Division and stiff, sometimes too militaristic, but so quick and smart that it was sometimes scary to remember that he was human; youthful Glynn Lunney, who'd been with me from the beginning, senior in time to all of them despite his years and a man I trusted to always do exactly the right thing; and Gerry Griffin, the newest Flight in the crowd and already a veteran who would stand up to any crisis and find the answer.

I sipped my tea and knew that I'd chosen well. Of all the men in the world, these were the ones to see that America's lunar landing came off as planned. On July 16, 1969, they were my men called Flight.

The day is both a blur and a bright, vivid memory. Time compressed. One moment, I was sitting at my back row console next to Bob Gilruth, with our headsets plugged in and thinking that the countdown would never reach zero. Then suddenly it was three hours later and Mike Collins was separating the CSM from the stack, turning around, and docking with the lunar module. They were beyond Earth's orbital pull and on the way to history.

This time the astronauts went patriotic with their spacecraft names.

The CSM was *Columbia*. The lunar module was *Eagle*. I'd gotten a lump in my throat when I first heard their choices. Now *Columbia* was docked with *Eagle*, Collins was backing the spacecraft away from the S-IVB rocket stage, and I got another lump just listening to the conversation.

All these years since Bob Gilruth put my name on his original Space Task Group roster in 1958 bounced around in my mind. It was forever and it was only yesterday. It almost didn't make sense. Scenes came in no order. Gus and Ed and Roger in their space suits, then in their coffins. That magnificently embarrassing Redstone that lifted off only four inches, then sat shrouded in its parachute. Atlas after Atlas exploding. An Angry Alligator. Those incredible photos of Ed White's space walk, and the moments of fright and helplessness we felt while Neil Armstrong and Dave Scott were alone up there with an out-of-control spacecraft.

And so many meetings. How many thousands of hours had we sat in meetings, in a dozen or more different cities and assembly plants and government offices? We'd planned, we'd argued, we'd fought for budget, for fixes to bad spacecraft, even for permission to simply get on with our jobs.

Another thought caught me. Working year after year with seventy- and eighty-hour weeks had made time expand. But it was only eight years and two months—or a week or so less than that—since Jack Kennedy stood before Congress and proclaimed the moon as an American goal. Eight years and here we are! *Good God!* I thought. *How did we do it?* And the only answer I had was that we did it because we had to do it. We focused some of the best of American brains on the goal, and we had the undivided attention and labor of thousands of people. Apollo was the peacetime equivalent of a nation at war. It may be—no, it was the best thing America did in the twentieth century.

I looked at Bob and saw how he'd aged. He'd always been balding. Now there was just a white fringe over his ears. The hard years of leadership took their toll, but his eyes were still young and missing nothing in mission control. In that moment, I've never respected or loved another man more. "Thank you," I said to him, and he just nodded. He knew what I was thinking.

It was seventy-five hours to the moon. The television networks said that more than a billion people had watched Apollo 11's liftoff. When I tried to get out of the Manned Spacecraft Center and on the road home, I thought that there must be that many right here in Clear Lake. Traffic was jammed for miles and it wasn't just reporters. Americans, Texans, tourists, they were all crowding in to be there.

History is seldom made by schedule, and so it's experienced by only a few. Jack Kennedy announced this schedule to the world, and we wanted the world to come in to see it. The world responded. As I inched through traffic and looked at the faces in those other cars and trucks, I knew we were doing a very good thing.

The Russians were out of this race, had been for years, but they made one last attempt to look good. They'd launched Luna 15 toward the moon a few days earlier, and I was worried that their operations and communication at the moon might interfere with us. We'd had other comm interruptions over the years when the Russians insisted on operating near and at our radio frequencies. They were breaching international agreements, but we never made a formal complaint. I talked about the Russian mission at a press conference, to put some public pressure on them over their flight.

Then I asked Frank Borman to go through channels at the White House and to use contacts he'd made with the Russians since his Apollo 8 flight. The next day I received a telegram from the Russians saying that I shouldn't be concerned; their operation wouldn't interfere with Apollo 11. They were true to their word. Their radio transmissions didn't bother us. And then they lost it all anyway when Luna 15 crashed into the moon instead of landing on July 21.

July 20, 1969. They were in orbit around the moon. I was in mission control well before dawn, in time to hear the wake-up call that roused Armstrong, Collins, and Aldrin for their big day. I assumed that they didn't sleep too well, full of anticipation and awe. I know that I didn't. Getting out of bed before the alarm rang was no problem at all.

George Low had an official seat at my console in the back row of mission control. I made sure that Bob Gilruth did, too. Every top manager from NASA was in the glass-walled viewing room, and I knew that Bob was in no mood to put up with their questions and conversations at this point in the mission. Bob belonged in the middle of the action, not in some executive observer's chair.

The hours passed so smoothly that it was more like a simulation than the real thing. Aldrin moved into *Eagle*, setting its switches and getting it ready to fly while Armstrong put on his space suit in the command module. Then they switched places, Armstrong finishing the power-up while Aldrin dressed for his day on the moon.

Mike Collins watched *Eagle* move away. An hour later, after the two spacecraft had gone another time around the moon, he saw *Eagle*'s descent engine glowing fire. It was the first big burn in the landing sequence, descent orbit insertion (DOI). Armstrong and Aldrin were getting their first in-flight feel for their machine, and they liked it. In mission control, heart rates and blood pressures started to creep up. So far, almost nothing was different from Apollo 10. That wouldn't last long and we felt the final moments coming.

They were behind the moon and we had no data. Then the plot boards and console screens came alive. *Columbia* and *Eagle* emerged into view of our big network radar dishes. The data was glitchy, but readable. Their orbit numbers were good and telemetry from *Eagle* said all was well.

Gene Kranz took it all in. "Go for powered descent," he said in a calm, strong voice.

On the capcom console, I saw astronaut Charlie Duke take a deep breath. This was it. He took a second breath, pressed his mike key, and said the words. "*Eagle*, Houston. If you read, you're a go for powered descent. Over."

There was silence. Then more silence. After an age of waiting, I heard Mike Collins. "*Eagle*, this is *Columbia*. They just gave you a go for powered descent."

Still silence. In the back rooms of mission control, communications technicians were scrambling to sort out the problem. Gilruth, Low, and I didn't look anywhere but straight ahead at the plot boards. What the hell was wrong?

The comm techs found it. *Eagle*'s antenna was misaligned, just a little. Duke passed quick instructions to Collins, who relayed them to *Eagle*. In moments, Armstrong yawed *Eagle* just enough. The signal-strength meters in mission control jumped and we had them back.

"*Eagle*, Houston. We read you now," Charlie Duke radioed. He sounded relieved. "You're go for PDI [powered descent initiation]."

The stress level in mission control dropped perceptibly. Controllers who'd been standing almost at attention, or sitting straight in their chairs, relaxed and slumped. I breathed easier and sat back to listen.

"Ignition."

The voice was Aldrin's, coming from the moon . He'd punched on the descent engine, and now Neil Armstrong was flying *Eagle* past the point where Stafford and Cernan had gone only two months earlier. *Eagle* was on its way down.

Landing on the moon isn't like they show it in the movies. *Eagle* didn't just come straight down from fifty thousand feet. It was so much different in real life, and so much more hair-raising, that I couldn't stop myself from clenching the arms of my chair and from starting to breathe in shorter and shorter pulses.

Eagle's legs were forward and slanted down. Armstrong and Aldrin were standing inside, now feeling some gravity from the engine's thrust, but they were facing away from the landing site and looking down. To them, the sensation was of flying backward into something they couldn't see.

The reason was simple. They needed to see landmarks, to know that they were on course. When they sighted a familiar crater, used over and over in training sessions, they knew exactly where they were. Armstrong rolled *Eagle* 180 degrees. Now they were more or less on their backs, or slightly sideways, and looking mostly up while slanting down toward the smooth site we'd picked in the Sea of Tranquillity. They still couldn't see where they were going. It was an instrument approach like no human had ever done.

The instruments worked, at least so far. Landing radar kicked in. There was a big error between the onboard guidance system and the radar's report of how high they were. Within thirty seconds, the landing radar had fully corrected the onboard computer. We'd expected that. The data were good. My mission rules conversation with Armstrong jumped into my head. Without the landing radar, even Neil couldn't have saved the mission. The original altitude errors were just too much.

Then I sat up straight, almost in shock. "Program alarm!" It was Armstrong's voice. "1202! 1202!"

Gilruth and I looked at each other. "What the hell is that?" I snapped. My heart was thumping and I could feel the skin on my face tightening.

The guidance officer on Kranz's team, Guido, was Steve Bales. He was twenty-seven years old that day, which made him a high school kid when *Sputnik* went up in 1957. I was thumbing through my papers looking for alarm codes when I heard Bales answer Kranz's unasked question.

"We're go on that, Flight," Bales said.

Charlie Duke didn't wait, either. "*Eagle*, Houston," he called. "We're go on that alarm."

I found the explanation almost at the same time that Bales explained it to Kranz. *Eagle*'s computer was having trouble completing its calculations. It was overloaded with data and had flashed an alarm code.

But that wasn't the end. That 1202 came up three more times. It was tense enough in mission control without that awful alarm. Later we figured out that *Eagle*'s rendezvous radar was on when it should have been off. It was an artifact from the Apollo 10 mission, and nobody had changed the required setting. So now *Eagle*'s computer was receiving streams of useless data and reacting with those sudden alarms. Steve Bales was quick to give Kranz a go, and Charlie Duke reassured Armstrong and Aldrin. I looked over the top of my console, finding Bales at his console down in the front row that we called the Trench. "Good kid," I said to Gilruth and Low. "Were any of us ever that young?"

"I wasn't," Gilruth chuckled. "But I guess you two were."

Those damned alarms must have given Armstrong and Aldrin big jolts of adrenaline. But they didn't show it and they didn't sound like it. Their radio reports were simple and direct, and now they were down to the point we called "high gate"—barely seventy-five hundred feet above the craters of the moon. *Eagle*'s guidance system blipped thrusters, they pitched almost upright, and the moon was there out the windows and coming up fast. I could visualize Armstrong's hands lightly on the controls, waiting to take over at a heartbeat's notice.

Kranz nodded at Charlie Duke, and Capcom said the words we'd been waiting to hear, almost forever: "You're go for landing."

Go for landing! I know that every heart and every set of lungs in mission control—and maybe around the world—stopped in that second and waited. There was nothing from Armstrong and Aldrin. They were almost down to "low gate," about thirty-five hundred feet, and acknowledging the obvious wasn't on their minds.

"Program alarm!" This time it was Aldrin. "1201! 1201!" Armstrong was looking out at a boulder field we didn't know was there. He didn't need more alarms.

It was similar to a 1202. I had my finger on the page while Bales snapped a fast "Go, Flight," and Charlie Duke said simply toward the moon, "We're go."

We could see on our displays that Armstrong was flying manually now and that fuel was getting low. He was skimming over that boulder field, flying *Eagle* like a helicopter. My worry level soared because he should have been descending. He needed an open spot to land, and my worst fear was that he couldn't find one. Buzz Aldrin did his job reading off altitude and forward movement. The descent started again. I could only hold on to my chair and wait. They were slowing down quickly. "Sixty feet. Down two and a half. Two forward . . . two forward." The numbers from Aldrin were in feet per second. They were close and slow.

Charlie Duke saw the fuel plot drop. "Sixty seconds," he radioed. One minute of fuel left, and everyone in mission control was standing. It wasn't supposed to be quite this close. Bob Gilruth squeezed his eyes shut, then opened them with a deep breath. I don't remember if I was breathing or not.

"Getting some dust here." Aldrin's voice. If the rocket thrust was kicking up dust, they had to be almost down.

"Contact light," Aldrin said. The metal probe sticking out from the bottom of the landing pad had touched the moon. My mind watched what my eyes couldn't see. *Eagle* crunched the moon, settled in, was down. Aldrin's voice again.

"Engine stop.

"Engine arm off."

The landing came at exactly 3:17:39.9 P.M. CST in the United States of America. Damned if that LLTV training hadn't paid off. There was silence from the moon. There was absolute silence in mission control. I had time for one quick inhale before Neil Armstrong's voice was in my ear loud and clear.

"Houston, Tranquillity Base here. The *Eagle* has landed."

"Roger, Tranquillity," Charlie Duke said instantly. "We copy you on the ground. You got a bunch of guys about to turn blue. We're breathing again. Thanks a lot."

On the ground. Is the moon "ground"? Let the philosophers decide that one. Neil Armstrong and Buzz Aldrin were safe, the first of our human species to land on another world. We'd done it. At least, we'd done it this far. And all hell broke out in mission control, cheering and waving small American flags, handshakes all around, complete pandemonium from people who'd held it in and did their jobs in the face of historic stress. Bob Gilruth brushed a tear from his cheek, and my own eyes were bleary. We shook hands and he clapped my shoulder. Neither of us could speak over the lumps in our throats. We didn't have to. The shouting and cheering in mission control was saying it for all of us.

Gene Kranz gave the revelry about a minute before he took charge. He had to yell over the commotion to get everyone's attention. On the moon, Armstrong and Aldrin were getting *Eagle* instantly ready for an emergency takeoff. This was new country to all of us, and nobody was taking any chances. Kranz ordered his controllers back to their consoles. The job was to see what shape *Eagle* was in. It was done in a very few minutes and Kranz polled his team. For the first time ever, the question wasn't go/no-go. It was stay/no-stay.

The control team was unanimous: "Stay, Flight."

Will the world ever forget that afternoon moment on July 20, 1969? No. Nor will it forget the hours just before midnight. Just as we'd discussed that night in the crew quarters, Armstrong and Aldrin were too hyped to follow the flight plan and get some sleep. Moving around in one-sixth g was easy and they were comfortable. They wanted to get on with it, to get out of *Eagle* and walk on the moon. I conferred with Kranz, Lunney, and Slayton. The moon walk would come on Lunney's shift. The flight surgeons had no objection. We made the decision. Go out early.

That long day that began for us before 4:00 A.M. now came to the next moment of truth at 9:30 P.M. Lunney gave the go for cabin depressurization. Bruce McCandless, now Capcom, passed the approval and we watched our displays as *Eagle*'s atmosphere evacuated to the moon.

I thought back to the intense and private discussions we'd had about who should be the first man on the moon. In all the early flight plans and timelines, it was the lunar module pilot. Buzz Aldrin desperately wanted that honor and wasn't quiet in letting it be known. Neil Armstrong said nothing. It wasn't his nature to push himself into any spotlight. If the spotlight came, so be it. Otherwise, he was much like Bob Gilruth, content to do the job and then go home.

I thought about it. The first man on the moon would be a legend, an American hero beyond Lucky Lindbergh, beyond any soldier or politician or inventor. It should be Neil Armstrong. I brought my ideas to Deke, and then to George Low. They thought so, too.

So now we were in another Gilruth-Low-Kraft-Slayton meeting, talking it through from every angle. Not once did anyone criticize Buzz for his strongly held positions or for his ambition. The unspoken feeling was that we admired him and that we wanted people to speak their mind. But did we think Buzz was the man who would be our best representative to the world, the man who would be legend?

We didn't. We had two men to choose from, and Neil Armstrong, reticent, soft-spoken, and heroic, was our only choice. It was unanimous. Bob Gilruth passed our decision to George Mueller and Sam Phillips, and Deke told the crew. Buzz Aldrin was crushed, but took it like a stoic. Neil Armstrong accepted his role with neither gloating nor surprise. He was the commander, and perhaps it should always have been the commander's assignment to go first onto the moon.

There was an engineering side to it all that we hadn't considered. I thought of it while Buzz took a couple of tries to get *Eagle's* hatch open. When he did, it blocked Aldrin's position. The only way for him to be first out would have been to switch places with Neil hours earlier, before they put on their space suits. It was a small thing now. Our decision was based on other things.

Every eye within view of a television set anywhere in the world watched. Neil wasn't visible yet, but he was pulling a D ring to open a panel in *Eagle's* side and release that black-and-white television camera. There was fuzz and snow on the screen. Then there was Neil Armstrong.

I thought it, but I didn't say it. *What will he say? C'mon, Neil, back down that ladder and do it.* And I wondered, too, what we'd be thinking without that TV. Nobody in mission control was speaking.

He was white and fuzzy and bulky in his pressurized space suit. Those images are history now and they're replayed faithfully on every appropriate occasion these decades later. But at the moment, watching live with the rest of world, I stood almost at attention and let each second record in my memory. I'd never forget.

He was at the bottom, standing in the big dish of Eagle's front leg.

"Okay, I'm going to step off the LM now."

For the second time in a matter of hours, not a muscle twitched, not a man or woman moved inside mission control. We watched. Neil's leg moved, out and away, and down. We listened.

"That's one small step for man . . . one giant leap for mankind."

We applauded and shouted and it was pandemonium in mission control. Gilruth slapped me on the back and Low tried to hug me—he never showed that kind of emotion. Our headset cords kept it from happening. We shook hands all around. Euphoria held us and the moment never seemed to end. Man was walking on the moon. An American. One of us. We did it.

The next three hours flew away. Aldrin came out, they put up the flag, they deployed a small package of science experiments, they collected rocks and soil samples. They both had fine Hasselblad 70mm cameras attached to their chests, but Aldrin took no pictures of Neil Armstrong. We see Neil in the videotapes, and now and then in a few frames of 16mm movie film from a window-mounted camera. *Life* magazine's next cover showed an astronaut on the moon, saluting the American flag. The astronaut was Buzz Aldrin. Reflected in his gold visor is Neil Armstrong, taking the picture.

After twenty-one and a half hours on the moon, the job was done and they left. I heard Aldrin's clipped report, "Liftoff," as the ascent stage shot upward toward its rendezvous with Mike Collins in *Columbia*. A sudden thought hit me, an irony that could never adequately be explained. It takes untold resources and the most intricate of apparatus to launch a rocket from the Cape. But from the moon, more than 240,000 miles away, we did it with two guys and a fragile descent stage for a platform. And it worked every time.

Apollo 11 came home to a world waiting with honors and to a place sealed forever in history. Richard Nixon was on the carrier to greet them, taking advantage of a moment in history that he mostly inherited. A few years later, he cut the hell out of our budget. But for this moment, it was pandemonium in mission control all over again. These skilled and cool controllers kept it tight, then let loose completely. Flags appeared from desk drawers and under consoles. Cigars were passed all around. We celebrated.

The astronauts spent their time in quarantine and nobody brought an alien virus to Earth. The rocks were unpacked and the scientific study of real moon material began. We'd be involved, but our jobs still lay in doing more. We'd met the final piece of the Kennedy Challenge: ". . . and returning him safely to Earth."

It was time to look ahead. And quite suddenly, we didn't like what we saw.

21

Inevitably, Apollo Changes

The wake-up was rude.

We'd been to the moon and done exactly what the nation demanded of us. Some made the ultimate sacrifice and died. The rest of us pushed our bodies and our minds and families past any reasonable limits. The divorce rate among aerospace people during the space-race years was frightening—and then got worse after Apollo 11. All of us paid a personal price for being part of this amazing adventure, and I never heard a single person regret it.

My own family sacrificed. Betty Anne and the kids didn't have me around, in the daily role of father and husband. I was in and out, more of a remote authority figure to Gordon and Kristi-Anne than a typical American father. They all had my mother to fill some of the void, but it wasn't the same. Still, I don't know what I could have done differently. My nation called. I answered. All of us did our jobs. Gordon and Kristi-Anne, too, though they'd be adults before they understood.

Apollo 11 was a beginning. We had nine more Saturn V's either built or on the way, we had spacecraft being assembled, and plans for nine more lunar missions to exploit our capability and to explore for the sake of science and for the future of mankind itself. We'd clearly won the space race with the Russians. In retrospect, it wasn't even close. We'd put the best of American technology and the best of its engineering and scientific brains to work on that race. John F. Kennedy set the goal, and for more than eight years, nobody seriously got in the way or tried to stop us. We had support from the White House, from Congress, and from Americans everywhere. NASA should have been in its strongest position ever to ask for and get approval to keep moving outward.

That's surely what George Mueller believed. With our help in Houston, and with engineering support from Bellcomm and the Cape, he put together a far-reaching space exploration plan and gave it to Tom Paine, who'd been appointed NASA administrator by President Nixon. Mueller was ambitious. He wanted a series of spacecraft that would explore near-Earth space out to geosynchronous altitudes where the communications satellites lived. He wanted a space station permanently in orbit by the end of the seventies, the start of a permanent moon base, and a commitment that the United States would send an expedition to Mars. Wernher von Braun was solidly with Mueller on that one. Von Braun still harbored the dream of having his own manned space program and of supplanting Bob Gilruth as the man in charge.

Gilruth was more conservative. A mini–space station called Skylab was in the works, using an empty S-IVB upper rocket stage as living and working quarters. Gilruth wanted to go directly to a bigger space station and had Max Faget working on its design. He also envisioned a winged, reusable spaceship that would make routine trips to and from the space station. By now we knew that a space station was uppermost in Russian minds. Gilruth pushed for it to be our next big goal, too.

We had the technology. We had the team. We had the experience. A space station should have been given top priority. And after that, there wasn't a nickel to be bet against the idea that Americans would be walking on Mars by 1995. We could have done it. So why the hell didn't we?

There are complex reasons and simple reasons. The complex ones involve Vietnam, a president who'd had early interest in space and now had other problems to face, an inward-looking awareness of poverty, and much, much more. Our own NASA administrator found himself besieged by internal factions who couldn't agree on what to do. NASA's budget was chopped annually by Lyndon Johnson, and Richard Nixon cut it more. Congress didn't intervene. Its vision and daring were gone, ground under the rowdy feet of rioters in the streets and body bags coming home from Vietnam. The best of times for America was simultaneously the worst of times.

The simplest reason was that the public lost interest. The attention span of the American mind proved to be the biggest disappointment of the century. We had the solar system, and maybe even the stars, in our hands. And the American public told us to forget it. Each time we went back to the moon, fewer people watched and cared. It was old hat.

A Spiro Agnew task group did try to bring some order to the future, calling for Mars landings in the coming decades. Nobody listened. And at all levels of NASA, with the handwriting on the wall, people bailed out.

Sam Phillips went back to the Air Force. George Mueller stuck around long enough to see that no good was coming in the future, then returned to the aerospace industry. Tom Paine asked George Low to come to Washington as deputy administrator. Bob Gilruth wanted Jim McDivitt's strong administrative skills in the Apollo program manager's chair in Houston; Deke and I concurred, and McDivitt took over George Low's old job. Another of Gilruth's deputies, George Trimble, took a job in industry, too.

We were about ready to fly Apollo 12 back to the moon when Gilruth called me to his office. It was just him and me.

"I'm going to stay here another year or two," he said, "and then somebody has to move into this office. George and I have talked it over—"

"George will be a great center director," I interrupted. Gilruth held up his hand for silence.

"—and we've agreed that you're the best man for the job."

I was stunned. I'm sure that my face was falling to the floor when Gilruth told me the rest.

"Nothing's ever certain in these things. It'll be up to Paine when the time comes, but George and I are going to do everything to make it happen. Collaborators, you might say, to make you the next director of the Manned Spacecraft Center."

"Bob, I never thought of anybody but George as your successor."

"Well, it's time to think about yourself. I want you to start thinking about your own successor in Flight Operations, and then in a few months, I want you over here as deputy center director. We'll give you plenty of time to work into it and get used to running things. I'll back out slowly."

I'd been working around, with, or for Bob Gilruth since I was twenty years old. I'd seen him lead America to the moon and back, almost always from just far enough behind the scenes that most people didn't even notice him. The idea that I could even think about replacing him was beyond my immediate understanding. We talked it out for the next hour, and then I told him that I'd do it.

"You're the boss, Bob," I said. "I'm just kind of overwhelmed that I'm sitting here having this conversation."

In truth, I'd never been happier with any job than running the Flight Ops Directorate at MSC. I was in the action every day, every mission, and that flight director's blood was in me to stay. Stepping up to the next level was a challenge that would take me away from that and put me in charge of engineering, operations, astronauts, administration, personnel, facilities, and all the inside and outside politicking that went with being a center director. It was a damned big bite to take. I hoped it wasn't more than I could chew.

We flew Apollo 12 in November 1969. Nobody could say that it wasn't exciting. An unmanned Surveyor spacecraft had landed on the moon a few years earlier, next to a big crater and some big rocks. The photos it took were awe-inspiring, but it didn't have the automatic labs that later landers would have on Mars to do much science. So our scientists wanted us to go there and see Surveyor Crater close up and bring home some samples.

Pete Conrad and Al Bean made a perfect landing, near the crater's edge and opposite that lonely Surveyor sitting across the way. They'd had some stark moments leaving the Cape, when lightning struck their Saturn V and all the alarms and caution lights went off in the spacecraft. Then their lights went out. The lightning had knocked their fuel cells off-line. On board Apollo 12, with that huge Saturn V firing beneath him and the g loads building up, Pete Conrad not only stayed cool, but kept up a running commentary for controllers.

In mission control, the experts sorted things out quickly. The capcom relayed switch-by-switch instructions for restarting the fuel cells, and in the dark up there, Conrad and his crew got things working again. There were tense moments, but Apollo 12 reached orbit safely. Gerry Griffin was Flight on that one. He'd come up the ladder in mission control after a career as an Air Force pilot, and I listened from my console in the back row while he kept his team focused and did everything right.

Now came the tough question. Should Griffin let them go to the moon? Or should he keep them in Earth orbit, then bring them down again? All we knew was that the spacecraft and rocket had been hit by lightning, not once, but several times. Was Apollo 12 still safe? Or was it severely damaged? Gerry Griffin had ninety minutes to decide.

The conferences started. Through that first turn around the Earth, Griffin and his controllers devised enough tests, and the crew performed them, to convince us that any damage was superficial. The command and service module was operating as it should, but nobody knew about the lunar module. It was still inside its adapter cover, and except for a few small heaters, it was inert. There was no way to check its systems without doing the docking maneuver and connecting it to the CSM. And that had to be done after leaving Earth's gravity and heading for the moon.

I listened while Griffin was briefed by his LM experts. The lunar module was probably unscathed, they told him. But nobody knew for sure. Go or nogo? It was a decision that only Flight could make. Gerry Griffin made it. Apollo 12 would fire the S-IVB stage in orbit and press on to the moon. It was one of the gutsiest decisions in all of Apollo, and I was proud of it then and now. That moment was a tribute to the engineers on the ground, to the astronauts, and to the maturity of Apollo. We had the confidence to continue.

Now Conrad and Bean were on the moon, and Dick Gordon was orbiting in the CSM. Maybe we were making it look too easy again. The lightning strikes got the public's attention for a few hours, and the news media played it for the danger it truly represented. But then danger was past, the translunar coast was routine, and so was the landing. Then when Al Bean, setting up the first color television camera on the moon, accidently aimed it at the sun, the camera's tube burned out. Conrad and Bean were about to make the longest and most strenuous walk on the moon yet, and nobody could watch. It just wasn't the same to listen on the radio. At least not for the general public.

Conrad was a premier practical joker, and he'd gotten help from *Life* magazine and Deke Slayton in sneaking a camera timer to the moon. He and Bean intended to take a timed photo of themselves standing by Surveyor. We didn't hear the story until much later. But they were too good at their real jobs. Conrad and Bean spent nearly two hours working their way around Surveyor Crater. They carefully picked up rocks and dirt samples, documenting each with their camera and with the voice reports back to Houston. The scientists were ecstatic; these guys were proving that NASA was serious about getting the most information from the moon and turning it over for research and examination back home. And they never did take that picture.

On day two, they deployed a much bigger science station around their LM, again leaving the often-critical science community more than happy with how they did their business. Nobody at NASA thought it should be any different. Going to the moon was daring and dangerous. We did some science on Apollo 11. We did more on Apollo 12. And with each mission out through Apollo 20, we'd do more and more and more. If we didn't go as fast as the scientists wanted or do as much, we still did our best and made honest judgments about what to include.

There was a disquieting side to Apollo 12. I sat at my console in mission control and toggled back and forth between the networks to see how they covered man's second and third walks on the moon. Mostly they didn't. PBS stayed with us, but NBC, CBS, and ABC all went back to their soap operas or evening programs. Without that television camera on the moon, they didn't give a damn. It was a terrible display of news media disinterest, and I knew that apathy was now our biggest enemy.

How many times will men walk on another planet? How many times can you listen to history as it happens? How many times does the phrase *What have you done for me lately?* overshadow a great event?

After Apollo 12 was home and the astronauts sat in another long and unnecessary quarantine, I sat in my office and brooded. Some kind of terrible handwriting was on the NASA wall, and the big plans being hatched for space exploration were made of wisps and threads.

Apollo 12 was my last hurrah as director of flight operations. A new reality was forming in America, and I'd have to face it squarely. I stood and looked out my window, at the Manned Spacecraft Center, and then up at the moon.

It was going to be my job to run this place, and to do what I could for the space program's future.

Bob Gilruth announced me as his deputy director in January 1970. Sig Sjoberg took over my old job, one of the best decisions I ever made. He was an old-line flight ops manager and had been at my side since we'd met in the Langley days. I didn't have to look over his shoulder and worry.

We'd been to the moon twice, and all of us wanted to keep going out through Apollo 20, though now at a slower and more thoughtful pace. Two missions a year, we thought, was about right. Even more, we wanted firm guidance from the White House and Congress about the future. The word *post-Apollo* entered our vocabulary. We didn't get much guidance.

President Nixon was loud in proclaiming that the United States would continue manned space flight. What he wasn't loud about was our budget. When the Bureau of the Budget went over Tom Paine's submission, it used a sharp red pencil. There would be no money for any of George Mueller's grand plans, no money for Bob Gilruth's space station, no money for much of anything, in fact. NASA went back to the drawing board, and what it came up was Gilruth's winged spacecraft. We started calling it the Space Shuttle.

There also was no money to fly moon missions out through Apollo 20. Something had to go. If we didn't handle it right, the science community would be irate. Despite our best intentions, they still didn't fully trust us to be anything more than rocket jockeys out to prove that our hardware worked. We needed a compromise, and the one we worked out fit the definition perfectly: Nobody was happy, but everybody agreed.

Apollos 18, 19, and 20 were scrapped. We'd fly Apollo 13 in April 1970, then four more missions through the end of 1972. Scientists had all but free rein to design experiments and instruments to be set up by astronauts and left on the moon. We gave them their limits in size and weight, then let the scientists and their committees argue it out. The final word was NASA's, but we knew that we'd bend every way we could to accommodate them.

That boundless future in space suddenly had very real bounds, after all. And we had a very real responsibility to get every bit of data and every piece of new knowledge that we could before our time ran out.

22

Near Failure and Great Successes

Al Shepard had been off flight status for years, suffering from an inner-ear disorder called Ménière's syndrome that affected his balance. It was supposed to be incurable. Then somebody told him about a doctor in Los Angeles, and Al took a chance. A small tube was inserted into his ear, draining off the fluid that caused the trouble. He walked into the flight surgeon's office at the Manned Spacecraft Center, asked for a flight physical, passed with flying colors, and then told Deke that he wanted to go to the moon.

There wasn't a naysayer around. America's first man in space deserved his chance. Deke gave him Apollo 13. But it was coming up too soon for Al to get ready, and in Washington, George Mueller was irate that Deke seemed to be playing favorites with an old friend. When Mueller ordered Gilruth to get involved, Deke calmed the situation by swapping the Apollo 13 and 14 crews. That decision changed a lot of lives and made for some new space legends.

If there's such a thing as a bad-luck flight, it was Apollo 13. The new crew was Jim Lovell, Ken Mattingly, and Fred Haise. Lovell would be the first man to go to the moon twice, and this time he and Haise would land in the scientifically and geologically interesting Fra Mauro region of the moon. I kept sitting in on meetings about the current missions, and in mid-March 1970 things started to go wrong. Charlie Duke on the backup crew came down with the rubella form of measles. He got it from his kids.

Chuck Berry brought the problem to us. Charlie had exposed the prime crew. Lovell and Haise were immune; they'd had rubella as children. Mattingly's blood came up negative for the immunity factors, and he couldn't remember if he'd had it or not. The incubation cycle was just right for Mattingly to develop measles during Apollo 13. He could develop a high fever and perhaps have permanent aftereffects.

We had two choices: delay the flight by twenty-eight days or replace the crew. Deke came up with a third choice. He didn't want to pull the whole crew so late in the training cycle. "Let me talk to Lovell and Haise," he said. "If they'll agree, I'll move Jack Swigert up from the backup crew to replace Ken."

"Let me know," I said, "and I'll recommend it to Gilruth and headquarters." So it happened. Ken Mattingly was most unhappy and vocal in his comments at the resulting press conference. I didn't blame him. He was weeks from the greatest moment of his life and got sucker punched by fate.

Fate continued to converge on Apollo 13. The press had been lobbying since 1965 to be allowed into mission control during our flights. I'd refused from the day we moved mission control to Houston, believing that reporters watching over our shoulders would be a distraction we didn't need. Public Affairs had its console in the control center and over the years did a consistently good job of telling our real-time story to the news media. The one concession I made was to let reporters into the viewing room during practice sessions and simulations. It would be an extraordinary education for them, and they'd be better equipped to write about the real thing. Only a few ever accepted that offer.

By Apollo 13, my flight controllers were solid and experienced. Most of them knew the major reporters by first name, had been through scores of press conferences, and were not the least intimidated by being watched. That viewing room filled up with NASA executives, politicians, and astronaut families anyway. On the floor of mission control, we'd learned to ignore them.

So we changed the policy. The new rule was to allow one print reporter and one television reporter, serving as a pool for the rest of the press, into mission control for key events. Someone asked about emergencies and we added a pool slot for that contingency. The world's press selected my young friend from Time-Life, Jim Schefter, and NBC's Roy Neal as the only reporters they trusted to watch an emergency unfold and report it accurately. Schefter was one of the few reporters who sat through our simulations and had actually read the mission rules. He and Neal would soon be tested.

Apollo 13 was two hundred thousand miles from Earth on April 14. I left mission control in late afternoon and was in the shower at home when Betty Anne rushed in.

"Gene Kranz is on the phone," she said in a rushed voice. "He needs you right away."

I took the phone dripping wet. "Chris, we've got some serious problems." His voice was calm, but typically clipped. "The spacecraft . . . Chris, you better get out here quick."

"I'm on my way," I said, and hung up. I didn't ask for details beyond Gene's saying that they had electrical problems. His tone told me that he had his hands full. I was still damp driving back to work, turning pages in my mind and thinking about how many different things could be happening out there.

Kranz was deep in conversation around the flight console. He's not often a cussing man, but he looked me in the eye and said, "Chris, we're in deep shit." I believed him. One of his assistants briefed me quickly. The crew had heard and felt a loud noise. Within seconds, Jack Swigert saw that oxygen pressure was dropping in the two big service module tanks, and when Lovell looked out the window, he saw a cloud of spray. I plugged into the comm loops to listen to Kranz and to the CSM systems controller. There was still confusion about what was happening.

Then I looked carefully at the color monitor displays on my console. Apollo 13's oxygen system was in a total downturn. It was obviously too late to save it. I also saw that Swigert had taken quick action to pull two fuel cells off-line. They sucked up a lot of oxygen as they generated electricity. Pressure in the third fuel cell was falling away. It was going to be lost, and the command module would have nothing but its batteries for power.

I heard Kranz issue a critical order. Three oxygen bottles were in the command modules, to be used during reentry. They were kept charged from the main oxygen tanks. "Seal 'em off!" he barked. If the lines were kept open, they'd bleed dry into space. It was one of his early inspired decisions and a wise move. That order saved three lives in space.

There was a stir on the Public Affairs console next to mine. They'd been reporting the problem, but now the television networks were carrying a flash from Reuters that the mission would be aborted. Roy Neal still wasn't there, but Time-Life's Schefter was, had heard the fuel cell failures, and knew the rule: no fuel cells, no moon landing. He'd filed a fast pool story, and now all of the wire services and networks were picking it up. There was nothing for PAO to do but get a quick confirmation from Kranz that the report was accurate.

So now the whole world was watching. I had Kranz and his people in my ear, saying that the lunar module was still in good shape. It should be opened up and activated immediately. Kranz looked at Capcom Jack Lousma and gave the order. For the rest of this mission, that LM would be a lifeboat. Its electrical power and oxygen had to make the difference. There was no other choice.

In all the what-if simulations before each moon mission, a disabled spacecraft was invented. The solution was usually to fire up the big SPS engine, change trajectory, and get back home by the fastest route. In some of the solutions, the lunar module became a lifeboat. But nobody thought of this problem, where an explosion wrecked the oxygen tanks, the fuel cells were gone, sensors and instruments were destroyed, and nobody could tell if the SPS engine was still good or not. Any firing could be disastrous. Kranz declared the SPS, and the attitude control rockets on the CSM, to be off-limits. He needed another solution.

If they did nothing, Apollo 13 was on a free-return trajectory. It would circle around the back of the moon and head home. But that trip would take four and a half days. The lunar module people ran some quick calculations. There wasn't enough oxygen or electricity to make it. They offered another idea: Fire the LM's descent engine on the backside of the moon to speed up the return trip. Apollo 13 could get home a full day quicker. But even then, it would be a close thing, and they'd need some severe restrictions on using electrical power. The high-powered electronics in the spacecraft did much to keep it warm. Now much of it would be turned off and the astronauts would have to come home chilly.

Kranz told me what he was thinking, and I went into the viewing room to brief the Washington brass. A few minutes later, Kranz went off-shift and turned the operation over to Glynn Lunney's control team. I told him to put the details together. It was a plan.

We met the press at midnight in a grueling session. All that apathy had disappeared. They wanted to know everything and we gave it to them. Watching a video of that conference years later, I could see how worried I looked. I was, but this was no time to candy-coat things.

I didn't tell them about my personal impressions of the mission control guys. The control teams and the backroom support technicians were on top of their game. Not once did I hear the words "If we get them back . . ." It was "When we get them back . . ." and "Here's what we have to do . . ." Maybe the youth in mission control had something to do with it. Gene Kranz likes to say that failure is not an option. But it was more than that. Failure was not possible in their minds. It was never a question of if. The only question was how. That confidence affected us all. I was worried, all right. But I didn't doubt.

Apollo 13 has been the subject of books, a movie, and a television series. All of them got important parts of the story wrong. The need for drama on the screen overwhelmed any desire for complete accuracy. The movie scenes on board Apollo 13 were terrific, but the arguments in mission control didn't happen, were only made up for good drama. Nor were the Lovell and Haise families—Swigert was a bachelor—as terrified and uninformed as the movie made them seem. We were smarter and better than that, and the families were important to us. We kept them informed, almost minute to minute, and their worries were normal, not exaggerated. Finally, none of the writers, including Jim Lovell, understood what had really happened in that oxygen tank explosion. Even later, when everything was clear, some perceptions remained clouded.

Of such are lore and legend made. Of all the Apollo missions, 11 and 13 are uppermost in American minds. Between April 15 and 18, 1970, ground crews devised innovation after innovation to keep the lunar module going. Lovell, Swigert, and Haise rigged cardboard tubes to ventilation systems and carbon dioxide removal canisters. They napped mostly in the LM, where it was only slightly warmer than in the command module. The descent engine burn was perfect on the backside of the moon, and they came home cold, tired, dehydrated, and safe.

But what really happened? After they separated from the service module, the crew got a good look at it and took an invaluable set of photographs. One of six compartments in the service module had been blown away. That fact alone confirmed Gene Kranz's decision to avoid using the SPS engine. If it had worked at all, the result might have been fatal.

George Low named his old friend and college roommate Edgar Cortright, now director at Langley, to head an investigation board. One of the first discoveries was that the oxygen tank had been dropped about two inches. That minor accident got much of the blame for Apollo 13's trouble in the dramas and books written later. It may have been a contributing factor, but the real problems were much worse.

From the beginning of Apollo, there were times when pad technicians couldn't flush the tank clean after tests and were forced to turn on the low-power internal heater to boil off the oxygen. That could take eight hours. So in the Block 2 spacecraft they increased the pad power supply from twenty-eight to sixty-five volts. This took a special external harness. North American and the Apollo Spacecraft Program Office ordered the Beech Aircraft Company, manufacturers of the tank, to install a new sixty-five-volt thermostat switch in all Block 2 tanks. Beech didn't do it and the program office didn't follow up. Apollo 13, and the earlier missions, too, flew with a twenty-eight-volt switch. A unique set of circumstances brought on disaster.

On Apollos 7 through 12, during pad tests, the normal detanking by pressurized nitrogen forced the liquid oxygen out of the tank and it was never necessary to use the heaters to pressurize the tanks. So the tanks continued to operate satisfactorily on all these flights.

But on Apollo 13, the tube in the tank wouldn't function properly; it was misaligned, and liquid oxygen leaked back into the tank instead of draining. The Cape engineers then used the heaters to warm the cryogenic oxygen, convert it to gaseous oxygen, and allow the tank contents to boil off. This was the first time this process had been used, and the tank was considered good for flight. Nobody recognized the damage that could occur.

Postflight tests in the laboratory showed the dramatic effects of sixty-five volts when applied to the switch. With no liquid oxygen surrounding the switch, its contacts literally melted and permanently closed the switch. Continuous power ran through the Teflon-insulated wires to the heater for the next eight hours. The wire temperature rose past eight hundred degrees Fahrenheit, the Teflon melted, and the wires were bare.

During the pad tests, instruments available to the ground technicians only went to one hundred degrees; they didn't see anything happen. Before launch, the tanks were filled again, and now the bare wires were under the liquid oxygen surface. There was no danger yet. Not until the switch was turned on again.

Outbound to the moon, the liquid oxygen quantity gradually dropped. At fifty-five hours and fifty-five minutes, controllers sent word to Swigert: "Turn on the tank heater to build up more pressure." But now the bare wires weren't covered with liquid; they were in a pure gaseous oxygen environment. Swigert flipped the switch, the wires sparked, a fire started, and with the sudden increase in pressure, the tank blew.

The astronauts were both lucky and unlucky. The explosive force ripped away a service module panel, almost as if it had been zippered in place. The rest of the SM structure was intact, and that proved critical in keeping the crew safe. But the explosion also destroyed the oxygen manifold. There was nothing to stop all of the oxygen tanks from emptying into space.

It was a stupid and preventable accident. If Beech had replaced the switch as ordered, it would never have happened. If the program office had been more diligent in cross-checking its own orders, it would never have happened. If any of us had remembered the difference in heat loads between direct and alternating current, it would never have happened. But it did.

Ed Cortright was now at the Houston center, working with Max Faget's engineers on understanding all of this. He hadn't bothered making his manners to Bob Gilruth, or even giving us an update on his conclusions. So Bob made the move and invited him up to the ninth floor.

We listened to him summarize what he knew. We'd heard it all from Max already, so there were no surprises. The obvious solution was to replace the switch as previously ordered.

"I'm recommending that we design and build a new tank," Cortright said, and we were astounded.

"Why would you do that?" Gilruth asked.

"That tank is perfectly sound," I said. "We've got hundreds of test hours and flight hours on it. A new tank loses us all that experience."

"I have my reputation to worry about," Cortright said. "I can't recommend anything less."

What he meant was that a simple solution was just too easy. If it didn't look difficult, he didn't look good. We argued, but Cortright wouldn't budge. Vanity overcame his common sense. In the next weeks, Gilruth and I fought Cortright all the way through NASA headquarters. The bosses took the political way out and ratified Cortright's recommendation.

That new design cost $40 million. Cortright faded into the woodwork and left us to work out the details. It was a difficult engineering job, and ground tests couldn't possibly provide the kind of testing and flight experience we were throwing away.

I sat on the edge of my seat for the next few missions, worrying about that undertested tank out there. But it worked. If Cortright was proud of it, he never bothered to say.

23

Good-bye to the Moon

We made tough choices. Four Apollo missions were left on the books, and the constant rumor was that we wanted to cancel them all in favor of spending our money on the Space Shuttle. Some congressmen bit and urged us to quit. Scientists worried about us in public meetings and in private phone calls, demanding that we fly missions through Apollo 17. The press asked us about the future in every way they could, looking for an angle and a story.

The only story was that we never intended to quit. We wanted to go back, to get the most from the moon that we could. In every conversation with Apollo scientists, I stressed that these missions were our last chances in the decade to explore the moon. If I'd had a clue that it would be forty or fifty more years before we sent men back there, I would have fought like hell to get Apollos 18 through 20 back into the schedule, and so would Gilruth, Low, Paine, and all the rest. None of us thought that America would turn into a nation of quitters and lose its will to lead an outward-bound manned exploration of our solar system. That just wasn't possible.

We made the remaining missions as aggressive as possible. Each would go to a different and increasingly interesting place on the moon. Scientists put together bigger and better packages of experiments to be set up and left on the moon. We added big handheld drills, looking something like a skinny jackhammer, to bore deep underground and extract core samples that would show how the moon's surface was layered. Apollo 14, with Al Shepard, Stu Roosa, and Ed Mitchell, went to the Fra Mauro site that had been on Lovell's agenda. They brought a ricksha to carry their hardware on long and grueling science walks. It was a help, but hauling it over the gritty and rolling surface was sweaty work. On the next three missions, the LM carried a lunar rover—an electric car with wire wheels that let the astronauts roam far from their landing site and do even more exploration.

We made operational changes, too, to give science every edge. On Apollo 14, that first descent burn down to fifty thousand feet was done by the SPS engine before the LM undocked. That left Shepard and Mitchell with plenty of fuel during their final landing maneuvers to take their time, pick out the best place to land—where they could see the most interesting features—and then make a pinpoint landing. It meant more geology training in the premission months for the crew, and they soaked up their rock studies so well that any of them could have written a master's thesis for a geology degree. When they hovered over that lunar landscape, they were truly experts who knew what to look for and recognized it when they saw it.

Al Shepard made his own piece of history on the moon. He was almost as avid for golf as I am and, with Deke's connivance, had a golf club head made for the general-purpose staff he carried. At the end of the second moon walk, in front the television camera standing outside the LM, he screwed that head onto his staff, dropped a ball into the lunar dust, and swung away. It wasn't a great hit, and the ball bounced away. Deke had just rushed over to me in mission control and warned me that it was going to happen.

"I think you'll like this," he said, but the look on his face told me that suddenly he wasn't so sure.

I watched Al on the big screen on the front wall and could only grin. "What am I supposed to do at this point, Deke? Tell him to stop?" Hell, even if I'd known, I would have gone along with the gag. How many times could somebody play golf on the moon?

Al dropped a second ball, hit it squarely, and said that it flew out of sight. In one-sixth g, he probably drove it eight hundred yards. I'd say more, but it's hard to get a full swing when you're wearing a pressurized space suit.

Dave Scott missed his space walk on Gemini 8, got to stick his head out the command module hatch on Apollo 9, and was now going to drive around on the moon on Apollo 15. The landing site had the scientists excited—Hadley Rille, a steep-sided rift in the lunar surface, and close to the base of the moon's Apennine Mountains. It was the trickiest landing site yet, and the most geologically impressive.

We'd improved the Saturn V engines, too, getting another seven thousand pounds of payload going to the moon. Some of that went into additional fuel for the CSM and LM. The lunar rover weighed four hundred and fifty pounds. We added a heavy instruments package to the service module that would measure the environment around the moon and do a high-resolution survey of the surface. And the LM itself got more batteries to increase its surface stay-time and also more science instruments for Scott and Jim Irwin to deploy. After years of worry that we'd give them less, scientists were finally happy with Apollo.

Dave Scott didn't need the extra fuel. He put his LM down within five hundred feet of the aiming point, moving slightly to get the best spot, and had 103 seconds of hover time left in his tanks. For the next sixty-seven hours, he and Irwin stayed on the moon. In three long excursions, one of them lasting more than seven hours, they got it all done including a long drive to the Apennine foothills where they were looking up at mountains rising fifteen thousand feet above them.

The lunar rover had its own television camera, operated remotely from mission control by Ed Fendell. We and the world—or as much of the world as the television networks would allow to see it—watched it all. After each drive, they came back to the lunar module and put out more of the science instruments. The area was an outdoor laboratory, wired for sound, seismic activity, and much more when they finished.

Then came the part I'd been waiting to see. Dave Scott drove the rover off a bit, parked it with the camera pointing toward the LM, and walked back to his ship. Ed Fendell would try to catch their liftoff from the moon on live TV. He got it, too. We listened to the countdown from the moon, and when Dave fired that ascent engine, it must have been a real kick up through the astronauts' legs. That ascent stage bolted upward and was gone. *Good God*, I thought. *I had no idea it went so fast.* We'd been told it was like that by the other moon crews, but seeing it for real was a thrilling shock.

Fendell panned the camera up, but it was too late to see more than a few seconds of the first-ever manned launch televised from the moon. He'd do better on the next missions, now that he had a benchmark to work from.

There were two unfortunate aftereffects of Apollo 15. When Scott and

Irwin were safely back in the command module, Dr. Berry waved me over to his console. He pointed at the EKG traces coming in from the pair, and I could see something was wrong. Irwin's heart was skipping every other beat, then beating twice rapidly. Chuck called it a bigeminy.

"It's serious," Berry said. "If he were on Earth, I'd have him in ICU being treated for a heart attack."

"What do we do?" I asked.

"Find out how he's feeling." Berry sent the query through the capcom and Irwin was honest. He was tired and felt a strange heart sensation. He was 240,000 miles and a high-g reentry from the nearest hospital. My own heart skipped a beat.

"In truth," he said, "he's already in an ICU. He's getting one hundred percent oxygen, he's being continuously monitored, and best of all, he's in zero g. Whatever strain his heart is under, well, we can't do better than zero g."

We stood at Berry's console and watched the EKG trace. Within minutes, it went back to normal. It stayed there all the way home, but no flight surgeon took his eyes off that trace for more than a moment. Irwin's postflight exam didn't turn up a thing. In every test, his heart was normal. But something was hiding there. A few months later, he had a heart attack. Then another. He had the best surgeries the times could offer. And then another heart attack killed him. He was still young.

A few months later, Bob Gilruth moved on to Washington. I was sitting in his old office, now director of the Manned Spacecraft Center, when the second Apollo 15 problem landed in my lap. A news story revealed that a dealer in Germany was selling stamps and envelopes carried to the moon and signed by Scott, Al Worden, and Irwin. I told Deke to check it out. His report was straightforward, and it opened a Pandora's box of troubles for the astronaut office and for NASA.

"They did it," Deke told me a day later. "There was no hiding. Dave just said sure, nothing wrong with it, right? Hell, Chris, I don't know. They each got a large sum of money in a Swiss bank account. Maybe it wasn't smart, but was it wrong? Or illegal?"

"I don't know. But it doesn't smell good." I passed the information to George Low at headquarters, and he was a stickler. He immediately got the NASA inspector general and the lawyers involved. In a matter of days, we had a full-scale internal scandal on our hands.

The German dealer was a friend of Kurt Debus's. Through this association a number of the astronauts had signed stamps and envelopes, some of which were carried in space and to the moon. Deke was unaware of the fact that almost everyone was responding to the dealer's request for signing the documents he was providing. Deke was furious and so was I.

Most of the astronauts took money from the dealer. Deke confronted them one by one. The common story was that they gave the money to charity. Tom Stafford proved it for himself; his pay went directly to a charity and never touched his hands. But when the IG supoenaed Jack Swigert's bank records, he found more money than an astronaut should have. He also found a predated check to the March of Dimes, rigged through a Swigert friend in Las Vegas. Swigert's reaction was simple: "It's none of your damn business!"

A few others joined him. They'd gotten legal help, and nobody could find a law they'd violated. Some of my deputies told me to drop it; we were likely to discover things we didn't want to know. That wasn't going to happen. Many of the astronauts still had envelopes and stamps. With the press notoriety, those collectibles were suddenly worth a lot of money.

We confiscated them, sometimes under duress. The inspector general turned the whole package over to the Nixon Department of Justice, and we waited for the other shoe to drop. Our own NASA lawyers didn't expect indictments, but those Justice hotshots knew the law better than anyone else. If there was a crime here, they'd find it. They didn't.

It took a while, but Justice decided they would not pursue an indictment. It was questionable that any law had been broken and they realized that dragging astronauts into court would not be a popular pastime. The stamps and envelopes went back to the astronauts, and whatever happened to them was kept quiet. Before that day, we extracted our own punishment. From Wally Schirra to Dave Scott, a number of astronauts had broken faith with us and ignored a standing order from Deke. They weren't crooks. But they were still wrong. George Low handed the problem back to me, insisting that I call each of them into my office and reprimand them in person.

I did it. And we did more, too. Jack Swigert was to be suspended as an astronaut. He took it badly, quit, and soon found a job as the minority staff director for the House Committee on Astronautics and Aeronautics. Then he did everything he could to make life miserable for George Low and me, even trying to get us fired from NASA. We stayed.

I had a list of fifteen more astronauts for suspension. By now Dave Scott and Al Worden had left of their own accord. Scott was named director of our Dryden Research Center in the Mojave Desert, a move that pissed off Deke to his eyebrows. The others took their suspensions, apologized, and lived through it. Some of them flew on Skylab in the midseventies.

The final indignity fell on George Low and me. A secret session of a Senate committee was called. Lowell Weicker of Connecticut was on the committee and asked us for a briefing. Low and I told him the whole story. The next day he killed us with the information, hammering us in the committee meeting without mercy. I was testifying when somebody brought a note in to the chairman, Senator Cannon of Nevada. Scott, Irwin, and Worden were outside.

Cannon called a recess and brought them in. The senators treated them like gods, shaking their hands, patting them on the back, falling all over themselves in adulation. Then Cannon reconvened the session, and Senator Margaret Chase Smith of Maine was the first to speak. She turned on me.

"Dr. Kraft," she actually snarled, "how dare you allow those fine young men to be put into a position where they can be tempted like that."

I gritted my teeth, clenched my fists, and kept my mouth shut. What do you say to a senator who's got it all wrong?

Deke was there and did speak up: "I'm responsible for making sure those things don't happen. Blame me." Deke was a stand-up guy.

George Low and I disagreed in the end. I thought we shot ourselves in the foot over something that should have been handled internally. The reprimands and suspensions were in order. The bad publicity wasn't. But George was a detail guy. "It was unethical," he said. "They deserved what they got."

I talked to Deke and we made a few more changes for future flights. Everything in an astronaut's personal pack had to be listed, sealed, then signed off by Slayton, me, and the NASA administrator. No more would be sneaked into space, at least not if we could prevent it. Finally I insisted that each astronaut sign an agreement: They could give away items from those personal packs, but they couldn't sell them. Everybody signed.

That left us with two missions to fly in 1972. John Young and Charlie Duke landed on the moon with Apollo 16 and did a helluva job for NASA and for the scientists. Ken Mattingly got his ride to the moon as command module pilot. He never did come down with measles.

Then Joe Engle got robbed along the way to the moon. And NASA lost a good program manager.

It was common knowledge in late 1971 that Gene Cernan and Joe Engle would fly the final lunar landing on Apollo 17. We had two classes of scientist-astronauts on board, men with Ph.D.'s whom we'd put through Air Force flight training as well. Most of them were damned good pilots, and all of them were smart to brilliant as scientists.

Deke presented the Cernan crew, including Ron Evans as command module pilot, to Gilruth, in his last months as center director, and he sent it on to headquarters. Suddenly the scientists howled. This was the last Apollo mission and they wanted a scientist on the crew. The word came back from Washington: Put a geologist on the crew. Deke was livid. He never trusted those scientists. They didn't have the test-pilot gene and he was blunt to their faces, and to everyone else's, about the chances that any of them would get a flight. Now he had to back down.

"Okay," he told me the next day. "It's Jack Schmitt."

Harrison H. "Jack" Schmitt was the only geologist in the first class of five scientist-astronauts. He was good with the scientists and had worked hard on picking landing sites and training astronauts in the finer points of geology, and everybody liked him except Deke, who didn't like any of the scientists.

"He's a good choice," I said, and after we discussed it, Gilruth passed it along to Washington. The answer came back fast. Jack Schmitt was going to the moon.

Gene Cernan was unhappy and Joe Engle was worse. He'd just lost the only chance he'd ever have to walk on the moon. Cernan came to me and demanded that this decision be reversed. It couldn't happen.

"Gene, here's the situation," I said. "If you want to be the commander on this flight, Jack Schmitt's going with you whether you like it or not. Your choice."

That lowered his temper quickly. Gene Cernan was no fool, and this was his only chance to walk on the moon, too. A week later we announced the crew. It came out of the Public Affairs Office at 10:00 A.M. At 10:30, I got a call from Jim McDivitt and he was raging. At eleven-thirty we were in Gilruth's office with Jim ranting about Cernan.

"Why didn't you ask me about this crew?" he demanded. "Cernan's not worthy of this assignment, he doesn't deserve it, he's not a very good pilot, he's liable to screw everything up, and I don't want him to fly."

I was shocked at how strongly Jim was reacting. "Why didn't you ask me?" he pleaded "Why didn't you ask me?" Then he shocked me further. "If you don't get rid of him, I'll quit."

Jim McDivitt was a damn good manager for the Apollo Spacecraft Program Office, and Gilruth and I had to agree that we should have had him in the loop on the crew selection. How he'd missed the rumors that it would be Cernan was something I never knew. He stormed from my office and I called Deke.

"Yeah," Deke said, "there's some bad blood there. McDivitt's never liked Cernan. He rants on about that helicopter crash Gene had over at the Cape, but it's no big deal. Gene's a good pilot and he'll do you a good job."

I accepted Deke at his word. If I'd known then what I know now, it might have turned out differently. I called McDivitt and told him that Cernan was staying. We'd announced it publicly and it was too late to change. I apologized for not keeping him more fully informed. It wasn't enough.

"Thank you," he said. "You'll have my resignation shortly."

Jim McDivitt stayed through Apollo 16, then quit, and it was a shame. His country needed him a lot more than the electrical power company in Michigan that made him a vice president.

Gene Cernan wrote a book in 1999, *Last Man on the Moon*. He sent me the galleys and asked for my comments. When I got to the part about that helicopter crash, it all came back to me. Gene was bluntly honest about what had happened. Deke had been less so with me.

Flying helicopters was part of the training for landing on the moon. As backup on a moon crew, Gene took a chopper out over the Indian River and spotted some pretty ladies sunbathing below. That river water is very clear. He was buzzing the mermaids when he misjudged his altitude, hit the water, flipped over, and was covered in flame. He took a deep breath, swam under the fire, and got free. His eyebrows were gone, his face was black, and he had some minor burns on his hands. An hour later, he was back at the Cape getting cleaned up. Deke came in. Here's how Gene tells it:

"So exactly when did the engine quit on you?" Deke asked.

"Deke, the engine didn't quit. I just flew the son of a bitch into the water."

"Maybe you didn't hear me right, Geno. Exactly when did the engine start to sputter?

Gene got the message, but he held on one more time. "Like I said, Deke. It didn't quit. I just screwed up."

I give credit to Cernan for taking it like a man, but Deke covered for him anyway. The official story was mechanical problems. And then Gene was going to the moon.

Would I have booted him from the crew if Deke told me the truth? That thought crossed my mind twenty-seven years too late to make it happen. Then I put it into perspective. Gene Cernan was the last man on the moon. He and Jack Schmitt flew an amazing mission, did everything they were asked to do, and brought more science home to Earth than any other crew. Apollo 17 was America's last hurrah at the moon, and Gene Cernan was flawless.

I'd like to think that I would have accepted Deke's recommendation, no matter what, and done just what I did—sent Cernan to the moon. It worked out just fine.

Cernan and I had another misunderstanding, and it still rankles me. It started over dinner a few nights before the Apollo 17 launch. Gene and Jack would aim for one of the most rugged lunar areas yet; the terrain was mountainous and their landing site was strewn with boulders ejected from a nearby crater. The hazards were obvious.

I remembered the earliest days of manned space flight and the lessons we learned in mission control. *The crew comes first.* It was ingrained now, part of my being. If something goes wrong, I told Gene, don't worry about landing. Getting the crew back is the most important thing. But I knew Gene and he was no different from all the rest. In the moment of decision, he'd go to extraordinary lengths to land on the moon.

"Gene," I said, "I'm going to be proud of you no matter how the mission comes out. Just don't take the 'white-scarf' approach."

He didn't. Apollo 17 was a tremendous success, from a perfect landing to a spectacular takeoff that we all watched on live television. Jack Schmitt's contributions as a geologist on the moon proved beyond any doubt that we need scientists in space. And Gene Cernan was smart in every move, up to leaving that last footprint before he climbed back into the lunar module.

The story should have ended there. Then Gene wrote about our dinner conversation and I was stunned. My "white-scarf" remark, Gene thought, meant that I was timid about going back to the moon. It was a sign that NASA management hadn't really wanted a lunar landing program.

Nothing was further from the truth. I was hurt that Gene could think we'd do all that we did but still be halfhearted in our efforts. We did Mercury, Gemini, and Apollo because we believed. I still believe.

Epilogue

I t was over.

I'd stay on as director of the Manned Spacecraft Center—renamed the Lyndon B. Johnson Space Center—until 1982, and the adventures with Skylab and developing the Space Shuttle were no small things. The Shuttle in particular has an ongoing and major impact on space exploration, and not all of that impact is positive.

But it's Apollo and programs that led to it that will be there among the shining moments when the last history of the world is written. Apollo opened the way for humanity to move off the home planet and to populate the universe. We will migrate outbound.

I learned so much. One of the most important lessons was that any apocalyptic prediction by a scientist would almost certainly be wrong. Tommy Gold is the classic example. While we sat in mission control waiting for Apollo 11 to begin its descent to the moon, Bob Gilruth and I chatted about Tommy. He was a scientist who'd gotten more than his share of press by predicting that the moon's dust layer was up to a mile deep and that the lunar module would simply sink out of sight when it attempted to land. The scary thing was that some people took him seriously.

We'd invited Tommy to Houston and met with him in Gilruth's office, showed him the evidence—including the unmanned Surveyor spacecraft that had landed on the moon and sent back astounding photos of craters and boulders—and tried to change his mind. He was adamant. It only took one crater filled with dust to prove him right, and Tommy Gold wasn't going to change his dire prediction.

"We'll know in a few hours," Bob laughed, while Neil and Buzz began to fly free in the lunar module *Eagle*. "If Tommy's right, you can handle the press conference."

I didn't bother drafting a prepared statement.

The fright-monger scientists won another of their battles. They forced us to spend millions on quarantine facilities and to put astronauts, doctors, technicians, and even public affairs men through long weeks of isolation after the early moon flights. Somehow they made a case for the absurd idea that moon germs might kill off all life on Earth. Instead they proved that fear is more powerful than common sense, but by then the money was spent and the public's attention was elsewhere. Now the apocalyptists are saying that any discovery of life on Mars should lead to an immediate cessation of planetary exploration. Hogwash.

We are an adventurous people and exploration is our way. I read Einstein and listen to the theories of noted mathematicians and astrophysicists. The speed of light, they say, is some kind of limit that will keep us from the universe. I choose to believe otherwise.

Earth and our moon are about 4.5 billion years old. We know that now from the lunar samples. When Bob Gilruth formed the Space Task Group in 1958, scientists guessed the age of the universe at between 5 and 8 billion years. New instruments, mostly carried by satellites, now say 15 billion, and each new generation of astronomy birds is likely to push that limit further out. So there could be civilizations out there that are literally billions of years ahead of ours. If we don't find them, eventually one of them will find us.

That will be part of Apollo's legacy. But I have more immediate thoughts about what we did in those amazing years, and who did it.

President John Kennedy, for reasons political or otherwise, had the courage to challenge America to send men to the moon. America responded.

James Webb, Hugh Dryden, and Robert Seamans assembled an elite technical management team and never wavered from their beliefs that the job could be done. I credit Webb for holding the program together in its most trying times, and for having the savvy and experience to work in the political arena.

No man of space did more or received less credit than Robert R. Gilruth. He led brilliantly and strongly, pushing us to design, to build, and to fly Mercury, Gemini, and Apollo. He made the tough decisions that Washington approved, and he forced us always to be better than we would have been without him.

Bob had a knack for finding the fundamental answer to the most complex problem, and because of that, he was consulted on every important issue that came up during the space race. His reputation among his peers was without equal. But Bob Gilruth was more than a technician. He recognized skill and leadership in others and developed an organization of great strength and character. I did my best to emulate him. Why no monument to Robert R. Gilruth yet exists is beyond my understanding.

Wernher von Braun and the men he held at Huntsville produced the gargantuan Saturn V rocket. In many ways, von Braun followed his own agenda and always seemed rankled that he couldn't run the whole show. But he was a rocket man, not a spacecraft designer or a mission leader, and that was where he excelled. The Saturn V's track record of successful launches remains a marvel of technology. In the twenty-first century, I still find it hard to believe that von Braun did so much in the 1960s. The world has nothing like a Saturn V today.

George Mueller was a brilliant engineer, manager, and at times tough to work for. The concept of "all-up" testing with every piece of Apollo flying on the Saturn V was his. He saved us hundreds of millions of dollars and cut months, if not years, from the time it took us to land on the moon. I continue to respect him even as I chafe from certain memories.

George Low was a leader. He took charge of Apollo after the terrible fire, then forced a bond between NASA and aerospace industry professionals that was essential to our success.

No designer had more impact on sending Americans into space than Max Faget. He is brilliant.

Deke Slayton's friendship, insights, and management skills carried me through many a day. When the doctors decided that he was fit, after all, to fly in space, I was overjoyed to sign off on his appointment to the crew that rendezvoused with the Russians on the Apollo-Soyuz Test Project in 1975. Of the Original Seven, Deke got his ride last. We owed him that.

North American Aviation owes its postfire revival to Frank Borman, and to the executives it moved into power, Dale Myers and George Jeffs. At Grumman, the lunar module became a marvel under the reinvigorated leadership of Lew Evans, Joe Gavin, and Tom Kelly.

During Mercury and Gemini, John Yardley and Walter Burke at Mc-Donnell pushed the limits of their industry to produce spacecraft that let us move from down here to up there, and to learn so much about this new art of flying in space. Yardley in particular is an outstanding engineer and still a close friend for whom I have nothing but high regard.

The academic and scientific world was more intimately involved with Apollo than with any program since the Manhattan Project and with any since. Stark Draper's software people at MIT were the computer heart of the command, service, and lunar modules. Leon Silver, of the U.S. Geological Service, and Eugene Shoemaker of USGS and Arizona State University, made lunar science understandable to the astronauts and accessible to all of us.

The Apollo fire was an avoidable tragedy and a necessary catalyst. We were confused before January 27, 1967, and completely focused thereafter. There was just so much wrong with the spacecraft, the rockets, and our own management that it took a catastrophic incident to wake us up. I felt great anguish over the deaths of Gus Grissom, Ed White, and Roger Chaffee, and still do. We humans are a tough bunch to teach.

Courage, skill, and tenacity from one group of men made so many differences in Mercury, Gemini, and Apollo. The astronauts individually and as a group deserve all the respect and admiration they got, and then still more. It was my job to protect them, to speak out for them, and to ensure that every decision—whether it involved hardware, software, or operational procedures—was made in their favor.

A few people in NASA resented the astronauts and thought they were prima donnas. Some were. The adulation they received was bound to affect them. Whose head won't be turned by a White House invitation, deference from Congress, the applause of millions? A few astronauts took advantage of their position. At the same time, they endured grueling schedules. Their work required intense physical and mental activity, and their private lives suffered. I believe they all performed their jobs extremely well.

My flight controllers are too often unsung heroes. No mission then or now could be flown without the dedication, professionalism, and raw intelligence of the men and women who work the consoles. They are an American treasure.

I was that boy from Phoebus who found open doors and walked through them. My parents encouraged adventurism. That day I jumped from the

seats and ran out to Lou Gehrig sticks in my mind. Some fathers would
have yelled at their son and been embarrassed. Mine handed me a pen
when I rushed back to him and sent me out again to get that wonderful
autograph.

My own children, Gordon and Kristi-Anne, are among my best friends
today. But it wasn't easy for them. I had some fame, or more accurately in
the eyes of a child, notoriety, when we moved to Texas. It had to affect Gordon that I was too busy to hand him that pen and encourage him to be his
own person. He turned rebellious, wanted no part of anything technical,
and in college went down the "hippie" road. His degree from Colorado College was in philosophy. He found his way in San Francisco and today runs
his own industrial supply company. We talk and visit often, and I'm proud
of all that he's done.

Kristi-Anne reacted to Gordon's rebellion by being the model daughter,
student, and wife. She's an honors graduate from the University of Texas
in interior design, worked in the business for years in Houston, home-
schooled two boys, and plays a damn good piano. She and her husband live
close to Houston, and we see them often. It's good to have a daughter.

For more than fifty years, Betty Anne has been my strength and my joy.
She endured it all, usually without complaint, and I doubt that I could
have walked through many of those doors without her encouragement.
More than anyone else, she let me become the man called Flight.

Until 1957, nothing man-made had been into orbital space. In 2001, the
common events of our lives are touched by space.

The most visible and dramatic technology advance was the communication satellite, developed by NASA. We see and hear the rest of the world
in real time, and we take it for granted.

Mercury, Gemini, and Apollo forced advances in digital computers that
would otherwise have come later, or even not at all. NASA revolutionized
electronics.

Materials developed for those programs—composites, resins, graphites,
fireproof fabrics, and more—surround us with lightweight strength and endurance. Teflon and Velcro, however, were not invented for NASA. But we
used a lot of both.

Remote sensing in medicine is a direct fallout of our need to monitor
astronauts' bodies and health in space. Today's doctors' offices, emergency
rooms, and intensive care units are filled with machines whose ancestors

were born of necessity in Mercury, Gemini, and Apollo. Health studies of the astronauts in space and on the ground gave medicine understandings of human physiology that could have arrived in no other way. The number of treatments for our frailties that came from space research cannot be counted.

There's still so much to be done out there. We don't use near-Earth space, out to geosynchronous orbit, to proper advantage. The U.S. government has lost much of its will to invest in space exploration, and private industry still doesn't flourish there. Every technical and scientific field can benefit from being in space. But government policies inhibit and even prevent private adventures out there. Commercial rockets launching commercial communications satellites are more common. But they need more encouragement from the White House, Congress, and NASA.

Perhaps they will see things differently in 2020, when the International Space Station is near the end of its projected life. But even the ISS is an example of politics intruding on technical reality. It went through so many redesigns to accommodate national and international politics that it is virtually a white elephant in space. Even before it was occupied, the ISS was encountering problems that needed extra Space Shuttle missions to fix. How it plays out will be interesting to watch.

Those who argue that all of this would have come by spending NASA money on Earth are wrong. We achieved through challenge what we could never achieve through our individual motivation. That challenge worked because it involved people in space, not machines and robots. It's hard to love a robot, or to care about its health.

Any argument for dropping or curtailing manned space flight is fallacious. Exploration requires emotional commitment. It even requires letting go for a moment and giving in to the sheer joy and fun of being there. We did that on Apollo 16 when, with all other work completed, we told John Young to let loose with a grand prix drive of his lunar rover on the surface of the moon. He whipped and wheeled that little car in tight turns, skidding through the lunar dirt, then opening it up to a full eighteen miles per hour and leaving a rooster tail of moon dust behind him. We watched it all, mouths agape and hearts racing, on live television, and I felt like I was sitting next to him out there, 240,000 miles away. I couldn't work up that emotion when a robot rover drove on Mars in 1997. It was a great feat, but it wasn't the same at all. And it won't be until an American astronaut, man or woman, drives on Mars.

Intellectual commitment is not enough to inspire a president, propel a

Congress, or draw intense support from a diverse public. Off-planet expeditions will certainly include robotic vehicles. But there must be man.

This nation can find no better investment in the health, safety, security, education, and overall well-being of the American public than for a visionary president to declare that Americans will land on Mars. And then make it happen.

I was lucky to be part of it. Growing up in forgotten Phoebus, Virginia, made me determined to excel. I saw my chance and I took it. I reached most of my goals. But I was just part of a crowd.

Scientists say there is no life on the moon. I look at the moon today, see the faces from NASA, industry, science, and academe who brilliantly sent Americans to that place, and I know differently. The people of Mercury, Gemini, and Apollo are blossoms on the moon. Their spirits will live there forever.

I was part of the crowd, then part of the leadership that opened space travel to human beings. We threw a narrow flash of light across our nation's history.

I was there at the best of times.

INDEX